Beaches and Coasts

Beaches and Coasts

SECOND EDITION

Richard A. Davis, Jr.
University of South Florida,
Texas, USA

Duncan M. FitzGerald
Boston University,
Massachusetts, USA

Registered Office(s)
John Wiley & Sons, Inc., 111 River Street, Hoboken, NJ 07030, USA
John Wiley & Sons Ltd, The Atrium, Southern Gate, Chichester, West Sussex, PO19 8SQ, UK

Editorial Office
9600 Garsington Road, Oxford, OX4 2DQ, UK

For details of our global editorial offices, customer services, and more information about Wiley products visit us at www.wiley.com.

Wiley also publishes its books in a variety of electronic formats and by print-on-demand. Some content that appears in standard print versions of this book may not be available in other formats.

Library of Congress Cataloging-in-Publication Data
Names: Davis, Richard A., Jr., 1937– author. | FitzGerald, Duncan M., author.
Title: Beaches and coasts / Richard A. Davis, Jr, Texas, US, Duncan M. FitzGerald, Massachusetts, US.
Description: Second Edition. | Hoboken, New Jersey : John Wiley & Sons, Inc., [2019] |
    "Wiley-Blackwell (1e, 2004)"–T.p. verso. | Includes index. |
Identifiers: LCCN 2018061448 (print) | LCCN 2019001886 (ebook) | ISBN 9781119334514
    (Adobe PDF) | ISBN 9781119334552 (ePub) | ISBN 9781119334484 (Cloth)
Subjects: LCSH: Beaches. | Coasts. | Coast changes.
Classification: LCC GB451.2 (ebook) | LCC GB451.2 .D385 2019 (print) | DDC 551.45/7–dc23
LC record available at https://lccn.loc.gov/2018061448

Cover Design: Wiley
Cover Image: Courtesy of Duncan M. Fitzgerald

Set in 10/12pt Warnock by SPi Global, Pondicherry, India
Printed and bound in Singapore by Markono Print Media Pte Ltd

10 9 8 7 6 5 4 3 2 1

Brief Contents

Contents

1

Coastline Variability and Functions in the Global Environment

The surface of the earth is covered by two contrasting media; land and sea. They meet at the coast. There are, of course, glaciers that span parts of both the land and sea such as in Greenland, Iceland, parts of Alaska and the Canadian Arctic, and on Antarctica. Each of these two surfaces may cover millions of square kilometers over continents and oceans or much less in the case of small oceanic islands or some lakes within continental masses. Nevertheless, a narrow coastal zone separates these two major parts of the earth's surface. We are defining the coastal zone as that combination of environments that is influenced by marine waters. Typically this includes barrier-inlet systems, anything that is subjected to astronomical tides, including wetlands and coastal bays.

The world coastline extends for about 440,000 km, but the *coastal zone* comprises less than 0.05% of the area of the landmasses combined. Because nearly half of the global population lives within less than 100 km of the coastline, the coastal zone has become arguably the most critical part of the earth's surface in terms of global economy and strategies, and management needs. Unfortunately, the coastal zone is also the most desirable property for human development. Much of our development of the coast has had negative impacts on various environments and has caused problems that are very expensive or even impossible to correct.

1.1 Coastal Settings

The coast is global in its distribution but it is limited in width. We cannot give an average width, an average character, or any other average category that adequately typifies the coast. It is much too varied and complicated in its characteristics. In some places the coastal zone might be only a few hundred meters wide, whereas in others it might be more than 100 km wide. Some coastal zones include a wide range of environments that separate the true ocean from the terrestrial environment. In other situations, a single coastal environment may define the land–sea boundary.

In this book we will consider the controlling factors that determine what type of coast develops. The processes that develop and maintain coastal environments, as well as those that destroy the coast, are discussed in order to convey the dynamic nature of all coastal environments. Each of the major environments will be considered in light of these controlling factors and processes. The impact of human activity along the coast has been enormous, especially over the past century. Many examples of this impact appear throughout the book. Most of the emphasis here is directed toward geologic and physical attributes of the coast, although organisms are not overlooked.

Some coasts are chronically eroding and others tend to be depositional and are

Beaches and Coasts, Second Edition. Richard A. Davis, Jr. and Duncan M. FitzGerald.
© 2020 John Wiley & Sons Ltd. Published 2020 by John Wiley & Sons Ltd.

accreting. Both situations can take place in any of the coastal environments and even in close proximity to each other. These changes may be cyclical, for instance seasonal, typically with erosion in the winter because of storms and deposition in the summer. There are also places where one or the other is an ongoing process. In addition, coasts are, very rarely, impacted by hugely destructive tsunami (Figure 1.1). The controlling factors tend to be one or more of three factors: energy imparted to the environment, sediment availability, or human impact.

Depositional coasts include a wide spectrum of systems such as river deltas, barrier island systems, strandplain coasts, and glaciated coasts. Each of these may contain numerous distinct environments. The variety of morphologic features and the complex interaction of coastal *morphodynamics* is emphasized in this book.

Climatic differences may cause a wide variety of coastal types in that temperature and rainfall exert a major influence on coastal development. Extreme climates such as those in the very high latitudes can cause coastal areas to be covered with ice; all the time in some places, and for only a few months in others. Parts of Greenland and the Antarctic coast are covered with ice continually whereas some of the coasts of Alaska, Canada, the Scandinavian countries, and Russia have ice cover at least a few months each year. Desert conditions can directly influence coastal environments as well. Few significant rivers and therefore few river deltas are produced from desert areas. Some coastal deserts are dominated by huge sand dunes such as along Namibia on the southwest coast of Africa. Along the Persian Gulf the arid, low-latitude environment produces extensive coastal environments called *sabkhas* that are nearly at sea level and have an almost horizontal surface dominated by chemically precipitated carbonates, salts and other minerals.

The tectonics of the earth's crust also produce a major influence on the coastal zone. Coasts that coincide with or that are near plate boundaries tend to have more relief and are narrow as compared to those that are away from plate boundaries. Collision coasts that are produced by colliding plates provide a particularly rugged coast such as we see along the Pacific side of both North and South America. The opposite situation exists

Figure 1.1 Photograph of an irregular coast with high relief in northern California. (*Source:* NOAA, https://en.wikipedia.org/wiki/National_Oceanic_and_Atmospheric_Administration.)

on trailing edge coasts where sediment is abundant and deposition dominates. Relationships between plate tectonics and coastal development will be treated in detail in Chapter 2.

1.2 Population and the Coast

The coast is many things to many people. Depending on where and how we live, work and recreate, our perception of it varies greatly. Large populations live on or near the coast because it is typically very beautiful and interesting. Many more visit the coast for the same reasons. A large number of people gain their livelihood directly or indirectly from the coast, and some have the task of protecting it from intruders or enemies.

1.2.1 History of Coastal Occupation

The ancient civilizations of the eastern Mediterranean Sea were largely associated with the coast, including the famous Greek, Roman and Phoenician settlements and fortifications of biblical times and before. Many of the great cities of the time were located on the natural harbors afforded by the geologic and physiographic conditions along the coast. These cities provided a setting that was conducive to trade and that could be defended against enemies.

Far to the north, Viking settlements in the Scandinavian countries of Norway, Sweden and Denmark were typically located along the coast as well. Here the great fjords provided shelter and fortification along with ready access to the sea, which was a primary food source and the main avenue of transportation; they were the sites of many battles. At about the same time, the northern coast of what is now Germany and the Netherlands was also occupied for similar reasons but in a very different coastal setting; one of lowlands and barrier islands.

Many centuries later, cities in the New World such as Boston, New York, Baltimore and San Francisco owe their location to the presence of a protected harbor. In their early stages of development many of the major civilizations of the world were either directly on the coast or had important interaction with it.

In the early civilizations, reasons for this extensive occupation of coastal areas were strictly pragmatic. Coasts were essential for harboring ships, a primary means of transporting goods, one of the major activities of the times. The adjacent sea was also a primary source of food. Similar reasons were the cause for the settlement of many of the great cities of Europe such as London, Amsterdam, Venice and Copenhagen. All were settled on the edge of the water because their location fostered commerce that depended on transportation over water.

1.3 General Coastal Conditions

Varied geologic conditions provide different settings for the coast, and give variety and beauty to that part of the earth's surface. As a consequence, some coasts are quite rugged with bedrock cliffs and irregular shorelines (Figure 1.2) whereas others are low-lying, almost featureless areas with long, smooth shorelines (Figure 1.3). To be sure, with time, any coast may change extensively but some important relationships continue through geologically significant periods of time, up to many millions of years.

Changes at a given part of the coast are typically slow and continuous, but they may be sporadic and rapid. Rocky cliffs tend to erode slowly but hurricanes can change beaches or deltas very quickly. Overprinted on this combination of slow and rapid processes of change is a fluctuation in sea level, a very slow change throughout time of about $3\,\text{mm year}^{-1}$. In the geologic past this rate has been both much faster and even slower. The

Figure 1.2 Photograph of a depositional coast on the Texas coast.

Figure 1.3 Fluvial delta that empties into the Atchafalaya Bay on the Louisiana coast. (*Source:* earthobservatory.nasa.gov.)

point is, as coastal processes work to shape the substrate and the adjacent land, the position of the shoreline changes as well. This translates the processes and their effects across the shallow continental shelf and the adjacent coastal zone, producing long, slow, but relatively steady, coastal change.

Each specific coastal setting, regardless of scale, is unique yet is quite similar to other coastal settings of the same type. Although each delta is different, a common set of features characterizes all of them. The general approach of this book will be to consider the general attributes of each of the various types of coastal environments. Numerous examples of each environmental type will provide some idea of the range available for each. Finally, the overprint of time will demonstrate the dynamic nature of all of these coastal elements.

1.4   Coastal Environments

Coastal environments vary widely. This section will briefly introduce each of the major environments to demonstrate this variety. All of these and more will be discussed in detail in the following chapters.

Rivers carry tremendous quantities of sediment to their mouths, where they deposit it. Much of the sediment is then entrained by waves and currents but commonly there is a net accumulation of sediment at the river

Figure 1.4 River Exe estuary in Devon on the coast of England. (*Source:* Steve Lee, https://www.flickr.com/photos/94466642@N00/178926998/. Licensed under CC BY-SA 2.0.)

mouth—a delta (Figure 1.4). In fact, most of the sediment along all types of depositional coasts owes its presence, at least indirectly, to a river. Deltas range widely in size and shape. Most are dominated by mud and sand but a few have gravel. The primary conditions for delta formation are a supply of sediment, a place for it to accumulate, and the inability of the open-water processes to rework and remove all of the sediment from the river mouth.

Sea level has risen considerably over the past several thousand years as the result of glaciers melting and a combination of other factors. This increase in sea level has flooded many parts of the land and developed extensive and numerous coastal bays. Streams feed most of these bays. These bays are called estuaries (Figure 1.5) and are commonly surrounded by some combination of wetlands; usually either salt marshes or mangrove mangals, and tidal flats.

Another common type of coastal bay is one that tends to parallel the coast and is protected from the open ocean by a barrier island. These elongate water bodies—lagoons—have no significant influx of fresh water or tidal exchange. Tidal flats and marshes are uncommon along this type of bay because of an absence of tidal flux. Other coastal embayments that cannot be considered as either an estuary or a lagoon are simply termed coastal bays.

Barrier islands are another important part of the scheme of coastal complexes. They are a seaward protection of the mainland, typically fronting lagoons and/or estuaries. The barriers include beaches, adjacent dunes, washover fans and other environments (Figure 1.6), and wetlands, especially salt marshes, are widespread on their landward side. Barrier islands may be dissected by tidal inlets (Figure 1.7), which are among the most dynamic of all coastal environments. They not only separate adjacent barrier islands, but also provide for the exchange of water, nutrients and fauna between the open-ocean and estuarine systems.

Strandplain coasts are low-relief coastal areas of a mainland that have many characteristics

Figure 1.5 Overview of a barrier island from open-water beaches to washover fans that extend into the estuary behind the barrier.

Figure 1.6 Oblique aerial photo of the North Carolina coast with Drum Inlet bisecting the barrier island on the Outer Banks. (*Source:* Courtesy of A.C. Hine.)

of the seaward side of a barrier island. They contain beaches and dunes but lack the coastal bay (Figure 1.8). Examples include Myrtle Beach, South Carolina and the Nayarit Coast of western Mexico.

The presence of rocky or headland coasts can be present as short isolated sections within extensive sandy depositional coasts, such as along parts of the east coast of Australia or the Pacific Northwest coast

Figure 1.7 Coast of South Carolina near Myrtle Beach which is a strand plain coast without any open water in the backbarrier. (*Source:* Data SIO, NOAA, U.S. Navy, NGA, GEBCO. Image c 2018 TerraMetrics.)

Figure 1.8 Cliff on erosional coast in Oregon with a narrow beach.

Figure 1.9 Large groin to protect the beach from erosion on the North Sea coast of the Netherlands. There is a small dike landward of the dry beach.

(Figure 1.9). Other geomorphically different coasts may have their origin in glacial deposits, with New England being a good example.

1.5 Historical Trends in Coastal Research

Early systematic efforts for studying the coast in the early twentieth century were made by *geomorphologists*, scientists who study the morphology or landforms of the earth. Geomorphologists also investigate mountains, deserts, rivers, and other earth features. Their studies produced various classifications, maps, and reports on coastal landforms. Some scientists focused on the evolution of coasts and the processes responsible for molding them. For example, Douglas W. Johnson, a professor at Columbia University, wrote a classic and pioneering book in 1919 entitled *Shore Processes and Shoreline Development*, a monograph that is still commonly referenced. Another more recent individual who has contributed a great deal to our knowledge and terminology of coastal dynamics is Dr. Miles O. Hayes, who

had spent his career in both academia and private enterprise.

Engineers have also given special attention to the coast over many centuries. Their interest has been directed toward construction of dikes, harbors, docks, and bridges, on one hand, and stabilization of the open coast on the other (Figure 1.10). Although geomorphologists and engineers direct their efforts toward different aspects of the coast, their interests overlap in many circumstances.

Ancient people recognized that the coast is potentially dangerous during storms and is continually changing due to processes associated with wind, waves and storms. They understood that the shoreline is one of the most dynamic areas on the earth. Erosion was a particularly important problem, and settlements were lost or threatened as the shoreline retreated. For centuries, dikes have been constructed along the North Sea coast of Holland and Germany, both for protection (Figure 1.10) and for land reclamation. In many other areas, however, construction on the open coast was designed to slow or prevent erosion. As a result, various types of structures were emplaced at critical locations along densely inhabited areas of the coast in

Figure 1.10 South end of the Galveston sea wall on the Texas coast showing how it is protecting the shoreline from erosion.

Figure 1.11 Allied troops landing on a steep and narrow beach at Gallipoli during the early stages of World War I. (*Source:* Courtesy of Australian War Memorial.)

attempts to stabilize the beach and prevent erosion (Figure 1.11).

For decades these activities represented the major efforts of science and technology to understand, and in some respects, to con- trol the response of the coast to natural pro- cesses. World War II was also an important period in furthering our understanding of the coast. Major war efforts took place along the coast, particularly the landing of troops

Figure 1.12 Photograph of the late Douglas L. Inman, famous professor at Scripps Institution of Oceanography and major contributor to research on the coast of southern California. (*Source:* Scripps Institution of Oceanography, UC San Diego. Reproduced with permission of Scripps Institution of Oceanography, UC San Diego.)

(Figure 1.12), supplies and equipment, whether on the European mainland or on Pacific islands (see Box 1.1). All branches of the military were involved in studying coastal geomorphology, coastal processes including waves, tides and currents, and the analysis of weather patterns along the coast. Much of the world's coast was mapped in detail during this period. The Beach Erosion Board, a research committee of the U.S. Army, Corps of Engineers, made many very important contributions to our knowledge of coasts. This group conducted extensive research on beaches, waves, erosion and other important aspects of the coast using both their own staff and academic researchers from many of the best universities. Francis P. Shepard and Douglas L. Inman (Figure 1.13) of the Scripps Institution of Oceanography were prominent contributors to the research programs of this group and later became among the most prominent coastal researchers in the world.

Box 1.1 Why is the Coast so Important?

This seems like a pretty stupid question but maybe there are some reasons that you have not considered before. Most of us think about the beach and swimming or surfing; maybe fishing. All of these are important but they pale in comparison to some of the huge economic benefits of the coast. All of our ports and harbors are on the coast. Many of the major cities of the world are on the coast. These account for many trillions of dollars in economic benefit. Coastal tourism is also a huge business in many countries, with the United States leading the pack.

Now think back into history and see if the coasts have always been important to civilization. In early historical times eastern Mediterranean countries such as Greece, Turkey, Egypt and Italy all had established

coastal cities that functioned both as important economic centers and as places to protect their countries from invaders. In those days invasion by vessel was the most common way to conduct such activities. Ships were small and required little water to float, so attackers could beach their transport ships to unload the invading armies. This type of activity was also common in the Scandinavian countries when the Vikings were invading the high-latitude parts of Europe.

As time marched on, coastal invasions continued but in somewhat different fashions. During World War I some invasions were launched large transport ships from which soldiers made their way to the coast in small landing craft. This left the invading troops exposed and vulnerable to well-placed defenders on land and, as a result, landing troops were often destroyed. More recently in World War II there were examples of how the coast, and the beach in particular, played an important role in famous military events. The first to be considered is the landing of US troops at Iwo Jima in the Pacific. We all know of the famous picture and statue of the marines raising the flag on the island. Here the landing craft and submarines could proceed right to the shore because of the steep nearshore bottom (Box Figure 1.1.1), meaning that deep water extended very close to the beach. Tides on Pacific islands are low; less than a meter. This steep beach of volcanic sand posed some difficulty for the landing troops but they managed. Eventually the US troops prevailed, which led to the raising of the flag, but more US military were killed in

(a)

(b)

Box Figure 1.1.1 (a) Boats carrying troops right up to the beach at Iwo Jima and (b) marines on the steep beach of volcanic sand. (*Source:* U.S. Department of Defense.)

this battle in a few hours than in more than a decade of the wars in the Middle East and Afghanistan in this century.

The other famous landing of World War II was the Normandy Invasion on the north coast of France (Box Figure 1.1.2). The beach here is quite the opposite of that on Iwo Jima. It is wide and flat with tides of near three meters in magnitude. Bad weather was predicted for the invasion, which included hundreds of aircraft as well as vessels. Tides also had to be considered because the Germans had placed various

(a)

(b)

Box Figure 1.1.2 (a) Aerial view of the huge task force landing on Normandy Beach, France (*Source:* Wikipedia, https://commons.wikimedia.org/wiki/File:NormandySupply_edit.jpg); and (b) troops making their way across the shallow nearshore to the wide, gently sloping beach. (*Source:* U.S. War Department/ National Archives, Washington, D.C.).

obstacles in the shallow water to hinder the landing craft. The weather cleared and the decision was mad to land at low tide which would help to see the obstacles. It was the largest seaborne invasion in history, with 150,000 participants. Fewer were killed than at Iwo Jima.

These are just a couple of examples of extreme activities that benefited from the military knowing the nature of the coast that they were invading. Knowledge of tides, wave climate, nearshore bottom and other environmental parameters helped in these invasions.

Figure 1.13 Photograph of the devastation near the coast of Japan where the 2011 tsunami came ashore.

Suggested Reading

Collier, M. (2009). *Over the Coast: An Aerial View of Geology*. New York: Mikaya Press.

Davidson-Arnott, R. (2010). *Introduction to Coastal Processes and Geomorphology*. Cambridge, UK: Cambridge University Press.

Davis, R.A. Jr. (1993). *The Evolving Coast*. New York: Scientific American Library.

Johnson, D.W. (1919). *Shorelines and Shoreline Development*. New York: Wiley.

Masselink, G., Hughes, M., and Knght, J. (2014). *Introduction to Coastal Processes and Geomorphology*, 2e. New York: Routledge (Kindle Edition).

Nordstrom, K. (2000). *Beaches and Dunes of Developed Coasts*. Cambridge, UK: Cambridge University Press.

Woodrofe, C.D. (2003). *Coasts: Form, Processes and Evolution*. Cambridge, UK: Cambridge University Press.

2

The Earth's Mobile Crust

2.1   Introduction

Coastlines of the world exhibit a wide range of morphologies and compositions in a variety of physical settings (Figure 2.1). There are sandy barrier island coasts such as those of the East and Gulf Coasts of the United States, deltaic coasts built by major rivers including those at the mouths of the Nile and Niger Rivers, glacial alluvial fan coasts of the Copper River in Alaska and the Skiedarsar Sandar coast of southeast Iceland, coastlines fronted by expansive tidal flats such as the southeast corner of the North Sea in Germany, volcanic coasts of the Hawaiian Islands, carbonate coasts of the Bahamas and the South Pacific atolls, gravel beaches of southern England, mangrove coasts of Malaysia and southwestern Florida, bedrock cliff and wave-cut platform coasts of the Alaskan Peninsula and southwest Victoria in Australia, and many other types of coastlines. The diversity of the world's coastlines is largely a product of the Earth's mobile crust. The eruptions of Mount Saint Helens in Washington State (1980) and Mount Pinatubo in the Philippines (1991) and the devastating earthquakes of in Mexico City (1985) and in Kobe, Japan (1995) are dramatic expressions of this mobility. The formation of pillow basalts and new oceanic crust at mid-ocean ridges as well as the presence of hydrothermal vents at these sites are also a manifestation of crustal movement. The theory that explains the mobility of the Earth's crust and the large-scale features of

continents and ocean basins, including the overall geological character of coastlines is known as **Plate Tectonics**

Plate tectonics theory has done for geology what the theory of evolution did for biology, the big bang theory for astronomy, the theory of relativity for physics, and the establishment of the periodic table for chemistry. Each of these advancements revolutionized their respective fields, explaining seemingly unrelated features and processes. For example, in geology the cause and distribution of earthquakes, the construction of mountain systems, the existence of deep ocean trenches, and the formation of ocean basins are all consequences of the unifying theory of plate tectonics. The germination of this theory began many centuries ago with scientists' and world explorers' interest in the distribution of continents and ocean basins. As early as the 1620s, Sir Francis Bacon recognized the jigsaw puzzle fit of the eastern outline of South America and the western outline of Africa. By 1858, Antonio Snider had published two maps illustrating how North and South America were joined with Africa and Europe during Carboniferous time (~300 million years ago) and how the continents had split apart to form the Atlantic Ocean. He reconstructed the positions of continents 300 million years ago to show why plant remains preserved in coal deposits of Europe are identical to those found in coal seams of eastern North America. Snider's maps were an important step in promoting a theory that later became known as

Beaches and Coasts, Second Edition. Richard A. Davis, Jr. and Duncan M. FitzGerald.
© 2020 John Wiley & Sons Ltd. Published 2020 by John Wiley & Sons Ltd.

(a)

(b)

Figure 2.1 Dissimilar tectonic settings produce very different types of coastlines. (a) The coastal plain setting of South Carolina fronted by barriers and tidal inlets is in sharp contrast to the (b) mountainous fjord coast of the Kenai, Alaska.

Continental Drift (the theory that envisions continents moving slowly across the surface of the Earth).

In the early 1900s, the idea of continental drift was popularized by Alfred Wegener, a German scientist at the University of Marburg (Figure 2.2). Wegener was a meteorologist, astronomer, geologist and polar explorer, and led several expeditions to Greenland. He was the first to present a sophisticated and well-researched theory of continental drift, which initially he did through a series of lectures to European scientific societies in 1912. Three years later he published his ideas in a book entitled: *Die Entstehung der Kontinente und Ozeane* (*The Origin of Continents and Oceans*). Wegener believed that all the continents were once joined together in a super continent he called **Pangea** (Greek for "all Earth") which was surrounded by a single world ocean he named **Panthalassa** (all ocean) (Figure 2.3). The northern portion of this super continent, encompassing North America and Eurasia, was called **Laurasia**, and the southern portion consisting of all the other continents was **Gondwanaland**. Partially separating these two landmasses was the **Tethys Sea**.

Figure 2.2 This picture of Alfred Wegener was taken when he was 30 years old, before one of his expeditions to Greenland. It was during this period of his life that he proposed his theory of "Continental Drift." (*Source*: Bildarchiv Foto Marburg, https://en.wikipedia.org/wiki/Alfred_Wegener#/media/File:Alfred_Wegener_ca.1924-30.jpg).

Wegener used a variety of supporting evidence to bolster his theory of a single continent, including the continuity of ancient mountain fold belts and other geological structures that extend across continents now separated by wide oceans. He noted that the coal deposits in frigid Antarctica and glacial sediments in what are now the tropical regions of South Africa, India, and Australia could only be logically explained by moving the continents to different latitudinal settings. He also demonstrated that identical fossils and rocks of similar age could not only be found on widely separated continents but actually plot side by side when Pangea is reconstructed. Using additional fossil evidence, he theorized that Pangea separated into a number of pieces approximately 200 million years ago, forming the Atlantic Ocean among other features. He reasoned that if continents could move, then the presence of glacial deposits in India meant that India had once been close to Antarctica and that following the breakup of Gondwanaland, its northward movement and ultimate collision with Asia caused the formation of the Himalayan mountains (Figure 2.4).

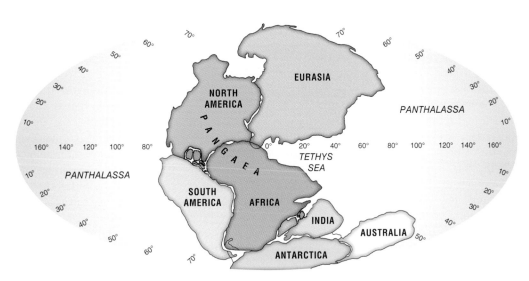

Figure 2.3 One of the compelling pieces of evidence that Wegener used to argue for his Continental Drift theory was the geometric fit of South America and Africa. In fact, he believed that about 200 million years ago all the continents were joined together in a super continent he called Pangea. (*Source*: D.J. Miller, United States Geological Survey.)

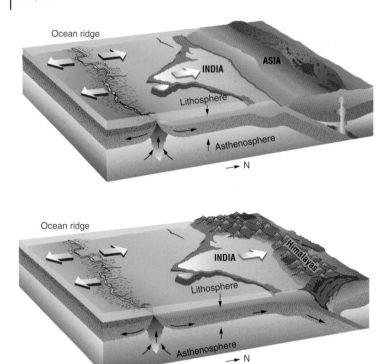

Figure 2.4 According to Wegener (and later verified) the Indian subcontinent traveled northward following the break-up of Gondwanaland and eventually collided with Asia, forming the Himalayan mountains.

Although Wegener's book and ideas were initially very popular, the geological community of that time ultimately scorned his new theory. His failure to convince the scientific community stemmed from his weak argument that the continents plowed their way through or slid over the oceanic crust. He believed that this movement was caused by the gravitational pull of the sun and moon (tidal force) acting with a differential force on the surface crust relative to the underlying mantle. Contemporary scientists showed that the tidal forces were much too weak to account for the drifting of continents and thus his theory was abandoned. Wegener did not live long enough to witness the revival and acceptance of many aspects of his theory as he perished in 1930 at age 50 during a fourth expedition to Greenland while on a rescue mission. Indeed, it was not until the late 1950s and early 1960s that scientists solved the puzzle of the moving continents and Wegener's theory was revived!

In this chapter we will demonstrate that the overall physical character of the edge of continents for several thousands of kilometers of coast, such as the West Coasts of North and South America, is a function of plate tectonic processes. Plate tectonics explains the seismicity of some coasts and the distribution of active volcanism, mountain ranges, a broad coastal plain, or something in between, and whether the coast is fronted by a narrow or wide continental shelf. Plate tectonics will also be shown to have an important influence on the supply of sediment to a coastline and the extent of depositional landforms such as deltas, barrier chains, marshes, and tidal flats.

2.2 Earth's Interior

In order to grasp why continents have "drifted" to new positions, why ocean basins have opened and closed, how mountain systems and ocean trenches have formed, as well as the dimensions of these systems, it is necessary to understand the composition, layering, and processes that occur within the Earth's interior. In its early beginning, the Earth was essentially a homogeneous mass consisting of an aggregate of material that was captured through gravitational collapse and meteoric bombardment. The heat generated by these processes, together with the decay of radioactive elements, produced at least a partial melting of the Earth's interior. This melting allowed the heavier elements, specially the metals of nickel and iron, to migrate toward the Earth's center while at the same time the lighter rocky constituents rose toward the surface. These processes, which have decreased considerably as the Earth has cooled, ultimately led to a layered Earth containing a **core**, **mantle** and **crust**. The layers differ from one another both chemically and physically (Figure 2.5). Most of our knowledge of the Earth's interior comes from the study of seismic waves that pass through the Earth and are modified by its different layers.

The center of the Earth is divided into two zones: a solid inner core (radius = 1250 km)

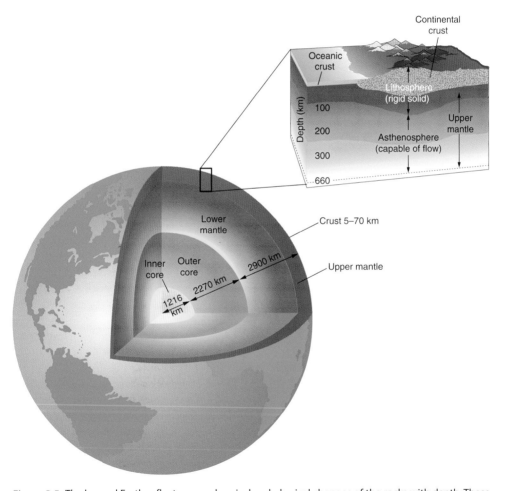

Figure 2.5 The layered Earth reflects gross chemical and physical changes of the rocks with depth. These changes (which are really gradual changes) are responsible for the major boundaries in Earth including the inner and outer core, lower and upper mantle, and thin crust. As the blow-up shows, the lithosphere is composed of oceanic and continental crust and the upper rigid portion of the mantle. The lithosphere is broken up in lithospheric plates that move across and descend into the semi-plastic region of the mantle called the asthenosphere.

having a density almost six times that of the crust, with a temperature comparable to the surface of the sun, and a viscous molten outer core (radius = 2220 km) some four times as dense as the crust. The mantle is 2880 km thick and contains over 80 % of the Earth's volume and over 60 % of its mass. It is composed of iron and magnesium-rich silicate (silica–oxygen structure) minerals, which we have observed at locations where material from the mantle has been intruded into the overlying crust and subsequently exposed at the Earth's surface. The chemical composition is consistent throughout the mantle; however, due to the increasing temperature and pressure the physical properties of the rocks change with increasing depth.

Compared to the inner layers, the outer skin of the Earth is cool, very thin, much less dense, and rigid. The crust is rich in the lighter elements that rose to the surface during the Earth's melting and differentiation stage, forming minerals such as quartz and feldspar. There are two types of crust: the relatively old (up to 4.0 billion years) and thick granitic crust of the continents (20–40 km thick), and the thin and geologically young (less than 200 million years) basaltic crust beneath the ocean basins (5–10 km thick). Continental crust can be up to 70 km thick at major mountain systems. The lower density and greater thickness of the continental crust (2.7 g cm^{-3}) as compared to the thinner, more dense (2.9 g cm^{-3}) ocean crust has important implications for the way that these two crusts have formed and the stability of the crusts.

The divisions described above are based mostly on the different chemical character of the layers.

However, in terms of plate tectonic processes, especially in understanding the movement of plates, the crust and mantle are reconfigured into three layers on the basis of physical changes in the nature of the rocks:

1) **Lithosphere** – This is the outer shell of Earth and it is composed of oceanic and continental crust and the underlying cooler, uppermost portion of the mantle (Figure 2.5).

The lithosphere is approximately 100 km thick and behaves as solid, rigid slab. In comparison to the diameter of the Earth, the lithosphere is very thin and stands comparison to the shell of an egg. The outer shell is broken up into eight major plates and numerous smaller plates. The Pacific plate is the largest plate, encompassing a large portion of the Pacific Ocean, whereas the Juan de Fuca plate off the coasts of Washington and Oregon is one of the smallest. The **tectonic or lithospheric plates,** as they are called, are dynamic and are continuously moving, although very slowly with an average rate varying from a few to several centimeters per year. Whereas once this movement was calculated through indirect means, such as age determinations of oceanic crust, rates can now be measured directly from satellites orbiting the Earth.

2) **Asthenosphere** – The lithospheric plates float on top of a semi-plastic region of the mantle called the asthenosphere, which extends to a depth of about 350 km. In this part of the mantle the high temperature and pressure causes the rock to partially melt, resulting in about 1–2 % liquid. The partially melted rock allows the asthenosphere to be deformed plastically when stress is applied. Geologists have compared this plasticity to cold taffy, hot tar, and red-hot steel. One way of illustrating this concept to yourself is by putting a chunk-sized piece of ice between your back teeth and applying slow constant pressure. You will see that the ice will deform without breaking, the asthenosphere behaves in a similar manner when stressed. In terms of plate tectonics, the semi-plastic nature of this layer allows the lithospheric plates to move.

3) **Mesosphere** – Below the asthenosphere is the mesosphere, which extends to the mantle–core boundary. Although this layer has higher temperatures than the asthenosphere, the greater pressure produces a rock with a different and more compact mineralogy. This portion of the mantle is mechanically strong.

2.3 Plate Boundaries

In a general sense, the edges of the eight major lithospheric plates as well as those of the smaller ones are defined by a number of tectonic processes and geological features (Figure 2.6). Most of the significant earthquakes and volcanic eruptions coincide with plate boundaries, as do major mountain systems, ocean trenches, and mid-ocean ridges. As we will discuss in more detail later, the movement of plates is a consequence of heat transferal from interior of the Earth toward the surface. Because the plates are rigid slabs, when they move over the surface of the Earth they interact with other plates. This produces three types of plate boundaries (Figure 2.7): plates that move apart from one another (Divergent Boundary), plates that move toward one another (Convergent Boundary), and plates that slide past one another (Transform Boundary). It should be noted that a single lithospheric plate can contain many different plate boundaries. For example, the Arabian plate is separating from the African plate along the Red Sea (divergent boundary), which is causing a convergent zone with the Eurasian plate at the Zagros Mountains (along northern Iraq and Iran). Finally, the northwestern edge of the Arabian plate is slipping past the African plate (transform boundary) along the Dead Sea Fault, which coincides the Gulf of Aqaba, Dead Sea, Sea of Galilee, and the border between Israel and Jordan.

2.3.1 Divergent Boundaries

This type of boundary, also called a spreading zone and a rift zone, occurs in the middle of ocean basins, or in the middle of continents (Figure 2.8). Today, most divergent

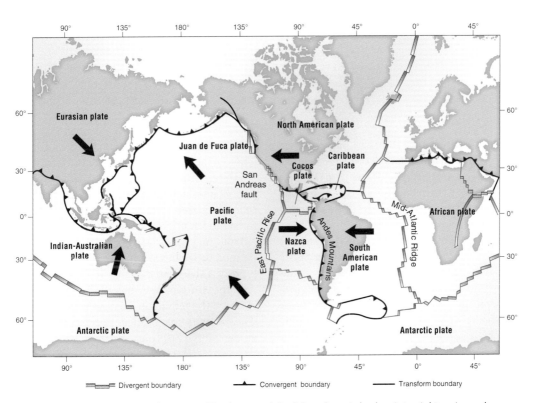

Figure 2.6 The outermost rigid portion of Earth, termed the lithosphere, is broken into eight major and several smaller lithospheric plates. Plates are separated from adjacent plates by divergent, convergent, and transform boundaries. Arrows indicate directions of plate movement.

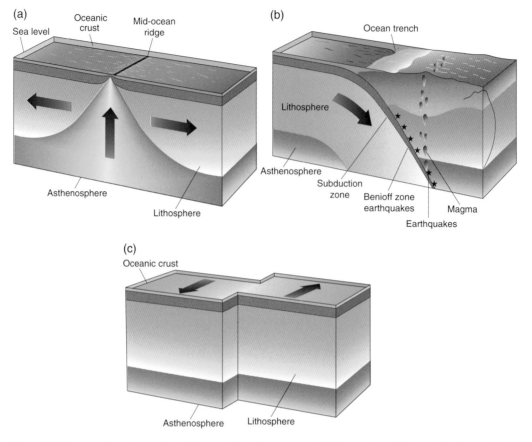

Figure 2.7 There are three types of plate boundaries: (a) convergent, (b) divergent, and (c) transform boundaries.

boundaries are found in ocean basins; however, 200 million years ago extensive rift zones on land produced the breakup of Pangea. In ocean basins, as the two lithospheric plates move apart, the asthenosphere wells up between the diverging plates forming new oceanic crust. The decrease in pressure of the upwelling mantle produces partial melting of the asthenosphere and the formation of molten rock called **magma**. Some of the magma ascends to the sea floor producing submarine volcanoes. The combination of volcanism, intrusion of magma in the overlying ocean crust, and the doming effect of the upwelling mantle, creates a submarine ridge that extends along the length of the divergent boundary. The ridge, which rises approximately 2 km (6600 ft) above the sea floor, rivals some mountain systems on land in size and stature. It is the longest continuous feature on the Earth's surface. The

Mid-Atlantic Ridge, which marks the location where North America and South America separated from Europe and Africa, is only part of an extensive mid-ocean ridge system that winds it way through the world's oceans for some 65,000 km (41,000 miles).

The exact boundary between two plates is defined by a central rift valley. A view of this type of valley can be seen in the northern Atlantic where the rift valley of the Mid-Atlantic Ridge is exposed in western Iceland. The valley is steep-sided and there is active volcanism. The volcanic rock comprising the large island of Iceland increases in age away from the rift valley. Likewise, the age of the oceanic crust on either side of the Mid-Atlantic Ridge gets older toward the bordering continents. This leads to greater sediment accumulation away from the ridge, resulting in a general smoothing of the once irregular, young seafloor and the

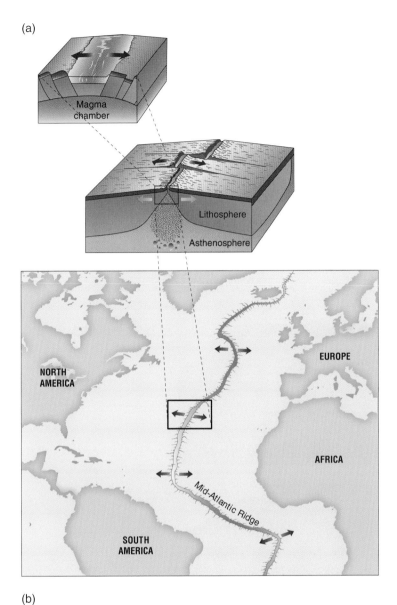

(a)

Magma
chamber

Lithosphere

Asthenosphere

NORTH
AMERICA

EUROPE

AFRICA

Mid-Atlantic Ridge

SOUTH
AMERICA

(b)

Figure 2.8 The divergent boundary of the Mid-Atlantic Ridge system is (a) conceptualized in cross-section and (b) shown as it appears on land where the divergent boundary moves onshore and bisects Iceland. (*Source*: Photo by Albert Hine, University of South Florida.)

formation of the relatively flat abyssal plains. Thus, divergent boundaries are like two conveyers moving newly formed oceanic crust away from a central ridge to cool, sink and become covered with sediment.

2.3.2 Convergent Boundaries

Lithospheric plates moving toward one another are composed of the upper rigid portion of the mantle and topped by either the dense basaltic oceanic crust or the lighter granitic continental crust. This leads to three types of plate convergences (Figure 2.9), each dominated by different tectonic processes and resulting landforms: an oceanic plate colliding with an oceanic plate; an oceanic plate colliding with a continental plate; and a continental plate colliding with a continental plate. The contact between the plates is not always head-on but rather the two plates commonly meet in an oblique convergence.

Ocean–ocean plate convergences occur throughout the margins of the northern and

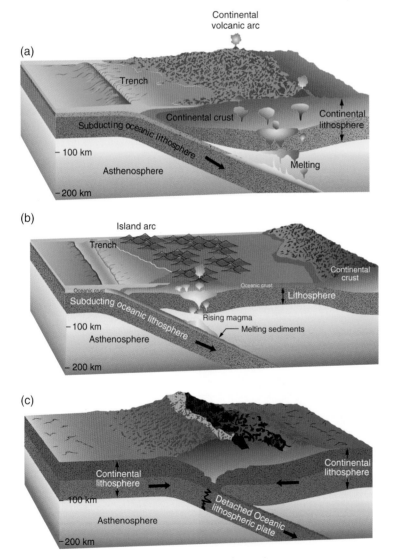

Figure 2.9 Convergent zones involve three types of lithospheric plate collisions: (a) Oceanic–continental plate collision (Nazca and South American plates). (b) Oceanic–oceanic plate collision (Pacific and Philippine plates). (c) Continental–continental collision (Arabian and Eurasian plates).

western Pacific Ocean where the Pacific oceanic plate, moving northwestward, collides with the oceanic crust of the Eurasian and North American plates. This type of plate boundary is also found along the Caribbean Islands and in the Southern Atlantic (South Sandwich Islands). When two oceanic lithospheric plates converge, the older and hence cooler and denser plate descends beneath the younger and more buoyant ocean plate. Thus, the Pacific plate is sliding under the Eurasian and North American lithospheric plates and is descending into the semi-plastic astheno-sphere. This process whereby plates that are created at mid-ocean ridges descend into the mantle and are consumed at convergent zones is called **subduction** (Figure 2.10). The depth to which the plate descends into the mantle and geometry of the down-going slab is known from the numerous earth-quakes that are produced during the subduction process.

One of the major features associated with subduction zones is deep **ocean trenches,** which are caused by the flexure of the downgo-ing plate. Trenches are the deepest regions in the oceans, being 8–11 km below sea-level or 3–5 km deeper than the surrounding ocean floor. They can be thousands of kilometers long. The deepest region on Earth is found in the Mariana Trench (11 km) in the western Pacific, which is more than 2 km deeper than Mount Everest (8.8 km) is high. Ocean trenches are relatively steep on the descending plate side whereas the over-riding plate margin has a shallow slope. As the oceanic plate is sub-ducted, much of the sediment that has accu-mulated on the ocean floor is scraped off and plastered against the adjacent plate margin. This produces an accretionary sedimentary prism that manifests itself as low submarine ridge along the length of the ocean trench.

Paralleling the overriding plate margin is a chain of volcanic islands referred to as an **island arc**. Examples include the Philippine Islands in the western Pacific and the Aleutian Islands that extend 2500 km westward from the Alaskan Peninsula (Figure 2.11). As the subducted plate descends into the mantle, water is released from the downgoing slab. This water lowers the melting point of the already hot sur-rounding rock facilitating the generation of magma at a depth of about 120 km in the overlying asthenosphere. Being less dense than the mantle rock, the magma rises toward the surface. Some of the magma intrudes and solidifies within the overlying ocean crust. A small portion of the magma reaches the surface and erupts on the ocean floor. As this process proceeds, the volcanic pile coupled with the thickening ocean crust produces the island arc. It should be noted

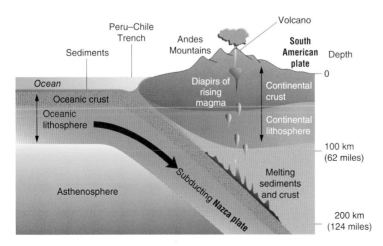

Figure 2.10 Subduction is the process whereby an oceanic lithospheric plate descends into the mantle at a zone of plate convergence. Earthquake activity, volcanism, mountain-building and formation of oceanic trenches characterize subduction zones.

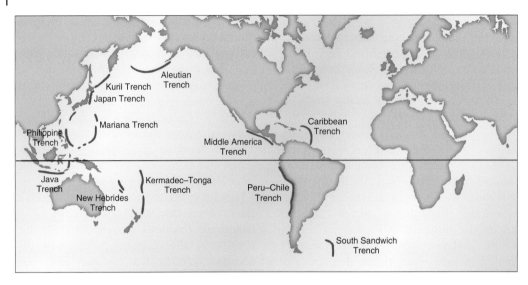

Figure 2.11 Distribution of oceanic trench systems. Trenches are sites of the greatest depths in the oceans.

that the volcanic islands are the surface expression of this arc and comprise only a small portion of the system. As the island arc increases in elevation, more and more sediment is shed from the volcanic arc and some of the sediment is transported to the ocean trench where it is metamorphosed and deformed by the compressive forces of the converging plates. Mature island arcs, such as the Japanese arc, consist of a complex mix of volcanic rocks, deformed sedimentary and metamorphic rocks, and intruded igneous rocks. Because of the varying angle of descending plate, the formation of the island arc along the adjacent plate margin occurs 50–200 km from the ocean trench.

Ocean–continent plate convergences occur where a lithospheric plate containing relatively thin and dense oceanic crust is subducted beneath a lighter, thicker continental crustal plate. As with the ocean–ocean plate convergences, the flexure of the downgoing plate causes a deep ocean trench offshore of the continent. Likewise, dewatering of the subducting slab produces partial melting of the asthenosphere. Although most of the rising magmatic plume solidifies within the overlying crust, some of the magma reaches the surface causing explosive volcanic activity. The combined processes of magma intrusions and volcanic eruptions produce thick continental crust and high mountain systems.

One example of this type of convergent boundary occurs where the Nazca plate, moving eastward, is being subducted beneath the South American plate moving westward. The convergent boundary is marked by the 5900 km-long Peru-Chile Trench. The thickening of the South American plate margin is evidenced by the immense Andes Mountains that reach over 6 km in elevation and are the site of frequent volcanic and earthquake activity. Another ocean–continent convergent boundary is found where the Juan de Fuca and Gorda Plates are descending beneath North America along northern California, Oregon, and Washington, forming the Cascade Mountain chain. The devastating eruption in Washington State on 18 May 1980 in which a cubic kilometer of rock was ejected, lowering the mountain by 410 m (14% of its height) is a product of this subduction process (Figure 2.12). During the initial eruption of Mount Saint Helens 59 people were killed. Mount Rainier, also located along the Cascade chain, beautifies the scenery of Tacoma, Washington (Figure 2.13) and will continue to do so until the mountain erupts, as it inevitably will, potentially sending a devastating wall of mud and volcanic debris toward the city.

In some ocean–continent plate convergences, such as the former western margin of North America 180–80 million years ago, the long-term subduction of oceanic crust led to

(a)

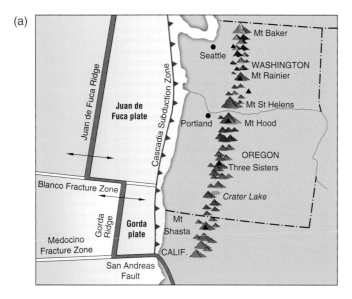

(b)

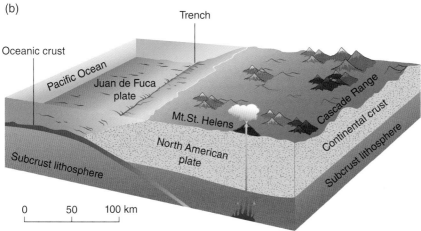

Figure 2.12 Subduction of the Juan de Fuca and Gorda Plates beneath the North American Plate has produced (a) the Cascade Range and is responsible for (b) the 1980 eruption of Mount St. Helens.

the "docking" of numerous micro-continents. These terrains, such as island arc systems, rode the ocean plate eastward toward North America but were not subducted into the mantle because they were too light. Instead, they are added to the continental margin, widening the continent by up to 1000 km.

Continent–continent plate convergences are responsible for the closure of ocean basins and the formation of majestic mountain systems. A geologically young and spectacular example of this type of convergence occurred as a result of the breakup of Pangea when India rifted away from Antarctica and rode a northward moving plate toward Eurasia. The leading edge of this plate consisted of oceanic crust. Several thousand kilometers of oceanic lithosphere were subducted beneath Eurasia before Eurasia was impacted by the subcontinent of India, approximately 45 million years ago. During the period of oceanic–continental convergence, the margin of Eurasia was greatly thickened by magmatic intrusion and volcanic accumulation. As the subcontinent of India approached and the ocean basin closed, the great pile of sediment, that had been shed by rivers draining Eurasia and deposited into the adjacent sea, was bulldozed, along with some of the oceanic crust, onto the continental margin (Figure 2.14). The formation of this

Figure 2.13 View of Mount Rainier from downtown Tacoma, Washington, which is only 60 miles away. Lahars due to volcanic activity could threaten Tacoma and its suburbs. (*Source*: https://volcanoes.usgs.gov/volcanoes/mount_rainier/mount_rainier_multimedia_gallery.html).

accretionary sedimentary prism in combination with the aforementioned crustal thickening produced the highest mountains on Earth, the Himalayas and the Tibetan Plateau. Although we know that the Indian–Australian plate continues to collide with Eurasia, geologists are uncertain what is happening to the down-going portion of the plate. Continental crust is too buoyant to be subducted and therefore most geologists believe that the oceanic part of the Indian–Australian plate has broken from the rest of the plate and continues to descend and be consumed into the mantle.

In addition to the formation of the Himalayas, other continental–continental convergences have occurred in the geological past including the collision between North America and northern Africa (during the formation of Pangea), which was responsible for the development of the Appalachian Mountains. The suture of Europe to Asia formed the north–south trending Ural Mountains.

2.3.3 Transform Boundaries

Transform boundaries occur where crustal plates shear past one another. Unlike the other plate boundaries, lithosphere is neither created or destroyed at transforms. The most common site of transforms is along mid-ocean ridges. The spreading that occurs at ridges is produced by upwelling magma from many different magma chambers. These magma sources are not aligned and may not even be connected. Consequently, the ocean ridges and spreading that occur at these sites are not aligned. A close inspection of mid-ocean ridges, such as the Mid-Atlantic Ridge, shows that while the ridge is continuous, it is offset along its axis. On either side of the offset the newly formed ocean crust moves in opposite directions, but only between the two ridges. Beyond the ridge axis the crust spreads in the same direction. These shear zones are called transforms because the motion along the boundary can terminate or change abruptly. They are sites of shallow earthquake activity.

Transforms not only lie between spreading ocean ridges but also between other types of plate boundaries. One of the most famous is the San Andreas fault, which is a transform boundary between the Pacific and North American plates (Figure 2.15). This transform connects the spreading zone of the Juan de Fuca ridge and a spreading zone centered in the Gulf of California. It is almost 1300 km

(a)

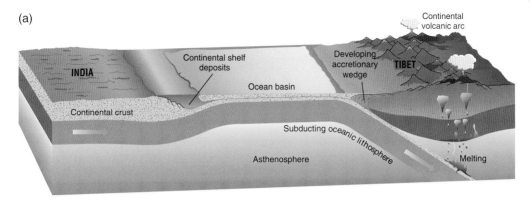

(b)

(c)

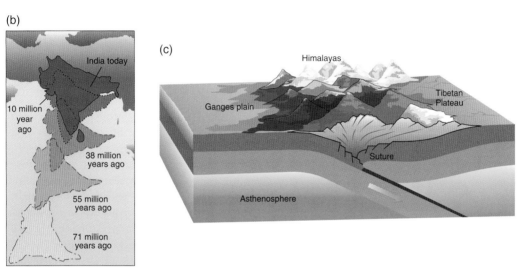

Figure 2.14 Evolutionary model of the Indian–Eurasian continental collision. (a) The actual collision of the two landmasses, which began approximately 45 million years ago, was preceded by the subduction of extensive oceanic lithosphere beneath Eurasia. (b) During this period of oceanic and continental convergence, the margin of Eurasia was greatly thickened through magmatic intrusions, thrust sheets, and an accreting volcanic arc. (c) Ultimately, the collision with India further uplifted the margin of Eurasia producing the majestic Himalayas and the Tibetan Plateau.

(780 miles) long and encompasses a large segment of western California. Because the boundary is located on a continent, the sideways grinding movement between the two plates involves a much greater thickness of crust than oceanic transforms. Hence, the energy released when the plates abruptly slide past one another can be large and catastrophic, such as the great San Francisco Earthquake of 1906. The San Andreas fault has been active for approximately the last 30 million years and lateral movement along the fault has amounted to more than 550 km (340 miles).

2.3.4 Plate Movement

While satellite technology can now accurately measure the exact rate at which North America and Northern Africa are moving apart from one another and the how fast the Hawaiian Islands are approaching the Aleutian Trench where they will be consumed, the mechanism that drives the lithospheric plates, and the continents and island chains that they carry, is much less well-known. Most scientists believe that plate movement is due to some type of convection process produced by the unequal distribution of heat within the Earth.

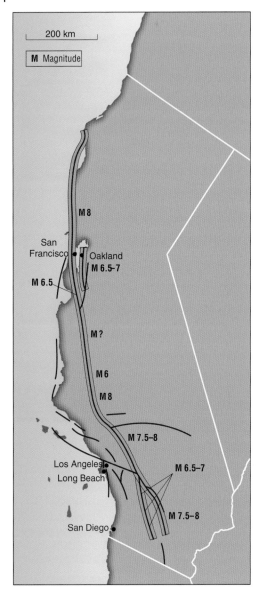

Figure 2.15 The San Andreas fault in western California has been site of some of the largest recorded earthquakes in the Northern Hemisphere. Land on the western side of the transform is part of the Pacific Plate moving in a northwesterly direction, whereas the land on the eastern side of the fault is attached to the North American Plate.

Although considerable heat has been given off since the Earth's original melting phase, additional heat is still being conveyed to the surface. This heat is derived from the molten core and from heat that is produced in the

mantle and crust through radioactive decay of unstable atoms as they change to more stable atomic configurations. The expulsion of this heat likely involves convection cells whereby hot material rises toward the surface and cool material descends (Figure 2.16).

Convection cells operate in a manner similar to the way in which a beaker full of liquid reacts to being heated at the bottom along a single side. As the liquid heats, it expands and rises upward. Upon reaching the surface, the warm liquid moves out laterally toward the opposite side of the beaker and cools. At the same, the void left by the upwelling heated liquid is replaced by cool liquid moving along the bottom. The circulation cell is completed as the moderately cool liquid at the surface descends along the other side of the beaker.

There is still considerable debate concerning the dimensions of convections cells within the Earth and the processes that produce them. Proposed models range from convection cells that are relatively shallow and circulate material within the upper mantle to cells that encompass the whole mantle down to the core boundary. One synthesis of various theories envisions large masses of very hot rock forming at the core–mantle boundary and ascending toward the lithosphere. Upon reaching the relatively cool lithosphere the **mantle plumes**, as they are called, are believed to mushroom and spread laterally. The doming process and shear force of the expanding plume ultimately lead to a rifting apart of the overlying plate. Once spreading (divergent zone) is initiated, movement of the plates is facilitated by a number of forces including:

1) *Gravitational sliding*: Spreading zones are hot elevated regions due to upwelling mantle material. The newly formed lithosphere at these sites is thin and hot. As the plate spreads from the ridge, it thickens, cools, and sinks as the doming effect is lost. Thus, the base of the newly forming lithosphere slopes away from the ridge. This slope may produce a condition

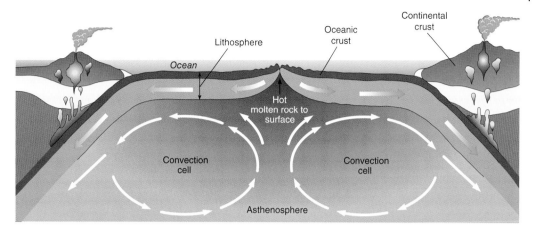

Figure 2.16 A simplified view of the convection process that drives plate motion and is involved with the ascension of magma at mid-ocean ridges and the descent of oceanic lithosphere at subduction zones.

whereby movement of the plate is caused by gravitational sliding.

2) *Slab pull*: Lithospheric plates float on top of the asthenosphere because they are less dense. As stated above, when a plate spreads away from the mid-ocean ridge where it was formed, it gradually cools and thickens. At some distance, the contraction of the lithosphere due to cooling produces a slab denser than the underlying asthenosphere. At this point, the plate descends into the mantle and a subduction zone is formed. It is thought that the subducting plate may pull the rest of the plate with it. These descending slabs have been traced by earthquake activity as deep as 700 km into the mantle.

In trying to assess the importance and viability of the convection processes described above, geologists have made a number of important observations that help to constrain the various models. The North American plate that is spreading westward from the Mid-Atlantic ridge contains no subduction zone, indicating that the slab pull is not a required force in the movement of all plates. Similarly, when ocean basins are first being formed, such as in the Red Sea, there is no topographically high mid-ocean ridge system to produce gravity sliding. Earthquake studies

have also shown that the magma chambers responsible for formation of new lithosphere at mid-ocean ridges appear to be no deeper than 350 km. Therefore, while deep mantle plumes may produce the rifting apart of continents, they may not sustain the spreading process along the entire length of the plate. Submersible investigation of the axial valleys of the mid-ocean ridges reveals evidence of giant cracks and fissures, suggesting that these sites are not a product of magma pushing the plates to the side, but rather the upwelling magma is a passive response to the plates moving apart by some other force. Thus, we are left with a working hypothesis that plates are formed, move laterally, and are subducted into the mantle as a result of some type of convection cell encompassing the lithosphere and mantle, but the details of the process are still being discovered!

2.4 Continental Margins

Continental margins are the edges of continents and the container sides of the deep ocean basins. Geologically, they represent a transition zone where thick granitic continental crust changes to thin basaltic oceanic crust. The margin includes the physiographic regions known as the continental shelf, continental slope, and continental rise (Figure 2.17).

Continental Shelves are the submerged, shallow extensions of continents stretching from the shoreline seaward to a break in slope of the seafloor. Beyond the shelf break, water depths increase precipitously. The average shelf break occurs at 130 m but this depth ranges widely. Continental shelves are underlain by granitic crust and covered with a wedge of sediment that has been delivered to the shore primarily by rivers where it has been redistributed by tides, waves, and shelf currents. In glaciated regions much of the sediment may have been derived through glacial deposition (Georges Bank east of Cape Cod). In equatorial areas the sediment cover may consist of calcium carbonate (main component of sea shells) that is biogenically derived (e.g. exoskeletons of various organisms including coral) or precipitated directly from seawater (particles called oolites, a calcium carbonate concretion).

Although all shelves are relatively flat, their gradients and widths vary considerably. The widest shelves occur in the region surrounding the Arctic Ocean and in a band extending from northern Australia northward toward Southeast Asia. Here the shelves may be more than 1000 km wide. At other sites, such as the eastern margin of the Pacific Ocean, continental shelves are comparatively narrow. The average shelf is 75 km wide and has a slope of 0°07′. This is equivalent to the

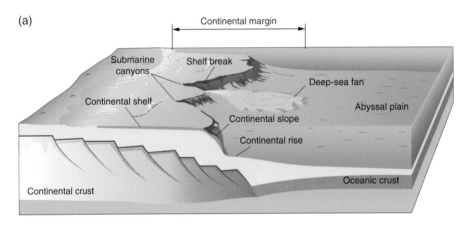

(a)

(b)

TYPICAL DIMENSIONS				
Feature	Width	Relief	Water depth	Bottom gradient
Continental shelf	<300 km	<20 m	<150 m	<1:1,000 (~0.5°)*
Continental slope	<150 km	locally >2 km	drops from 100+−2000+ m	~1:40 (3–6°)
Continental rise	<300 km	<40 km	1.5–5 km	1:1,000–1:700 (0.5–1°)
Submarine canyon	1–15 km	20–2,000 m	20–2,000 m	<1:40 (3–6°)
Deep-sea trench	30–100 km	>2 km	5,000–12,000 m	–
Abyssal hills	100–100,000 m (100 km)	1–1,000 m	variable	–
Seamounts	2–100 km	>1,000 m	variable	–
Abyssal plains	1–1,000 km	0	>3 km	1:1,000–1:10,000 (>0.5°)
Midocean ridge flank	500–1,500 km	<1 km	>3 km	–
Midocean ridge crest	500–1,000 km	<2 km	2–4 km	–

* A bottom gradient of 1:1,000 means that the slope rises 1 m vertically across a horizontal distance of 1,000 m.

Figure 2.17 (a) Physiographic provinces of continental margins including the continental shelf, continental slope, and continental rise. (b) Depths and dimensions of continental margins and other ocean regions. (*Source*: Adapted from B.C. Heezan and L. Wilson, "Submarine geomorphology," in *Encyclopedia of Geomorphology*, R.W. Fairbridge, ed. (New York: Reinhold, 1968); and C.D. Ollier, *Tectonics and Landforms* (Harlow: Longman, 1981).)

slope of a football field (100 yards long) in which one goal line is 6 in. (15 cm) higher than the other goal line; such slopes appear flat to the human eye. However, the topographic expression of continental shelves is not always flat. There can be tens of meters of relief in the form of valleys, hills, sand ridges and other features that can be attributed to the erosional and depositional processes associated with glaciers, rivers, tidal currents and storms.

Continental Slopes mark the edge of the continental shelves. Here the sea floor gradient steepens dramatically as the continental crust thins and is replaced by oceanic crust. Continental slopes descend to depths ranging from 1500 to 4000 m but may extend much deeper at ocean trenches. They are commonly only 20 km in width. Slopes are composed of sediment that forms the outer portion of the tilted sedimentary layers comprising the continental shelf. In regions where oceanic lithospheric plates are being subducted offshore of a continent (e.g. the west coast of South America) or an island arc (the Aleutian Islands off Alaska) the slope may also consist of oceanic sediment that is scraped off the downgoing slab. The gradient of continental slopes varies greatly from 1° to 25°. The average gradient is slightly steeper along the margin of the Pacific Ocean (5°) than in the Atlantic or Indian Oceans (3°) due to the number and extent of deep ocean trenches that are associated with the Pacific margin's subduction zones.

Continental Rises are formed from the transport of sediment down continental slopes and its accumulation in an apron-like fashion at the base of the slope. The thickness of these deposits can be more than several kilometers. The slope of the rise decreases toward the flat abyssal plains and is typically less than 1°. Continental rise widths vary greatly but average a few hundred kilometers. Continental rises are not found where the continental borderlands coincide with subduction zones. In these regions, continental rises are replaced by deep-sea trenches (e.g. the Mariana Trench).

Submarine Canyons are the conduits through which sediment is delivered to continental rises (Figure 2.18). As the names suggests, they are V-shaped, usually steep-walled valleys, and they resemble river-cut canyons on land in both size and relief. Most commonly, they are incised into the edge of

(a)

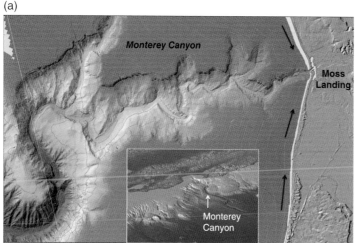

(b)

Figure 2.18 (a) Physiographic drawing of Monterey Canyon off the California coast. Note that canyons are erosional features that form along the shelf–slope break and are found ubiquitously on continental margins throughout the world. (*Source*: USGS, https://www.usgs.gov/media/images/bathymetry-monterey-canyon-and-soquel-canyon-tributary. Inset- *Source*: Brian Romans.) (b) Photograph of sand cascading down the head of a submarine canyon. (*Source*: Courtesy of Robert Dill).

continental shelves and extend down the continental slope to the rise. They occur ubiquitously throughout the world's oceans. Their origin has been the subject of controversy since they were discovered and studied by the late Francis P. Shepard (1898–1985) of Scripps Institute of Oceanography in California. Because of their likeness to river canyons and the fact that some submarine canyons extend across portions of the continental shelf toward major rivers, they were once believed to have formed by river erosion when sea level was much lower than it is today and the shelf edge was exposed. However, some canyons begin at water depths much lower than sea level is ever thought to have dropped. Additionally, the fact that canyons extend across continental slopes requires some mechanism of submarine erosion. It is important to note that the relationship between rivers and submarine canyons cannot be totally discounted. For example, Hudson Canyon off the eastern seaboard of the United States can be traced to the mouth of the Hudson River.

It is now generally accepted that submarine canyons are formed due to erosion by turbidity currents. A **turbidity current** is a sediment-laden torrent of water moving downslope under the influence of gravity. Turbidity currents are initiated along the shelf edge by sediment being put into suspension by a turbulent event such as an earthquake or underwater landslide. As the suspended sediment moves downslope, the sedimentary particles entrain water, forming the turbidity current. In a manner identical to rivers, the moving sedimentary particles, including sand and silt, act as abrading agents and erode the valley, producing submarine canyons. At the base of the continental slope the sediment carried by turbidity currents through canyons is deposited, creating a submarine fan. It is the coalescence of these fans that forms the continental rise. The largest submarine fans in the world coincide with some of the greatest sediment-discharge rivers of the world. The Bengal Fan (2500 km long) and Indus Fan (1600 km long) are a direct product of the Ganges-Brahmaputra and Indus Rivers, respectively, draining huge quantities of sediment from the Himalayan Mountains and the gradual transport of this sediment into deep water by turbidity currents. The Amazon Cone is another very large submarine fan forming from a high sediment-discharge river.

2.4.1 Tectonic Evolution of Continental Margins

Although the morphological character of margins ranges widely throughout the world, much of the variability can be traced to different stages of two evolutionary tectonic styles; the Atlantic and Pacific models (Figure 2.19).

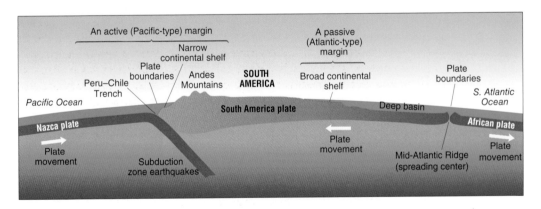

Figure 2.19 Cross-sectional view of Atlantic (passive) and Pacific (active) continental margins using South America as a model.

Atlantic Margin – This type of margin, which is also referred to as a **Passive Margin**, occurs where the edge of a continent coincides with the middle of a lithospheric plate, hence there is no tectonic plate interaction. Because of this location, there is little seismic activity and no volcanism. As the name implies, the eastern margin of North and South America and the western margin of Africa and Europe are examples of this margin type. The margin surrounding India and most of Australia fits into this category too.

During its early history, the evolution of Atlantic-type margins involves the rifting apart of large landmasses such as the breakup of Pangea. In the initial phase the continental crust is bulged upward by a rising mantle plume, perhaps ascending from the core–mantle boundary as discussed earlier in this chapter or from elsewhere within the mantle. This stage may be represented by Yellowstone National Park in northwestern Wyoming and the Rio Grande region of Colorado and New Mexico, where high heat flow values have been measured, and uplift and crustal thinning are occurring. As the plume mushrooms and spreads laterally, the overlying crust is stretched and thinned until it ruptures producing down-dropped continental crustal blocks that form a series of elongated basins. The central basin of this system is called the rift valley, which marks the separating landmasses. In the early stages, basaltic lava derived from the mantle plume may flow onto the valley floor. As the central valley widens and deepens, the ocean invades, forming a linear sea. The Red Sea and Gulf of California are examples of this stage of Atlantic margin development. Continued spreading of the landmasses through the creation of oceanic crust leads to the formation of a new deep-ocean basin and mid-ocean ridge system.

As the ocean basin evolves, so too do the continental margins. The linear basins that were formed by the down-dropped continental blocks during the initial rifting phase become depocenters for sediment delivered chiefly by river systems. As the underlying granitic crust subsides due to a deflation of the mantle plume, the rift valley is ultimately transformed into a juvenile ocean. Shallow marine sediments then cover the land-based, river-lain deposits. The incipient shelf continues to develop and the sedimentary wedge slowly thickens as sediment is shed from the adjacent continent and dispersed by waves, tidal currents, and storm processes. During this phase the slowly subsiding shelf provides accommodation space for continued sedimentation while maintaining a shallow marine platform. Subsidence of the shelf results from cooling and contraction of the mantle plume as well as from the weight of the accumulating sedimentary deposits. In warm water environments, coral reef formation along the edge of the shelf aids the trapping of sediment by providing a barricade to the deep sea. The end product of these sedimentation processes is the formation of a broad, shallow, flat continental shelf.

Along the eastern margin of the United States, oceanographic studies have revealed that some of the sedimentary basins are more than 10 km thick and their average thickness is 4–5 km. The outermost portion of the shelf from Florida to New England contains the framework of a buried coral reef that is several kilometers in height. It appears that this extensive coral reef system lived in relatively shallow water (10–30 m) and was able to grow vertically as the shelf platform slowly subsided. Sometime, approximately 100 million years ago, the coral reefs died and were subsequently buried by shallow marine sands, silts, and clays. Not all Atlantic-type margins are identical to that of the east coast of the United States. The widths and gradients vary greatly, as do the sediment thicknesses and the topography of the shelf. For example, along northwestern margin of Africa the continental shelf is very narrow (30 km wide) only about one sixth the width of the eastern United States. The variability of Atlantic margins is related to the original processes of rifting, the extent of crustal subsidence, the supply of sediment to the shelf, climatic factors (e.g. controlling reef development), the strength of oceanic currents, and other factors.

Pacific Margin – This type of continental margin occurs at the edge of lithospheric plates and thus it is also called an **Active Margin**. These margins are confined primarily to the rim of the Pacific Ocean where oceanic plates are being subducted beneath continental plates. Because they coincide with subduction zones, they are tectonically active and are characterized by earthquake activity and onshore volcanism. Pacific margins have narrower continental shelves and steeper continental slopes than Atlantic margins. Continental slopes of Pacific margins descend deep into adjacent oceanic trenches and thus continental rises are usually absent. An exception to this trend occurs along much of the west coast of North America, south of Alaska, where there is no subduction zone today but where one existed 25 million years ago. Here, sediment transported across the narrow shelf drains through submarine canyons and is building large sediment fans on the floor of the deep ocean.

Differences in the dimensions and morphology of Pacific versus Atlantic margins are the result of contrasting tectonic histories. Whereas Atlantic margins develop wide continental shelves due to rifting, the slow subsidence of broken up continental blocks, and the accumulation of great thicknesses of sedimentary deposits (several kilometers), Pacific margins are narrow and are a product of the compressive forces of an oceanic plate being subducted beneath a continent. As the oceanic plate flexes downward, forming the seaward margin of the ocean trench, the deep sea sediment that overlies the basaltic ocean crust is scraped off and plastered onto the adjacent continental margin. The forces involved in this process chemically alter the sedimentary layers and physically disrupt the sedimentary prism through folding and faulting. The shelf region receives some sediment from material that is eroded from adjacent continental hinterlands and delivered to the coast via river systems. However, the sedimentary wedge comprising Pacific shelf margins is relatively thin when compared to most Atlantic margins.

2.5 Tectonic Coastline Classification

In the early 1970s, plate tectonic theory, which had served to enlighten and transform many subdisciplines of geology, was applied to the field of coastal geology. Two scientists from Scripps Institute of Oceanography in California, Douglas Inman and Carl Nordstrom, produced their now classic work: *On the Tectonic and Morphologic Classification of Coasts*. This scheme provides a first-order characterization of the morphology and tectonic processes of 1000 km-long stretches of continental margin, including not only a description of the coastline but also the continental shelf and the uplands bordering the coast. The classification is based primarily on the tectonic setting of the coast (Pacific- versus Atlantic-type margins). Secondary factors dictating coastline sub-classes include: tectonic setting of the opposite side of the continent; geological age of the coast; and exposure of the coast to open ocean conditions.

It is important to note that this classification is meant as a first-order characterization of a coastline along the length of a continent. There will be many exceptions to the general trends presented here due to secondary factors such as the presence of a major river, the effects of glaciation, or climatic influences, which may have widespread effects too.

The Inman and Nordstrom classification consists of the following (Figure 2.20):

I) Collision Coasts
   A) Continental Collision Coasts
   B) Island Arcs Collision Coasts
II) Trailing Edge Coasts
   A) Neo-trailing Edge Coasts
   B) Afro-trailing Edge Coasts
   C) Amero-trailing Edge Coasts
III) Marginal Sea Coasts

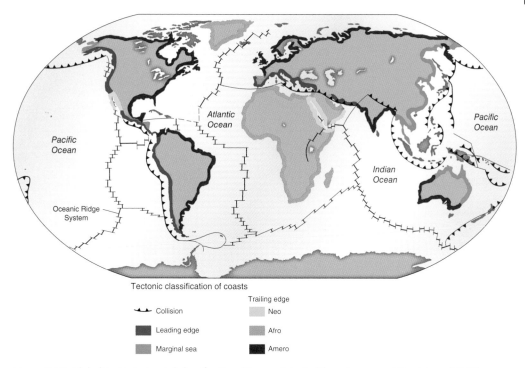

Figure 2.20 Global tectonic coastal classification. (*Source*: Adapted from Inman and Nordstrom (1971)).

2.5.1 Collision Coasts

A. *Continental Collision Coasts* coincide with convergent, Pacific-type margins where oceanic lithospheric plates are being subducted into the mantle beneath the continent. Characteristically, these coasts have narrow continental shelves that terminate next to deep ocean trenches. The coastlines are backed by high mountain systems such as the Andes, a system containing many peaks over 6000 m (Figure 2.21). Mountain-building is a product of crustal thickening and volcanism caused by upwelling magma emanating from the subducted slab. In some locations, including the Coast Ranges along northern California, the mountains adjacent to the coast are relatively low in elevation because they consist of ocean sediments and pieces of ocean crust that were scraped off the subducted plate and thrust onto the continental margin. Located much further inland are the considerably higher Sierra Nevadas (>6000 m), which are similar in origin to the Andes but now consist primarily of granite.

Because the subduction process along most of California ceased tens of millions of years ago, most of the Sierra's volcanic pile has been removed through erosion, exposing the underlying granite plutons. These rocks are evidence of the magma that solidified within the crust during the subduction process.

Continental shelf widths along collision coasts are a function of how steeply the ocean plate descends into the mantle. Along Chile and Peru there is almost no continental shelf because the Nazca Plate dips very steeply beneath South America. In contrast, the continental shelf along the coasts of Oregon and Washington is comparatively wide because the Juan de Fuca Plate descends at a shallow angle beneath North America and material has scraped off the Juan de Fuca Plate to form a moderately narrow shelf.

Collision coasts tend to be rocky due to the scarcity of sediment; where sediment does exist, gravel-sized material is often a major component. This is particularly true at high-latitude coasts due to the effects of glaciation

Figure 2.21 Alaska. Mountainous coast along a tectonically active continental margin.

and frost weathering. The lack of major rivers leads to localized sediment sources that produce isolated accumulation forms such as small barrier spits and pocket beaches. Extensive barrier development is normally absent. Exceptions to this trend occur in regions where glacial meltwater streams transport large quantities of sand and gravel to the coast or near the mouths of moderate to large rivers that exist along collision coasts. The south-central coast of Alaska and the mouth of the Columbia River, respectively, are examples of these conditions.

Collision coasts are not only majestic and ruggedly beautiful, they also experience some of the Earth's most dramatic processes. The southern coast of Alaska exhibits many of the features and processes that typify collision coasts. The Pacific plate is being subducted beneath North American along the south-central margin of Alaska continuing along the Aleutian Islands. This tectonic setting produces volcanism, frequent earthquakes, and large crustal displacements. The coastal mountains of the region are young and their uplift has been rapid. Coastal Alaska is the site of some of the world's largest earthquakes and volcanic eruptions. On March 27, 1964, at the head of Prince William Sound (130 km west of Anchorage) the largest earthquake ever recorded in the

northern hemisphere, registering 8.6 on the Richter scale,[1] shook much of the central Alaskan coast along an 800 km-long tract (Figure 2.22). The effect on Anchorage and many other coastal communities was one of complete devastation, including the loss of 131 lives. The primary shock lasted from three to four minutes and produced large-scale slumps, landslides, and avalanches. The earthquake also created tsunamis (seismic sea waves) that completely wiped out native communities in Prince William Sound and large sections of several seaports including Valdez and Seward. These giant waves are not only be triggered by movements of the ocean floor, but in case of some Alaskan fjords, they also may be caused by earthquake-induced gigantic rock falls that crash into the heads of deep, elongated water bodies, sending walls of water toward the mouth of the fjords (see Box 2.1). The great 1964 Alaskan earthquake produced permanent changes along much of the central coast due

---

1 Scientists have re-evaluated the 1964 Alaskan earthquake and determined that it was a magnitude 9.2, second only to the 9.5 Chilean earthquake of 1960. The new calculations are used for very large earthquakes whose signal saturates the recording instruments. The moment magnitude, as it is called, is based on the amount of fault slippage and surface area of the plates involved.

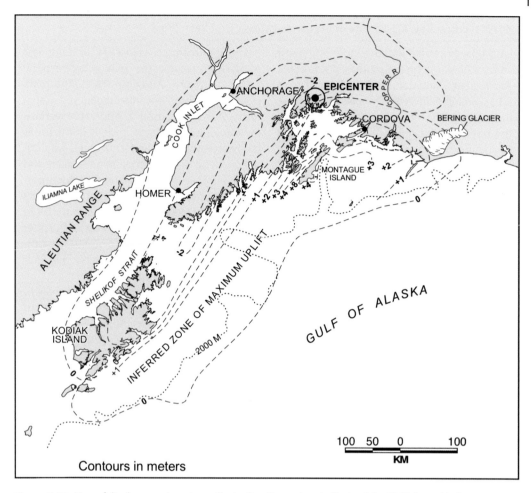

Figure 2.22 Map of displacement contours illustrating the regional effects of the 1964 Great Alaskan Earthquake. Note that portions of the Alaskan coast east of Anchorage experienced uplift, whereas areas southwest of Kodiak Island underwent subsidence.

to uplift and subsidence (down-faulting). In the Prince William Sound region, the Island of Montague rose as much as 9 m, and there are many other areas along the coast in which results of the uplift can be seen today in the form of raised beaches and intertidal platforms. Contrastingly, 200 km southwest of Anchorage, the earthquake caused a general subsidence of 2–3 m along the Kenai Peninsula, resulting in drowned forests.

In addition to being earthquake prone, coastal Alaska experiences continual volcanism, particularly along the Alaska Peninsula where at least 17 volcanoes have erupted during the past 10,000 years. Some of these volcanoes have erupted with a tremendous ferocity. In 1912, Mount Katmai erupted, ejecting 15 km^3 of volcanic debris. Much of the ash was blown southwestward blanketing portions of Kodiak Island, 150 km away, with up to a half meter of ash. This enormous eruption is the third largest during recorded history with only Tambora and Krakatoa in Indonesia being larger. Truly Alaska is a very active collision coast.

B. *Island Arc Collision Coasts* are similar to continental collision coasts and are active tectonically with characteristic volcanism and earthquakes. They differ from continental collision coasts because the convergence

Box 2.1 Lituya Bay, Alaska: Site of the Largest Waves Ever Seen (Box)

Lituya Bay is located on the rugged southeast coast of Alaska along the southwest side of the Fairweather Mountain Range (Box Figure 2.1.1). Frenchman Jean Francois La Perouse first explored the region in 1788. He noted that the bay provided good harborage although navigating the strong currents at its mouth was hazardous. He was unaware, however, of the apocalyptic waves that frequent this site. From the coast the bay extends 13 km inland, where it is bordered by mountains and glacier-filled valleys (Box Figure 2.1.2). In his study of the region Don J. Miller of the U.S. Geological Survey concluded that the bay was once occupied by the confluence of at least two major glaciers, which had deepened the valley and deposited a moraine at its mouth. Slight reworking of the terminal moraine created a spit-like landmass, La Chaussee Spit, which narrows the bay's entrance, forming a tidal inlet about 300 m across. Two inlets, Gilbert Inlet and Crillon Inlet, open onto the head of Lituya Bay. Gilbert and Crillon Inlets are currently enlarging as Lituya Glacier retreats to the northwest and North Crillon Glacier recedes to the southeast, respectively (Box Figure 2.1.3).

The mountainous landscape of southeast Alaska is the result of the collision between the Pacific and North American Plates. The plate boundary in the Lituya Bay region is a **transform** that has caused considerable vertical motion as well horizontal slip. It is known as the Fairweather fault and it coincides with the overall trend of Lituya and Crillon Glaciers and the inlets that form the upper T-shaped portion of the bay. On the evening of July 9, 1958 at 10:16 p.m. a section of the fault ruptured approximately 23-km southeast of upper Lituya Bay producing an 8.3 magnitude earthquake. Measurements taken along the fault revealed vertical ground displacements of 1.05 m and horizontal slip of 6.5 m. The earthquake was felt throughout southeastern Alaska as far south as Seattle, Washington and as far east as Whitehorse, Yukon Territory, Canada, covering an area of over 1,260,000 km^2.

On the evening of July 9, 1958, three fishing boats were anchored at the entrance to Lituya Bay. The boatsmen were enjoying the late evening sunlight that accompanies Alaskan summers. Only four people lived to tell the story of a giant wave that filled the bay. At the onset of the earthquake the ground around

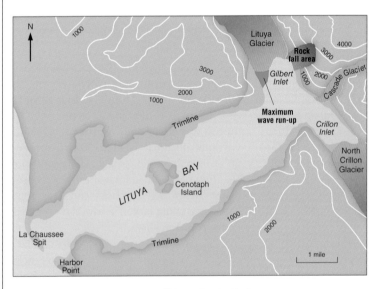

Box Figure 2.1.1 Location map of Lituya Bay in Alaska.

Box Figure 2.1.3 Aerial view of trimline cut by the seismic wave. The northern end of Cenotaph Island is in the foreground. (*Source*: Photo by D.J. Miller.)

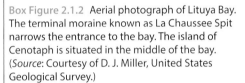

Box Figure 2.1.2 Aerial photograph of Lituya Bay. The terminal moraine known as La Chaussee Spit narrows the entrance to the bay. The island of Cenotaph is situated in the middle of the bay. (*Source*: Courtesy of D. J. Miller, United States Geological Survey.)

Lituya Bay shook violently and continued doing so for about a minute. Eyewitnesses from the fishing vessels indicate that about 1–2.5 minutes after the earthquake began there was a tremendous explosion in the upper bay. The noise they heard came from the mountain slope at the head of Lituya Bay giving way, sending 30 million cubic meters of rock into northeast side of Gilbert Inlet (Box Figure 2.1.4). Failure of the mountainside was caused by ground tremors, which weakened the high-angle bedrock slope that had been oversteepened by glacial erosion. Due to the steepness of the mountain slope (75–80°) the slab of rock fell almost directly into the inlet from an average height of about 610 m (2000 ft). The huge mass of rock plunged through the water column cratering the sediment bottom of the bay. The displaced water produced a massive wave that propagated across Gilbert Inlet traveling up the opposite cliff-side to a maximum height of 530 m (1740 ft) (Figure 2.1.4). Here the force of the

wave stripped away the vegetation and the underlying soil down to the bedrock surface. Where trees were rooted deeply in bedrock, the trunks, some of which were more than a meter in diameter, were sheared off as if they were small twigs. Scientists have theorized that generation of a wave of this magnitude would be similar to that caused by the impact of a small asteroid.

After carving away much of the southeast mountainside of Gilbert Inlet the giant wave swept down Lituya Bay with a speed of between 156 and 209 km h^{-1} (97–130 mi h^{-1}) advancing toward the fishing boats anchored at the mouth. As the wave approached the first fishing boat, Mr. Ulrich and his young son described the onslaught of a black wall of water that crested over a 30 m (100 ft). The giant wave snapped the boat's anchor chain and propelled the craft and occupants toward the bay's south shore where they were saved by the wave's backwash, which carried the boat and passengers back to the middle of the bay. There they encountered several smaller but still terrifying waves, some reaching over 6 m (20 ft) in height. Miraculously, the Ulriches sailed out Lituya Bay next day under the boat's own power. A second boat with Mr. and Mrs. Swanson aboard anchored closer to the

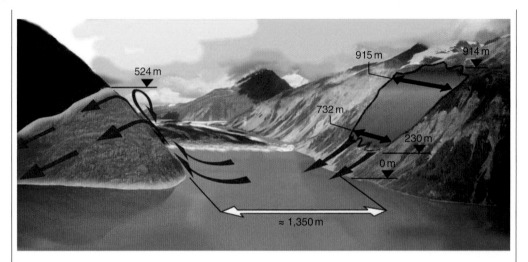

915 m

914 m

524 m

732 m

230 m

0 m

≈ 1,350 m

Box Figure 2.1.4 Oblique aerial photograph looking into Gilbert Inlet (fjord) illustrating where the slab of rock fell into the bay (shown in red) producing a wave that trimmed trees to an elevation of 524 m. (*Source*: Heller 2014, http://www.mdpi.com/2077-1312/2/2/400/htm. Licensed under CC BY 3.0.)

northern bay shoreline was not so lucky. The Swanson's fishing trawler was picked up by the giant wave and carried out the bay like a swimmer on a surfboard. By the Swanson's account as the trawler surfed over La Chaussee Spit the tops of the trees were 25 m (82 ft) below them. Once deposited in the ocean their boat immediately began to sink. Fortunately, they were able to climb aboard their dinghy and were picked up by a passing fishing boat a few hours later. The third boat was engulfed by the monstrous wave and neither crew nor boat was ever found.

The immense wave of July 9, 1958 was not the first extraordinarily large wave to sweep through Lituya Bay. On October 27, 1936 four people witnessed three waves traveling down the bay ranging in height from 15 to 30 m (50–100 ft). These waves cut a neat trim line along the tree-covered slopes surrounding the bay. Other trim lines at even higher elevations have been reported by Don J. Miller of the U.S. Geological Survey, indicating that giant waves repeatedly crashed along the bay shoreline. Trimlines have been found at elevations of 60–150 m (200–490 ft) (Box Figure 2.1.3). It is interesting to note that the 1936 waves were not related to an earthquake and no bedrock scar corresponding to a rockfall was ever found. Scientists have suggested that other mechanisms such as the drainage of a subglacial lake or the frontal collapse of a glacier may also trigger giant wave formation. Lituya Bay is a beautiful anchorage site. However, boaters beware of the roar in the upper bay that portends the coming of the great wave!

is between two oceanic plates. These coasts are backed by low to moderately high mountains and are fronted by narrow continental shelves that are bordered by deep ocean trenches. Most island arc collision coasts are located in the northern and northwestern Pacific and include the ocean coasts of the Aleutian Islands and the Philippine Islands. Japan is another collision island arc coast,

which experiences major earthquakes (Kobe, Japan) and volcanic eruptions.

2.5.2 Trailing Edge Coasts

A. *Neo-Trailing Edge Coasts* are geologically young coasts (<30 million years old) that have formed as a consequence of continental rifting (Figure 2.23). Examples include the

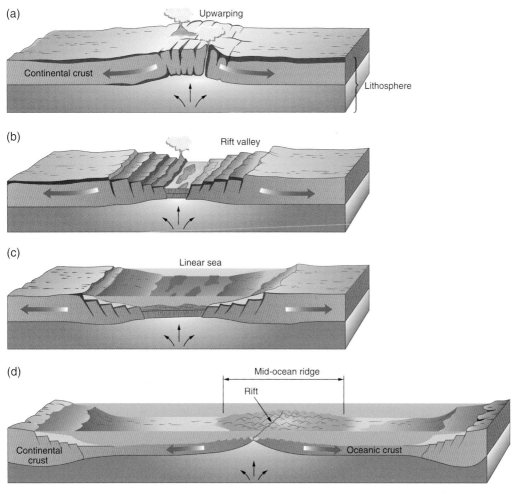

Figure 2.23 Four-stage model of rifting of a continent and establishment of an ocean basin: (a) Doming and stretching of lithosphere produced by upwelling magma; (b) Spreading causes rupture of lithosphere and creation of a rift valley due to block faulting; (c) Continued spreading, subsidence and ocean crust formation causes flooding of the rift valley and creation of a linear sea; (d) Long-term spreading creates an ocean basin.

coasts surrounding the Red Sea, Gulf of Aden and Gulf of California. These *linear seas* are a stage in the evolution of new ocean basins and are summarized below:

Stage #1a. Mantle plume causes doming and stretching of overlying continental crust.

Stage #1b. Alternatively, plate motion causes passive rifting and ensuing mantle upwelling.

Stage #2. Thinning crust ruptures and down-faulted blocks create a topographically low rift valley.

Stage #3. Subsidence and widening of rift valley causes invasion of ocean and formation of a linear sea.

Stage #4. Long-term sea floor spreading leads to ocean basin formation.

Neo-trailing edge coasts are rugged and coastal borderlands tend to have narrow to non-existent coastal plains with adjacent mountains. Sediment tends to be scarce along these coasts and there is little barrier development (Figure 2.24). Where sand is locally abundant, narrow mainland beaches or pocket

Figure 2.24 Baja California. Neo-trailing edge coasts tend to be rocky and mountainous with few beaches and barriers. (*Source*: Courtesy of Miles Hayes.)

beaches form. A major exception to this trend occurs in the Gulf of California along the mainland coast of Mexico where there are extensive coastal lowlands and several barrier island chains. Neo-trailing edge coasts experience frequent low-magnitude earthquakes due to crustal adjustment associated with past rifting. Volcanism is absent.

The Gulf of Suez is a relatively narrow body of water which connects the northwestern end of the Red Sea to the Suez Canal and the Mediterranean Sea. The Sinai Peninsula, which forms its northeast shoreline, is typical of neo-trailing edge coasts. The coast is bordered by a narrow hilly region, which gives way to high mountains. Little sediment reaches the coast and thus depositional landforms are mostly absent. The arid climate of the region and the lack of rivers contribute to this condition. In the Ras Mohammed area, at the southern tip of the Sinai, the barren landscape along the coast belies the abundant sea life immediately offshore. Not more than a 100 m from the shoreline, a robust coral reef community provides some of the best SCUBA diving in the world.

B. *Afro-Trailing Edge Coasts* differ from Amero-trailing edge coasts in that the opposite side of the continent is not a collision zone. Rather, the opposite continental margin coincides with the middle of a plate and thus it is also a trailing edge coast. Afro-trailing edge types are found along the east and west coasts of Africa, the southwestern half of the Australian coast, and the coast of Greenland. The lack of a collision zone along the east or west coast of Africa means that there is no large-scale organization of river drainage within the continent. Consequently, sediment delivery to the margin of Africa, as well as other Afro-trailing edges, has only been local and generally does not compare in magnitude to the quantity that has been transported to Amero-trailing edges. This condition is also a function of differences in climate. Similar to Amero-trailing edges, Afro-trailing edges are tectonically inactive with minor earthquake activity and no volcanism.

Due to the sedimentation history of Afro-trailing edges, they exhibit a great deal of morphological variability. For example, in some regions, such as most of the northeast coast of Africa, the continental shelf is quite narrow (<25 km wide), whereas along much of its southwest coast, shelf widths approach 100 km. Coastal borderlands along Africa range from coastal plains and hilly settings to cliffs and low mountains. Likewise, there are estuarine, barrier and deltaic coasts where sediment is abundant and other long stretches of coast that are rocky and barren of sediment.

The southern portion of South Africa is a good example of the variability of Afro-trailing edge coastlines. The port of Cape Town is a mixture of lowland areas surrounded by several flat-topped low mountains composed of layered sedimentary rocks. The coastline south of Cape Town extending to the Cape of Good Hope is rugged, with high cliffs and low mountains. The Indian and Atlantic Oceans meet at this site and their unlimited fetches combine to produce huge 6 m and higher waves that crash upon the rocks sending salt spray 40 m high, to the top of the cliffs (Figure 2.25). East of this region is False Bay where wave abrasion of sandstones has led to extensive sand accumulation. At the western end of the inner bay the beach is 300 m wide and fronted by a very wide surf zone. The abundance of sand at this locality is most clearly demonstrated by the extensive dune system backing the beach where individual

Figure 2.25 The coast of South Africa where large waves carve away at sandstone cliffs. In this region beaches only occur in embayments where sediment can accumulate under high wave-energy conditions.

dunes reach 10 m in height and the dune field extends several kilometers inland. Contrastingly, a few tens of kilometers from this site the coast consists of steep rocky cliffs devoid of any sediment.

C. *Amero-Trailing Edge Coasts* occur along passive, Atlantic-type margins in which the opposite side of the continent is a collision coast. The mountain chains associated with collision coasts organize the drainage of the continent such that the major rivers flow from the mountains and away from the collision coast, across the continent, and discharge their loads along the passive margins. Amero-trailing edge coasts include the eastern margins of North and South America, the Atlantic coast of Europe, and the coast of India. These are geologically old margins, in which long-term deposition of sediment has led to the development of wide, low-profile coastal plains and wide continental shelves. The boundary between the coastal plain and shelf is simply a function of sea level. This boundary has changed dramatically during the past two million years as sea level has fluctuated in response to the growth and decay of the continental ice sheets. When the ice sheets advance and sea level falls, the coastal plains expand; contrariwise, when the ice sheets shrink and sea level rises, the continental shelf widens landward.

Amero-trailing edge coasts tend to be tectonically inactive with relatively few major earthquakes; the ones that do occur are of low magnitude (<4 on the Richter Scale). The Charleston, South Carolina Earthquake of 1886, magnitude 7.3, is a major exception to this trend. These coasts also lack volcanic activity.

One of the primary characteristics of Amero-trailing edge coasts is their depositional landforms including barrier island chains, broad sediment-filled lagoons, marsh systems, tidal flats, and river deltas. For example, almost the entire east coast of the United States, south of glaciated New England, is fronted by barrier chains interrupted by only a few major re-entrants, such as Delaware and Chesapeake Bay

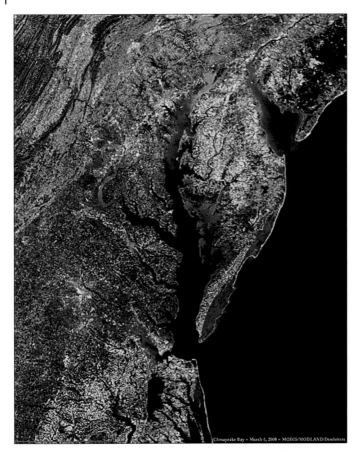

Figure 2.26 Much of the east coast of the United States is characterized by a coastal plain setting fronted by lagoons, marsh systems, and barrier chains. The drowned river valleys of Delaware (top) and Chesapeake Bay (bottom) are major exceptions to trend.

(Figure 2.26) and several mainland beaches, including Myrtle Beach, South Carolina and Rehoboth Beach, Delaware. Some of these barrier systems, such as the Outer Banks of North Carolina, are separated from the mainland by wide shallow bays (Albemarle Sound and Pamlico Sound), whereas other barrier chains, including those along much of South Carolina and Georgia, are backed by extensive marsh and tidal creeks. Although there are no active river deltas along the East Coast today, many existed in the past including the Santee River delta in South Carolina. This delta is no longer building and is now eroding because much of its river discharge, and hence its sediment source, was diverted into Charleston Harbor in the late 1800s.

The East and West Coasts of the United States illustrate well the sharp contrasts in morphology that different tectonic settings produce (Table 2.1).

2.5.3   Marginal Sea Coasts

Along much of the western and northern Pacific Ocean, a series of island arcs, including the Aleutians, Kuril Islands, Japan, and Philippine Islands, separates the edge of continents from the open ocean. These protected shorelines of Alaska and Asia are defined as marginal seacoasts (Figure 2.27). Characteristically, they are fronted by shallow water bodies such as the Bering Sea, Sea of Okhotsk, Sea of Japan, East China Sea, and South China Sea. In the Atlantic Ocean, the

Table 2.1 Comparison of the East and West Coasts of the United States.

Feature or Process	East Coast	West Coast
1) Tectonic Class	Amero-trailing edge coast	Collision coast
2) Earthquakes	Small-magnitude earthquakes	Site of large earthquakes
3) Volcanism	None	1980 eruption of Mount St. Helens. Other mountains in Cascade Range also active
4) Shelf width	Wide and flat	Narrow and steep
5) Coastal borderland	Coastal plain	Mountains, some very high
6) Sediment supply-	Long-term erosion of Appalachian Mountains provided sediment to the coast through numerous moderately sized rivers.	Coming mostly from short steep gradient rivers
7) Coastal morphology-extensive	Depositional coasts with extensive barrier island development. Numerous bays, lagoons, marshes, tidal flats tidal deltas, and estuaries.	Rocky coast with mostly pocket beaches and small spits.
8) Wave energy-	Low (0.7–0.8 m)	Moderate (1.5–1.7 m)
9) Tidal range-	Micro- to macro-tidal (1–6 m)	Micro- to macro-tidal (1–3 m)
10) Description of car trip along the seaboard	South of New York exceedingly boring due to flatness of coastal plain. Only excitement is Pedro's "South of the border" signs.	Magnificent coastal vistas and view of mountain ranges.

Gulf Coast and east coast of Central America are also marginal seacoasts and are protected by the Caribbean island arc as well as by the landmasses of Florida and Cuba.

Marginal seacoasts exhibit considerable variability depending largely upon the geological history of that portion of the continent, but overall, they tend to be more similar to Amero-trailing edges than any other class of coast. For example, many of the world's largest sediment-discharge rivers occur along these coasts, including the Mississippi and Magdalena rivers, which empty into the Gulf of Mexico and the Caribbean Sea and the Huang Ho, Yangtze and Mekong rivers, which discharge along the Asian continent. Interestingly, despite this apparent abundant supply of sediment, long barrier chains are sparse along the marginal seacoast of Asia. This condition may be related to the lack of a coastal plain and low-slope inner continental shelf, which would provide a continuous platform upon which barrier island chains could develop and migrate onshore during the period of rising sea level. Contrastingly, the Gulf coast of the US contains a wide, flat inner shelf and coastal plain, and barriers occur ubiquitously in this region. Along the marginal seacoasts of Asia, much of the sediment may go into filling valleys rather than building barriers.

Generally, marginal seacoasts experience low-to-moderate wave conditions due to their semi-protected nature afforded by the shallow seas and offshore island arcs. However, these coasts can suffer damaging short-term, high wave-energy conditions during the passage of major storms, such as hurricanes.

2.6 Tectonic Effects on Coastal Sediment Supply

The tectonic history of a continent strongly influences the distribution and abundance of sediment along coastlines, which in turn controls the size and extent of depositional features, such as barriers, deltas, marshes,

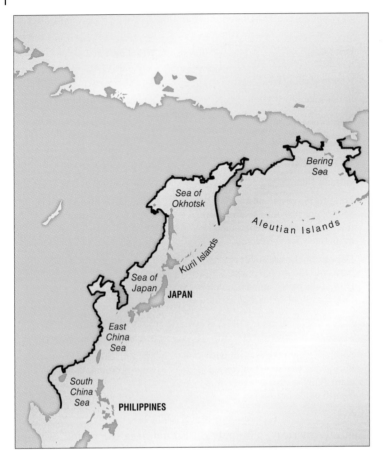

Figure 2.27 Most of the marginal sea coasts of the world are found along the borders of the northern and western Pacific Ocean where the Pacific Plate is being subducted beneath North American and Eurasian Plates.

and tidal flats. These influences can be on a scale of continental-wide drainage and the widths of continental shelves or they can be more regional such as dictating the position of a river mouth.

2.6.1 Continental Drainage

Drainage of a continent is determined by the distribution of major mountain ranges, which is a product of the continent's tectonic history. For example, the collision between the eastward-moving Nazca plate and the westward-moving South American plate produced the Andes Mountains, which span the entire length of its Pacific margin. The eastern margin of South America has experienced little mountain-building activity since the opening of the South Atlantic, approximately 130 million years ago. Thus, the Andes mountain range is the major drainage divide that runs the length of South America separating rivers that flow east from those that drain west (Figure 2.28). Eastward-flowing rivers, including the Amazon, Parana, Orinoco, and others, drain more than 90% of South America. Their headwaters begin along the eastern flank of the Andes and flow long distances across the continent before discharging large quantities of water and sediment along the Atlantic margin. Conversely, the west coast of South America receives relatively small volumes of sediment delivered by short, steep-gradient rivers. The diminutive size of the rivers reflects their small drainage basins resulting

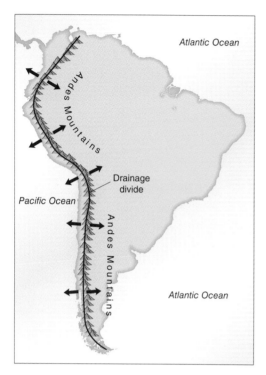

Figure 2.28 Physiographic map of South America illustrating that the Andes Mountains produce a drainage divide along the entire length of the continent. This asymmetric drainage pattern results in all the major rivers discharging their sediment loads along the Atlantic Ocean and Caribbean Sea, whereas only relatively small rivers empty into the Pacific Ocean.

from the proximity of the adjacent Andes Mountain drainage divide.

The African continent illustrates a very different drainage condition. Because most of Africa is in the middle of a plate, there is little active mountain-building and most of the existing mountain systems are geologically very old. Unlike South America, there is no large-scale organization of river drainage in Africa. Rather, Africa has moderate to small-sized rivers that discharge along its Mediterranean coast (e.g. Nile River), Atlantic coast (e.g. Niger and Congo Rivers), and Indian coast (e.g. Zambezi River). Due to the relatively small volume of sediment reaching the African coast, there are fewer depositional features along this coast as compared to the East Coast of South America. Certainly

the effects of climate and the overall lesser amounts of precipitation falling on Africa than South America contribute to this condition.

River drainage in the United States is strongly affected by the tectonic history of North America. Similar to South America, the western margin of the United States was a collision zone up until 30 million years ago resulting in the formation of the Rocky Mountains, Sierra Nevada Mountains, and Basin and Range. It is the Rocky Mountains that form the western drainage divide of the Mississippi River, whose drainage basin comprises two thirds of the United States. The eastern divide is defined by the Appalachian Mountains that were formed when the landmasses forming Pangea were being assembled.

As demonstrated for three different continents, the delivery of sediment to the coast is governed by the size of the rivers discharging along the coast. This, in turn, is a function of climatic factors and drainage basin size, which is largely controlled by the tectonic history of the continent (Figure 2.29). One other factor that influences the supply of sediment to a coast is the geology of the drainage basin. For example, the collision between the Indian and Eurasian plates created the Himalayan Mountains and the Tibetan Plateau, some of the highest landscapes on the surface of the Earth. The elevation of the Himalayas coupled with ongoing glaciation of this region produce huge quantities of sediment. The rivers draining the Himalayas including the Indus, Ganges, and Brahmaputra Rivers, rank as some of the largest sediment-discharge rivers in the world.

2.6.2 Location of Rivers

In addition to the quantity of sediment delivered to the coast, tectonic processes dictate the location of many large river systems. As described previously, the break-up of a continent involves several plumes of magma rising from the mantle that dome, stretch, and eventually rift apart the overlying lithosphere. If we consider a single mantle plume, the doming and rifting process produces a three-arm

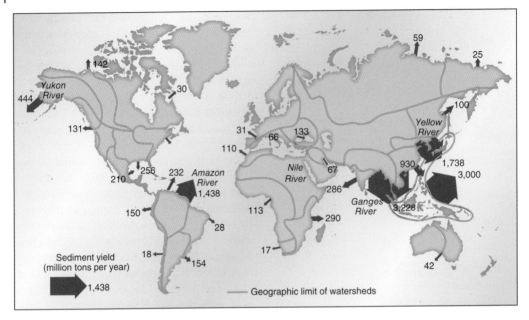

Figure 2.29 Pattern of sediment-discharge to the world's oceans. These estimates are based on suspended sediment loads and not contribution by bedload. Note that most of the suspended sediment discharged to the ocean by rivers occurs along Amero-trailing edge and marginal sea coasts. (*Source*: Adapted from Milliman and Meade, 1983.)

tear in the lithosphere (Figure 2.30). As the mantle plume spreads laterally beneath the lithosphere, two of the rift valleys widen and deepen while the third arm of the rift becomes inactive. The third arm fails because the continued spreading of the other two arms of the rift relieves the pressure of the upwelling mantle plume. An example of this process is seen where the Arabian Peninsula has rifted away from northeast Africa. The Indian Ocean has invaded the two active arms of the rift forming the Red Sea and Gulf of Aden. The *failed third arm* of this system is the East African Rift that extends into the African continent. Most of this region is a low-lying and parts of the valley are below sea level. The valley is bordered by escarpments and is still tectonically active with infrequent earthquakes and volcanism.

Many mantle plumes were involved in the break-up of Pangea, each having its own three-arm rift system. The two active arms of each rift widened, elongated and connected to one another eventually producing the separation of entire continental-sized

landmasses such as North America separating from Africa. In many cases the failed third arms of these systems coincide with sites where the internal drainage of a continent empties into the sea. It is believed that because failed third arms are topographic lows that extend into continents, major rivers either evolve at these locations or existing rivers migrate to these sites. The Niger River valley in western Africa and the Amazon River valley in eastern South America are examples of these developments. The Mississippi River valley may also be the location of a geologically very old rift valley.

Another, unique form of tectonic control of river drainage and sediment supply to a coast occurs in the Indian Ocean. The triangular shape of the Indian sub-continent dictated that when it collided with Eurasia 50 million years ago, not only was a sizable mountain system produced with a large potential sediment supply, but also the shortest route for drainage to the sea was on either side of India. Thus, the high sediment-discharge rivers of

Figure 2.30 Four stage model in the development of internal drainage of a continent: Time 1. Doming and thinning of lithosphere due to upwelling magma; Time 2. Rupture of lithosphere and development of a three-arm rift system; Time 3. Linear sea development while third arm of rift system becomes inactive; Time 4. Evolution of ocean basin, interior of continent drains through old rift valley.

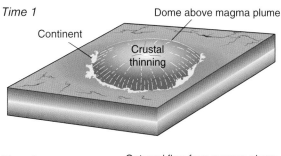

Time 1

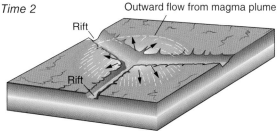

Time 2

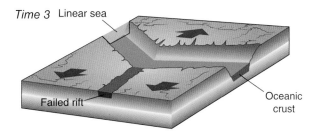

Time 3

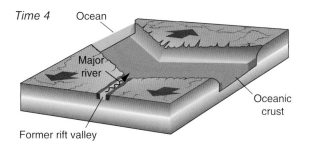

Time 4

the Indus draining into the Arabian Sea and the Ganges-Brahmaputra emptying into the Bay of Bengal is a predictable outcome.

2.6.3 Continental Shelf Width

To a large extent, plate tectonic setting dictates the width and slope of the continental shelf, with Pacific (active) margins tending to have narrow, steep shelves and Atlantic (passive) margins having relatively wide, shallow gradient shelves (Figure 2.31). Shelf width and slope, in turn, influence the effects of storms, wave energy and sediment dispersal, as well as cliff abrasion rates along rocky coasts. Along collision coasts, such as the west coasts of North and South America, the narrow shelf widths result in large wave energies and large potential rates of sediment transport. Remember, too, that these margins have localized sediment supplies

Figure 2.31 Physiographic map of southern South America illustrating the difference in shelf width between the active western margin and the passive eastern margin. Wide continental shelves lead to lower wave energy and submarine canyons being located further offshore as compared to narrow shelves.

Continental shelf width also influences sediment supplies along continental margins by dictating the proximity of submarine canyons to the coast. For example, along the coast of California the narrow shelf has led to canyons being located only several hundred meters seaward of the beach. During storms and other high wave-energy events, sand is eroded from beaches at these sites and transported offshore. Some of this sand makes its way into the heads of submarine canyons, such as La Jolla and Scripps Canyons, where eventually it is transported to deep sedimentary basins. In contrast, this process does not occur along the eastern margin of North America because the submarine canyons of this region are too far from shore to capture inner shelf sands.

2.7   Summary

Plate tectonics provides a means for explaining the overall features and major processes on the surface of the Earth including the character of the world's continental margins and its coastal settings. We are able to understand why coastal regions on either side of a continent can be so markedly different and why the West Coast of the United States is bordered by mountains and experiences volcanic eruptions and frequent earthquakes whereas the tectonically inactive East Coast abuts a flat coastal plain and is fronted by an almost uninterrupted chain of barriers. It has also been shown that the tectonic history combined with climatic factors determines the drainage of a continent and how much sediment is delivered to a particular coast. Even the location of individual major rivers has been shown to be a function of plate tectonic processes. In subsequent chapters it will be shown that coastlines are molded by numerous physical, chemical, and biological processes, but all these operate on the backbone of the plate tectonic setting of the region.

derived by relatively small river systems. Thus, sediment accumulation forms along these types of settings are dependent on nearby sediment input and embayments where sediment can collect.

A very different pattern exists along Amero-trailing edges and marginal sea coasts that are fronted by wide continental shelves. In these regions, much of the deep-water wave energy is attenuated by friction imparted by the shallow gradient continental shelf. Longshore and offshore sediment transport along these coasts occurs during the passage of major storms. Accretionary landforms are common along these coasts and are coincident with abundant supplies of sediment.

Reference

Inman, D.L. and Nordstrom, C.E. (1971). On the tectonic and morphologic classification of coasts. *J. Geol.* 79: 1–21.

Suggested Reading

Bird, E.C.F. (2009). *Coastal Geomorphology*. Chichester, UK: Wiley.

Cox, A. and Hart, R.B. (1986). *Plate Tectonics: How it Works*. Palo Alto, CA: Blackwell Science.

Davies, J.L. (1980). *Geographical Variation in Coastal Development*. Harlow, UK: Longman.

Davis, R.A. (1994). *Geology of Holocene Barrier Island Systems*. London: Springer-Verlag.

Glaeser, J.D. (1978). Global distribution of barrier islands in terms of tectonic setting. *J. Geol.* 86: 283–297.

Hayes, M.O. and Kana, T. (1978). *Terrigenous Clastic Depositional Environments, University of South Carolina, Technical Report No. IICRD*. Columbia, SC: Dept. of Geology, University of South Carolina.

Masselink, G. and Gehrels, R. (2014). *Coastal Environments and Global Change*. Chichester, UK: Wiley and American Geophysical Union.

Moores, E.M. (ed.) (1990). *Shaping the Earth: Tectonics of Continents and Oceans*. New York: McGraw-Hill.

Woodroffe, C.D. (2003). *Coasts: Form, Process, and Evolution*. Cambridge, UK: Cambridge University Press.

3

Sediments and Rocks

Materials of Coastal Environments

We learned in the previous chapter that the earth's crust has a wide range in composition, that is, it comprises many different rock types. Most of the volume of the crust is igneous rock, but metamorphic rocks and sedimentary rocks are also important. Most of the earth's surface is sedimentary rocks and sediments. All of these rock types and their contained assemblages of minerals may occur in the coastal zone. Rocky coasts are widespread along the shorelines of the world and are generally characterized by erosion (see Chapter 18). In this book, we are more interested in the depositional coastal environments where various types of sediment accumulate. The sediments that characterize these depositional environments, such as beaches, dunes, estuaries and deltas, are products of the destruction of bedrock through various physical and chemical processes we call *weathering*.

These weathering products will be transported on or near the earth's surface, eventually coming to rest as an accumulation of particles which either settle due to a change in surface conditions, or are precipitated from solution (physicochemical action) or by an organism (biochemical action). Even in the coastal zone, some of these particles become buried and are cemented into sedimentary rocks, but most remain as unconsolidated particles of various sizes while at or near the earth surface.

In this chapter both the textural and compositional aspects of sediments are discussed, along with a general treatment of the rocks from which they came.

3.1   Rock Types

The *igneous rocks* that comprise most of the earth's crust form from magma that originates in the mantle. They are composed primarily of what are called silicate minerals. Silicates are compounds of a variety of cations of various elements combined with silicon and oxygen; common elements in silicate compounds are iron, magnesium, potassium, calcium, sodium, and aluminum, although there may be others. Depending upon where these minerals form within the earth's crust they may result in granite, diorite, gabbro, basalt or other igneous rock types.

Weathering of igneous rocks can produce sediment particles of varying sizes that eventually become buried and cemented to form sedimentary rocks such as arkose, quartz sandstone or mudstone. Heat and pressure may be applied to these rock types to produce *metamorphic rocks* or so-called "changed rocks." These include gneiss, schist, slate and others.

In some cases *sedimentary rocks* can form from shells and other skeletal material, typically composed of calcium carbonate: the

Beaches and Coasts, Second Edition. Richard A. Davis, Jr. and Duncan M. FitzGerald.
© 2020 John Wiley & Sons Ltd. Published 2020 by John Wiley & Sons Ltd.

minerals calcite, aragonite and/or hi-Mg calcite. This produces limestone, which, when metamorphosed, becomes marble.

All of these rocks, as well as other, less common, types may occur in various combinations to produce rocky coasts. More importantly to the development of depositional coastal environments, they produce sediment as the result of physical and chemical weathering and abrasion. These sediment products, regardless of where they form, many eventually find their way to the coast where they accumulate in various coastal environments. The bulk of our discussion in this chapter will focus on these sediments; their size, composition, and arrangement within various coastal settings.

3.2   Sediment Texture

The production of sedimentary particles during erosion can be related to many phenomena including climate, waves, organisms, gravity processes such as rock falls or landslides, stream transport or others. In this discussion no attention will be paid to the composition of the sediment particles; instead the focus is on their size, shape and distribution as they accumulate in various coastal environments.

3.2.1   Grain Size

When determining the grain size of sediment particles it is usual to examine a sample of many grains. If we think about how we are going to try to measure so many grains we can see problems. Though there are multiple ways whereby one might assign a size to a sediment particle there are three common measures: particle diameter, volume or mass. Mass is rather easy to deal with because shape is not a factor and it can easily be related to the density of the material. However, because of density differences, mass is not practical as a parameter of grain size. Measurement of the volume of large number of grains is both difficult and time-consuming because it requires us to treat each grain individually. The diameter of a particle is far easier to measure and, although there is an infinite range of particle shapes, ranging from a sphere to a flat disc, and though relatively few perfect spheres are produced as erosion products, many sediment particles exhibit a rough approximation to a sphere.

The range in grain size of sedimentary particles is almost infinite, from less than a micron to a meter in diameter to boulders as large as an automobile. Obviously, more than one technique must be utilized to measure such a wide spectrum of grain sizes, especially because grain size analysis involves extremely large numbers of grains. This places further constraints on size-analysis techniques. Commonly, grains larger than a couple of centimeters or so in diameter are measured individually with calipers or some similar method. Grains ranging from a centimeter down to about 50 µm in diameter are commonly measured by sieving them through as sequence of several sieves that separate the grains into rather narrow size classes arranged in order of sieve opening from largest on top to smallest at the bottom. Another method is by allowing the grains to settle through a column of a fluid, usually water. The rate at which particles settle is proportional to their size, shape, and density, although certain assumptions have to be made. Settling tubes in which this type of analysis is conducted have become common in recent years and may be quite sophisticated, with direct connection to computers for rapid data processing. Sediment particles smaller than 50 µm may also be measured by settling rates, although such methods are time-consuming and not very accurate.

Currently, instruments utilizing photoelectric sensors, lasers, or x-ray beams have been developed for measuring grains ranging from tens of microns to a few millimeters in diameter.

Because sedimentary particles display such a wide range of sizes, it is practical to utilize a geometric or logarithmic scale for size

Table 3.1 Wentworth grain size scale, including Krumbein's phi scale.

	Millimeters (mm)	Micrometers (µm)	Phi (φ)	Wentworth size class		Rock type
	4096		−12	Boulder	Gravel	Conglomerate/ Breccia
	256		−8	Cobble		
	64		−6			
	32		−5	Pebble		
	4		−2			
	2		−1	Granule		
	1		0	Very coarse sand	Sand	Sandstone
1/2	0.50	500	1	Coarse sand		
1/4	0.25	250	2	Medium sand		
1/8	0.125	125	3	Fine sand		
1/16	0.0625	63	4	Very fine sand		
1/32	0.031	31	5	Coarse silt	Silt	Siltstone
1/64	0.0156	15.6	6	Medium silt		
1/128	0.0078	7.8	7	Fine silt		
1/256	0.0039	3.9	8	Very fine silt		
	0.00006	0.06	14	Clay	Mud	Claystone

classification. Almost 100 years ago a scale was devised by William Wentworth based on a factor of 2, such that as one moves from unity there is a multiplier or divisor of 2; one category is either twice or half the diameter of the next one. Later modifications resulted in the grain size scale as it is used today in which the values used are the $-\log_2$ (Table 3.1). W.C. Krumbein of Northwestern University came up with this conversion in 1937. He designated the simple numbers produced by this formula as *phi units (ø)*. Each size class possesses a name, e.g. fine sand, as well as a size range.

3.2.1.1   Statistical Analysis of Grain Size Data

After sediment has been subjected to one of the standard methods for measuring grain size, the measurements are then used to calculate statistical parameters which describe the population of particles in the sediment sample. Data obtained from the size analysis are usually in the form of weight percentages for each size class. These raw grain-size data may be plotted as a histogram or as a frequency curve (Figure 3.1). A cumulative curve (Figure 3.2) is also used so that specific percentage values can be taken from it and used to calculate sediment parameters that characterize individual samples.

The distribution of grain sizes in a sediment population typically follows or approaches a lognormal distribution, resulting in a normal or bell-shaped curve (Figure 3.3a). In a symmetrical curve for sediment distribution the mean, mode and median are all the same value; 50% or the perfect middle. The *mean grain size* is the average, the *median* is the statistical middle grain size and the *mode* is the most abundant grain size. Most sediments do not have a perfectly symmetrical grain size distribution but are "skewed." The *skewness* refers to the overall shape of the distribution curve. Those samples that have more coarse grains than fine ones are called skewed to the left and those with more fines than coarse grains are right-skewed (Figure 3.3b).

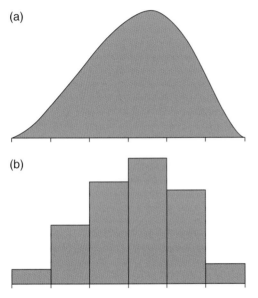

Figure 3.1 Graphs showing (a) the distribution curve and (b) a histogram format for the grain size distribution of a sediment sample.

Two statistical parameters are commonly used to describe grain size distribution: the *mean grain size* and the *standard deviation*. Such parameters are easily derived from the raw data by statistics software. The mean is the statistical average, determined in just the same way as a teacher determines the average score on an examination. The sizes of every grain are summed and the total is divided by the number of grains in the population. The *standard deviation* is the statistical determination of the deviation from the mean. In other words, were all of the values bunched together (a small standard deviation) or were they spread out along a wide range (a large standard deviation)? When dealing with sediments we generally refer to this measurement as the *sorting*. A well-sorted sediment has most of the grains about the same size, whereas a poorly sorted sediment has a wide range in particle sizes that comprise the population or sample (Figure 3.4). Quantitative values for sorting can be obtained using the key percentage parameters for a cumulative curve. These values also have verbal designations (Figure 3.5)

Another statistical condition that commonly occurs in natural accumulations of sediment is the mixing of populations. Examples include the combination of quartz

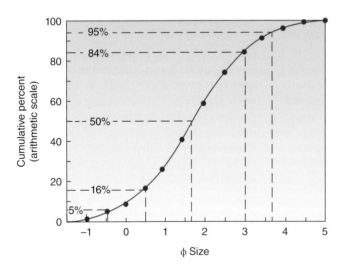

Figure 3.2 A cumulative curve for a sediment sample with the percentage values that are used in calculating statistical data for the samples.

(a)

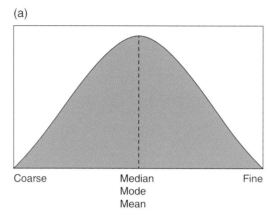

Coarse   Median   Fine
                  Mode
                  Mean

(b)

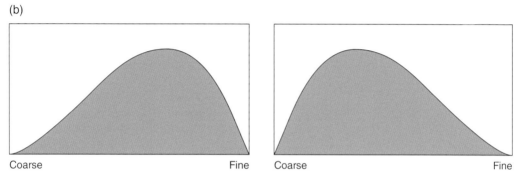

Coarse   Fine   Coarse   Fine

Figure 3.3 (a) A symmetrical curve for a sediment with the mean grain size, the median, the mean and the mode all falling at the 50% position, and (b) simple plots showing both fine-skewed (A) and coarse-skewed (B) grain size distribution.

sand with shells or pebbles (Figure 3.6) which is common on many beaches. Each mode is derived from different sources and each has a very different grain size; the quartz is sand and the shells are gravel. Such sediment is said to be *bimodal* (Figure 3.7). Each population represents a mode, that is, a large number of particles of about the same size as the shells—the coarser less abundant mode, and the sand—the finer, more abundant mode (Figure 3.6).

3.2.2 Grain Shape

Sediment particles display a great variety of geometries in their shape. This variation is due to a combination of the internal structure of minerals that comprise the particles plus the origin and history of the particle.

Some have simple and symmetrical shapes, whereas others are extremely complex.

The *roundness* of a particle refers to the sharpness or smoothness of its edges and corners. Both physical abrasion and chemical reactions contribute to this characteristic, although abrasion is generally the most important of the two. Descriptive names for this characteristic range from "very angular" to "well rounded" (Figure 3.8). Figure 3.9 shows a well-rounded, moderately sorted gravel deposit.

The term *sphericity* refers to the degree to which a particle approaches a sphere. Although many ways of determining sphericity are available, it is most common to compare the lengths of three mutually perpendicular axes. The closer this ratio is to unity(1) the more spherical is the particle

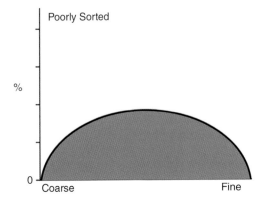

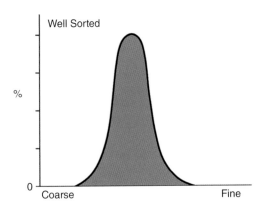

Figure 3.4 Plots of a poorly sorted and a well-sorted sediment.

relatively thin skin near the surface of this crust. Although there is a wide variety of minerals present in these rocks, they contain primarily four mineral types: feldspars, quartz, clay minerals, and carbonates, with feldspars being the least abundant of the four.

This difference in abundance between the crust as a whole (mostly igneous) and sedimentary rocks is largely a reflection of the relative chemical stability of the minerals present. Iron–magnesium silicate minerals weather readily and are not common in sedimentary rocks. Most feldspars are relatively stable, and they weather to produce clay minerals. Quartz is quite stable chemically and is physically very durable (Figure 3.12). Thus, during the cycle of weathering and transport, quartz tends to persist because it is hard and chemically inert at surface conditions. *Carbonate minerals* are dominantly the result of physicochemical or biochemical precipitation from solution at or near the surface of the earth. Most of the carbonate material that we see in coastal sediments is in the form of skeletal material; typically shells (Figure 3.13).

The result is that sediments and sedimentary rocks largely comprise a few minerals that are the relatively stable products of erosion.

(Figure 3.10). Sphericity is more strongly influenced by the origin of the particle than is roundness. Some grains are inherently elongate or flat because of their crystallographic or biogenic makeup (Figure 3.11). Examples are such minerals as mica, which tends to be flat or disc-shaped, and many shell types, such as bivalves or branching corals.

3.3 Mineralogy

Although the crust of the earth is composed of a wide variety of minerals, relatively few are present in great abundance. Those that are include the feldspars, iron–magnesium silicates such as pyroxenes and hornblendes, and quartz, with feldspars being most abundant. Sedimentary rocks are present as only a

3.4 General Origin and Distribution of Sediments

Having discussed the broad aspects of the composition of the earth's crust and some aspects of sediment textures, we can now turn our attention to where sediments on the coast come from and what controls them in their specific environment of deposition. There are several factors that determine how sediments are distributed throughout the coastal zone. Some of the important ones are:

- composition of coastal sediments is a direct reflection of the composition of the source materials that produced them;
- dispersal of sediments is related to the processes that transport them from their source and throughout the coastal zone; and

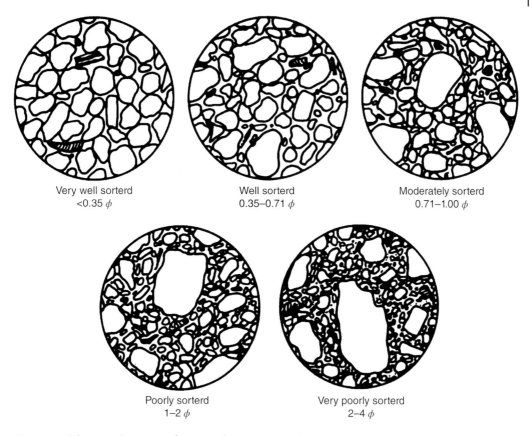

Very well sorterd
<0.35 ϕ

Well sorterd
0.35–0.71 ϕ

Moderately sorterd
0.71–1.00 ϕ

Poorly sorterd
1–2 ϕ

Very poorly sorterd
2–4 ϕ

Figure 3.5 Schematic depictions of a range of sorting categories.

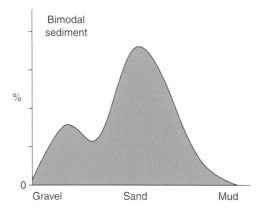

Figure 3.6 Plot showing a bimodal grain size distribution.

- grain size and sorting of the sediments are a reflection of the rigor of the processes acting in the environment in which they accumulate and the length of time involved.

3.4.1 Composition

Minerals and rocks that comprise the sediments of the coastal zone reflect the composition of the rocks and sediments that are in the source area from which these sediments come. On islands there is no significant fluvial system to distribute sediment. The high relief and short distance to the coast permits sediment particles of unstable composition to be carried downslope to the beach, as in the Pacific Islands (Figure 3.14). In most situations, sediments are produced by some type of weathering and erosion, then they are transported by a river to the coastal zone where they may end up in a river delta, are carried to the open coast then distributed by currents and waves, or are dumped into an embayment along the coast. Fine sediments are commonly dominated by clay minerals, sand is generally mostly quartz but might also have feldspar,

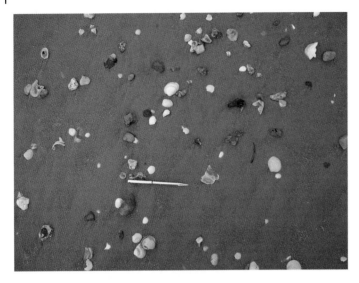

Figure 3.7 Beach sediment with a coarse shell mode and a fine sand mode.

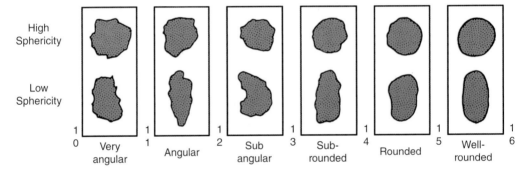

High Sphericity

Low Sphericity

1 0   Very angular   1 1   Angular   1 2   Sub angular   1 3   Sub-rounded   1 4   Rounded   1 5   Well-rounded   1 6

Figure 3.8 Plot of a spectrum of grain roundness appearances and values.

Figure 3.9 Example of a very well rounded gravel that is moderately sorted.

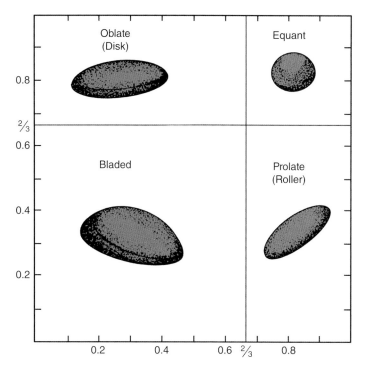

Figure 3.10 Grain shape diagram showing the four main categories as developed by Zingg. (*Source:* Adapted from Zingg (1935).)

Figure 3.11 Photograph of very spherical quartz grains.

Figure 3.12 Photograph showing a pure quartz sample.

rock fragments and other minerals, and gravel could be similar to sand in composition although rock fragments tend to be the most common. Coasts in glaciated areas show a similar match to source materials. The glacial sediments of these areas contain a wide range of grain sizes and sediment particle compositions (Figure 3.15). The coastal sediments derived from these materials also have this spectrum of textures and compositions. The New England coast is a good example of this situation (see Chapter 17).

Figure 3.13 Photograph of a carbonate shell accumulation with a wide range of shapes. (*Source:* Courtesy of Alex Simms.)

Regardless of how they originate or are transported to the coast, all of the coastal environments of the Bahamas are composed of a single mineral; $CaCO_3$ (calcium carbonate). This is because there is nothing but calcium carbonate in the entire Bahamas Platform; hence we can be certain that the islands are isolated from any landmass that could contribute other mineral and rock compositions.

3.4.2 Texture

The processes that dominate a particular environment in the coastal zone are responsible for the grain size of the sediments that

Figure 3.14 Tahiti. A Boulder and cobble beach made of volcanic material.

Figure 3.15 Exposure of Pleistocene glacial sediment composed of a wide range of grain sizes and compositions.

accumulate there and, to an extent, the grain shape (Figure 3.16). Mud carried by rivers ends up in the river delta, in bays where the rivers discharge their load or it may be carried offshore onto the continental shelf and beyond. In general, mud and other fine sediments tend to accumulate in places where waves and currents are absent or weak. Sand tends to accumulate where wind, waves and currents are relatively strong (Figure 3.17). This includes beaches, dunes, tidal inlets and other open coastal environments. In some places this sand is brought to the coast by rivers, or it may erode from rocks along the

coast, or it may be carried to the beach from offshore by waves.

In general, we can say that the grain size of the sediment along the coast is directly proportional to the rigor of the processes operating in the environment of sediment deposition.

3.5 Summary

The materials that make up the coast are fundamental in determining the nature of the coast. Obviously, one of the most important factors is whether the coast is dominated by bedrock or by sediment. Because sediments are transported by a wide variety of processes, and because there is a wide range in the intensity of these processes, they are important factors in determining the morphology and scale of the various coastal environments. Low-energy processes tend to accumulate sediments that are fine grained and/or those that are not well sorted. More energetic conditions result in sand and gravel that tends to be well sorted. Regardless of the nature or intensity of the processes operating along a particular coastal location, sediment is the product of the source materials in that vicinity.

Figure 3.16 Coarse beach gravel with a wide range of grain shapes and compositions.

Figure 3.17 Beach on the coast of Maine composed of sorted sand derived from shell debris. (*Source:* Courtesy of Joe Kelley.)

Box 3.1 The Late Robert G. Dean

The coastal world lost one of its greats in early 2015 with the passing of, Dr. Robert Dean, Graduate Research Professor Emeritus at the University of Florida (Box Figure 3.1.1). Bob Dean was born in Wyoming and lived there and in Colorado as a boy. He and his family moved to Long Beach, California when he was a teen, and after high school he attended a local community college. He received his BS in Civil Engineering from the University of California – Berkeley. His graduate work resulted in a MS in Oceanography from Texas A&M University and a PhD in Civil Engineering from the Massachusetts Institute of Technology.

Most of Bob's career was spent on the faculties of the University of Florida and the University of Delaware where he was a major player in the development of two of the most prominent coastal engineering programs in the United States. He was named a Distinguished Research Professor at Florida in 1992. Bob supervised numerous graduate students while at both of these institutions, many of whom are now leaders in the coastal research community. He was also very eager to help people at other institutions, both faculty and students.

Bob was a member of the prestigious National Academy of Engineering in recognition of the many contributions to the field of coastal engineering. He also served on the Beach Erosion Board of the U.S. Army, Corps of Engineers, and was a member of multiple committees of the National Academy of Sciences. Over the years he consulted for more than 100 firms and agencies all over the world. He was a prolific author on a range of coastal topics. These include the books titled *Coastal Processes with Engineering Applications*, and *Wave Mechanics for Engineers and Scientists*, both written with former student Dr. Robert A. Dalrymple.

Unlike many of his coastal engineering colleagues, Bob was interested in sediment and he recognized that geologists could make a contribution to coastal research. One of his important contributions, based largely on his work on the Florida coast, was the impact that tidal inlets made on beaches. He frequently stated that at least half of the erosion problems experienced by beaches could be related to tidal inlets.

Box Figure 3.1.1 The late Robert G. Dean.

Reference

Zingg, T. (1935). Beitrage zur Schotteranalyse: Min. *Petrog. Mitt. Schweiz.* 15: 39–140.

Suggested Reading

Boggs, S.N. (2011). *Principles of Sedimentation and Stratigraphy*, 5e. Englewood Cliffs, NJ: Prentice-Hall, Inc.

Davis, R.A. (1992). *Depositional Systems; an Introduction to Sedimentology and Stratigraphy*, 2e. Englewood Cliffs, NJ: Prentice-Hall, Inc.

Manaan, M. and Robin, M. (2015). *Sediment Fluxes in Coastal Areas*. Dordrecht: Springer.

Middleton, G.V. (ed.) (2005). *Encyclopedia of Sediment and Sedimentary Rocks*, 2e. Dordrecht: Springer.

4

Sea-Level Change and Coastal Environments

When sea level changes, for any reason, the coast responds with a change of its own. Some changes in sea level are very abrupt and can be recognized easily. Others are shown by sophisticated instruments. There are sea-level changes that take place globally but there are also those that are local or regional in extent. This chapter will consider the causes of sea-level change, the extent over which they occur, and also the rates at which they take place.

Sea level is changing throughout the world but it is doing so very slowly as a consequence primarily of climate change and plate tectonics. It is also changing locally by a wide variety of means. In just a few moments, an earthquake can lower sea level by lifting the earth at the shore, or can raise it by causing the ground to sink (Figure 4.1). All of these phenomena cause the position of sea level to change in comparison to patterns that shift over the seasons and as a result of the shifting positions of ocean currents. The El Niño phenomenon also causes changes in current patterns that can change sea level seasonally. In 2016 this occurred and influenced the Gulf Stream causing sea level to rise substantially on the southeast coast of the United States near Miami.

Slow and long-term changes in sea level on scales of decades to millennia take place as glaciers enlarge and subsequently melt in response to global changes in climate. The removal of the great mass of these glaciers then permits the crust to rise due to the removal of the large ice mass, thus producing a lowering of sea level (Figure 4.2). The isostatic adjustment of removing the immense ice mass from the land mass results in rebounding of the crust, which is still occurring in parts of Scandinavia. Such changes in elevation result in old shorelines being many meters above present sea level. A good example of this is on the shoreline of Hudson Bay in Canada (Figure 4.3). Here, numerous shorelines chronicle the rebound of this area after the ice sheet melted.

These kinds of sea-level change are all relative, that is the position of sea level has changed relative to the surface of the crust. They also occur locally or regionally as opposed to globally. These local relative changes may be limited in extent to hundreds of meters but they may be regional, extending over a thousand kilometers or more.

A eustatic sea-level change takes place throughout the world. For such a change to occur there must be a change in the volume of water in the world ocean or there has to be a change in the overall size, or shape of the ocean container, i.e. the crustal basins in which the oceans occur. For example, the melting of glaciers due to global warming will add water to the ocean system and thereby cause sea level to rise globally. This has been going on since the melting of continental ice sheets began about 18,000 years ago (Figure 4.4). In addition, the warmer water resulting from climate change takes up more space than colder water, again causing relative sea level to rise. Similarly, sea-floor spreading will change the shape of the ocean

Beaches and Coasts, Second Edition. Richard A. Davis, Jr. and Duncan M. FitzGerald.
© 2020 John Wiley & Sons Ltd. Published 2020 by John Wiley & Sons Ltd.

(a)

(b)

Figure 4.1 (a) Photograph of the coast in the Anchorage, Alaska area showing uplift and (b) property damage in the city associated with the earthquake of March 27, 1964. (*Sources:* (a) USGS, https://earthquake.usgs.gov/earthquakes/events/alaska1964/1964pics.php (b) U.S. Army, https://en.wikipedia.org/wiki/1964_Alaska_earthquake#/media/File:AlaskaQuake-FourthAve.jpg).

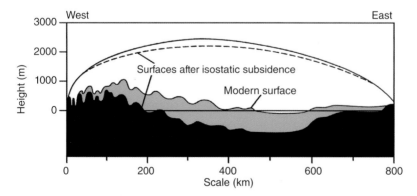

Figure 4.2 Diagram showing the pre- and post-melting of an ice sheet with the isostatic rebound that results from this change. (*Source:* Adapted from A. Bloom, 1978).

Figure 4.3 Shorelines on Hudson Bay, Canada preserved as sea level dropped due to isostatic rebound of the this area after the glaciers melted. (*Source:* Courtesy of Philippe Hequette).

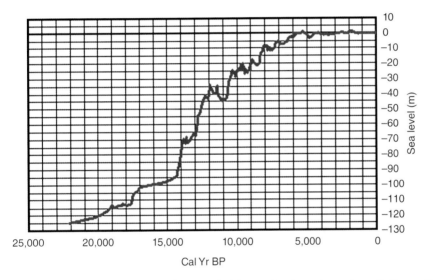

Figure 4.4 Sea-level curve during the post-glacial era as ice sheets began to melt. (*Source:* Courtesy of the Florida Department of Environmental Protection's Florida Geological Survey).

basins and thus cause sea level to change everywhere. The melting of the ice masses takes place over centuries to millennia but the size of the ocean basins takes many millions of years to change significantly.

Details of these and many other causes of sea-level change will be considered in the following sections. The regular and predictable change in sea level caused by astronomical conditions that we call tides is not treated here because they do not result in any net change. They will be discussed in a separate chapter.

4.1 Changing the Size and Shape of the Container

4.1.1 Tectonic Causes

Movement of the earth's crust may take place over a wide range of rates and scales; from meters per hour to millimeters per year and from local events such as volcanic eruptions to plate movements that extend through entire ocean basins. Some of these changes take place at or near the coast and cause the crust to shift, and thereby causing sea level to change. Others are so large that they change the size of an entire ocean basin or even multiple ocean basins.

4.1.1.1 Local and Regional Changes

Seismic events that result in crustal movement have taken place at or near a coast many times, especially when a coast is at or near the leading edge of a plate boundary.

A severe earthquake took place on the coast of the northeastern Pacific Ocean in Alaska on March 27, 1964. Here, the Pacific and North American crustal plates are colliding, and the movement along this crustal boundary produced some spectacular changes in local, relative sea level positions (see Figure 4.1). The earthquake had its epicenter along the north shore of Prince William Sound in Alaska, between Anchorage and Valdez, the area made famous by the Exxon Valdez oil spill disaster. The event registered about 8.3 on the Richter scale, making it one of the most powerful ever recorded in North America. Numerous locations along the shoreline displayed shifts of several meters upward and downward relative to sea level. Extreme examples showed movements of 6 m upwards, which caused harbors and docks to rise well above sea level and their docks to become useless. Even more extreme examples took place offshore, where islands were uplifted more than 10 m and a portion

of the sea floor on the adjacent continental shelf was lowered 15 m. It is important to remember that these sea-level changes took place in mere minutes and they were quite local in their extent.

Other local sea-level changes can be associated with volcanic eruptions. These are common along plate boundaries and will result in significant uplift of coastal areas affected by volcanic eruptions that move the shoreline seaward (Figure 4.5) on many islands. Some subsidence may also take place. Any of these conditions will cause local sea level to change; in some cases by several meters.

These extreme, event-related changes in sea level are not rare along collision coasts at plate boundaries. Such conditions are present along the west coasts of both North and South America, on many of the volcanic island arcs of the Pacific Ocean and in parts of the eastern Mediterranean such as Italy, Turkey and Cyprus.

4.1.1.2 Eustatic Tectonic Changes

Sea floor spreading and the evolution of crustal material at spreading centers on the ocean floor can also have an influence on sea level. As new crust moves up from the asthenosphere the older crust above moves away from the oceanic ridge, causing the plates to diverge (Figure 4.6a). Subduction takes place

Figure 4.5 Volcanic eruption of Kiluea on Hawaii depositing new rock and moving sea level seaward locally.

(a)

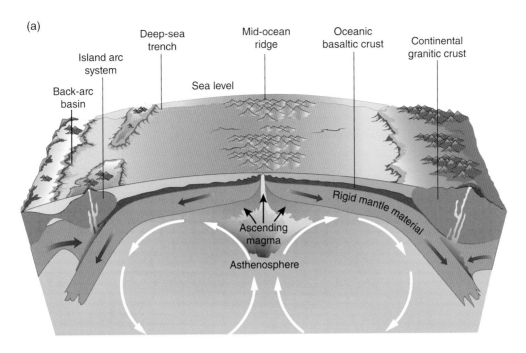

(b)

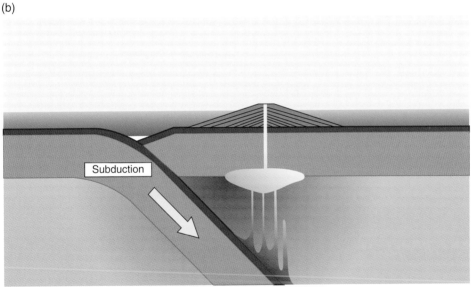

Figure 4.6 (a) Spreading of the sea floor caused by convection in the mantle forms the oceanic ridge system, and (b) subduction of crust producing volcanic eruptions. (*Sources:* (a) Courtesy of VCU (Virginia Commonwealth University), (b) Steven Dutch, https://www.uwgb.edu/dutchs/CostaRica2008/CRAccretion.HTM).

at collision zones, causing the descent of one plate under the other (Figure 4.6b). This combination of slow uplift and slow downwarping on the plate margins, along with movement of land masses on the conveyor belt called the crust, produces changes in ocean basins that affect their volume, resulting in sea-level change on a global scale. The rate of change, however, is small, on the order of millimeters per decade.

We therefore have both local and eustatic sea-level changes that can be produced by tectonic activity. Some are very rapid such as those resulting from volcanic activity or earthquakes, and others are very slow, caused by sea-floor spreading. We can generalize by saying that local changes in sea level are generally more rapid and greater in magnitude than eustatic changes.

4.2 Climate and Sea-Level Change

Sea level is affected by changes in climate. These changes may be due to seasonal or other short-term fluctuations in climate or to long-term changes; some changes are cyclic but many are not.

4.2.1 Seasonal Changes

Mean sea level typically shows a seasonal difference ranging from 10 to 30 cm depending upon location (Figure 4.7). Such differences are due to changes in wind patterns and velocities, and persistent, large-scale high- or low-pressure systems. For example, the Bermuda High, a large high-pressure system in the central Atlantic Ocean depresses sea

level. Atlantic sea level is typically lowest in the spring and highest in the fall, the rise and fall alternating between the northern and southern hemispheres as seasonal wind patterns change. In low latitudes the seasons are less marked and as a consequence, mean sea level shows less seasonal change than in the mid-latitudes.

The most basic and most easily predictable seasonal sea-level fluctuation is a consequence of weather patterns as the sun and the earth shift in their positions relative to each other. As the sun moves through the latitudes with the seasons there are changes in water temperatures leading to changes in wind patterns and velocities. A sea-level change of 10–30 cm is produced as a response to these wind differences; water is pushed in different directions as the result of differences in atmospheric pressure and wind direction (see Figure 4.7).

Seasonal temperature differences change the volume of water in the sea, thereby producing changes in sea level because water expands as it warms. However, the change resulting from this phenomenon is so small that it is difficult to measure. In some situations, however, this change coincides with seasonal change produced by wind to show typical annual patterns of sea-level change.

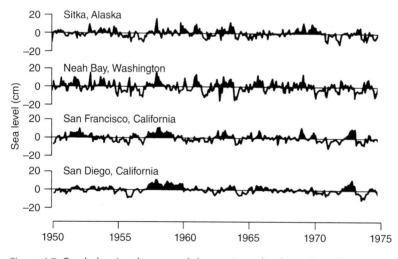

Figure 4.7 Graph showing the seasonal changes in sea level at various places around the continental United States. (*Source:* Komar and Enfield (1987)).

Sea level is highest in the summertime on the east coast of the United States because the warm, expanding Gulf Stream flows very near the coast at that time of year. On the Pacific side of the continent, sea level is at its lowest on the Washington coast during the summertime because of the cold coastal currents that flow from the north at that time of year. Highest mean sea level on this coast is in winter, when Arctic wind generates coastal storms that blow onshore and pile up water.

On the southeast coast of Florida climate change is causing abnormally high sea levels compared to other areas in the southeast. The warmer water in the Gulf Stream in this area, combined with the movement of water in the current toward the coast causes some flooding in the South Beach area of Miami. The phenomenon was particularly marked in 2016 due in part to the strong El Niño condition.

4.2.2 Non-seasonal Cyclic Changes

Probably the most famous of the weather-related causes of sea-level change is the El Niño phenomenon. This climatic change, which influences most of the world, is caused by the warm current produced by changes in wind conditions off the west coast of Peru and occurs over most of the South Pacific Ocean every four to seven years. Fishermen named the current El Niño—the child—referring to the birth of Christ, because it tends to begin in late December. Normally, during this middle part of the Southern Hemisphere summer, the Peru Current, is cold and flows to the north along the west coast of South America. As the current flows to the north, the Coriolis effect produced by the earth's rotation (explained in the next chapter) causes a deflection of the current to the left or west, resulting in upwelling of even colder, nutrient-rich deep water. This upwelling provides the nutrients that support a huge population of plankton, which is the food for one of the world's most important fishing grounds.

At four- to seven-year intervals, the west-to-east blowing trade winds diminish in their intensity, allowing a warm current to move south and force the Peru Current into the Pacific at a higher than normal latitude. This relatively warm current is the result of the El Niño conditions (Figure 4.8). When the

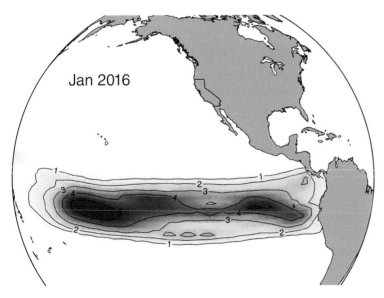

Figure 4.8 Map showing the high temperatures in the Pacific Ocean during January 2016 as caused by El Niño. The numbers are the amount of temperature increase in degrees Celsius above normal. (*Source:* Courtesy of youtube.com).

current enters South American coastal waters, its thick, surface layer of warm water inhibits upwelling of the cold, nutrient-laden water below. This keeps the plankton population from reaching its typical high density and so the fish populations are smaller than normal. The current may stay to the south for a year or more, causing fish populations to either migrate to other feeding grounds or die, thereby disrupting the economics of the adjacent coastal areas. The warmer water of the sea warms the air above it so that the rate of evaporation increases. This moist air mass moves landward, bringing torrential storms that produce flooding and coastal erosion in some areas and a large influx of sediments in others. Changes in other wind and current patterns are also associated with these air masses.

There is a rise in sea level associated with El Niño along the entire western South American coast and extending up to California, due to both the wind effects and the fact that the warm water takes up more space than the colder Peru Current. The El Niño condition also causes anomalous weather conditions, including storms on the South American side of the basin and some drought conditions in places like Indonesia on the western side.

4.2.3   Long-term Climatic Effects

Various explanations have been offered for long-term changes in climate, including sun spots, changes in the relative positions of celestial bodies, and orientations of those celestial bodies.

Climate-induced global or eustatic sea-level change can occur in two ways: by increasing or reducing the amount of water in the entire world ocean, or more subtly, by changing the temperature of the world ocean which causes its volume to increase or decrease as water expands or contracts. In either case, when the volume of water changes it causes an absolute, or eustatic, change in sea level. A worldwide and long-term change in climate can bring about both of these conditions simultaneously. A cold climate results

in the growth of glaciers, which incorporate huge volumes of water, while at the same time lowering the temperature of the ocean. Both of these conditions cause the volume of the world ocean to be reduced. A warmer climate melts the glaciers and frees the water to return to the ocean; simultaneously it raises the ocean's temperature causing an increase in ocean volume and a rise in sea level.

Drastic temperature changes are not required to produce major changes in sea level. A rise or fall in the mean annual global temperature of only 2–3 C has a profound effect on both the mass of ice retained on the surface of the earth and the volume of the water in the world ocean. During the last ice age, thick ice sheets covered much of the land masses of the Northern Hemisphere. The volume of water in the world ocean was greatly reduced; water was sequestered in ice and that remaining in the ocean chilled and contracted in volume. This combination resulted in the exposure of nearly all of the continental shelf and was produced with a reduction in global mean annual temperature of only about 2–3 °C as compared to the present. If the current trend toward a warmer global climate continues and the mean temperature increases by only a few degrees, the entire process will reverse—the ice sheets will melt and the ocean will encroach upon the continents until many of the port cities are at least partly under water.

Insufficient data have made it difficult to assess the pace and direction of global climate and sea-level changes over the past few decades. Accurate records of sea level have been kept for little more than a century and weather records in most parts of the world do not extend back any further. Currently, the International Panel on Climate Change (IPCC) has met many times and has produced several lengthy reports. Predictions are difficult at best and theirs extend to the end of the twenty-first century. The recent version of their predictions on sea level ranges from a 27 cm increase to more than 90 cm over this period (Figure 4.9).

The great sea-level changes of the past, as recorded in the layers of sediment and ice

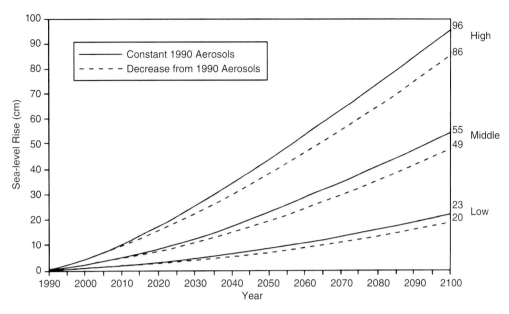

Figure 4.9 Plot of predicted sea-level rise during the twenty-first century according to the report of IPCC 2007.

accumulation, occurred in cycles of 10–30 thousand years or so. During the 2.6 million years of the Quaternary Period there have been 100 cycles (Figure 4.10). The most recent period of higher sea level than the present condition took place about 120,000 years before present – (BP) during oxygen isotope Stage 5e (Figure 4.11). Our hundred-year-old records, therefore, cannot be superimposed on past changes to make any sort of valid prediction. Nevertheless, the recent rise in global temperature has forced us to take note of a possible human-generated cause for the worldwide increase in the rate of sea-level rise taking place today. The still-accelerating release into the atmosphere of carbon dioxide and other greenhouse gasses has some climatologists projecting a global warming of 3 °C by the year 2030. This increase in mean annual temperature could melt a large portion of the ice cover in Greenland and Antarctica; enough to raise global sea level by as much as 5 m in only a few centuries. Too long for a person to worry about but still a short time in terms of human occupation of the coast as we know it. Even here, however, we may be trying to superimpose a

short-term data base on very long periods of cyclicity. The current increase in carbon dioxide might only be part of a longer cycle that pre-dates civilization and will decline by itself as the cycle proceeds. It is difficult, however, to discount the obvious contribution to global warming being made by our current high levels of combustion and our destruction of photosynthesizing plants that take up huge amounts of carbon dioxide and release large quantities of oxygen. It is important for us to examine the details of the warming and cooling patterns of the past and their effects on global sea level to help understand the global warming pattern and sea-level changes that we are presently experiencing.

4.3 Sea-level Rise due to Sediment Compaction and Fluid Withdrawal

In some of the coastal areas associated with river deltas sea level is rising locally because the land itself is sinking, or subsiding. This is

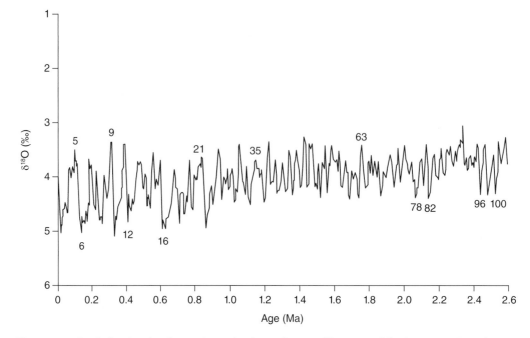

Figure 4.10 Graph showing the changes in sea level over the 2.6 million years of the Quaternary Period. (*Source:* Shackleton et al. (1990); reproduced courtesy of The Royal Society of Edinburgh and W. R. Peltier).

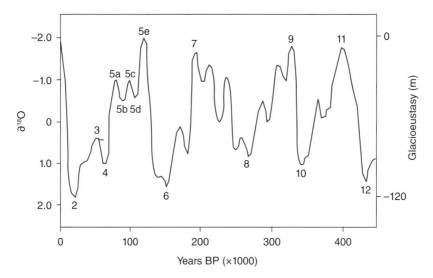

Figure 4.11 Graph of the past few hundred thousand years showing the position of the most recent sea-level high stand during Stage 5e, 120,000 years before present. (*Source:* Shackleton et al. (1990); reproduced courtesy of The Royal Society of Edinburgh and W. R. Peltier).

particularly apparent along east Texas and in the Mississippi Delta in Louisiana. Here, local sea level is rising much faster than the average global rate. The highest rates of local sea level change in the United States are in the Mississippi River Delta area where it is rising by 9–10 mm year^{-1}, about four times the global average (Figure 4.12). As much as 6–7 mm of that rise is due to land subsidence caused by a combination of compaction of

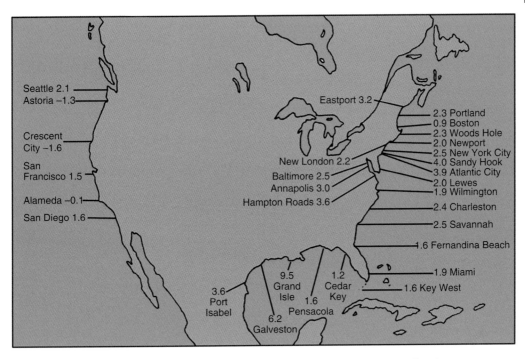

Figure 4.12 Map of the United States showing the annual rates of sea level rise at many locations. (*Source:* Adapted from National Academy of Sciences, 1987).

muddy delta sediments and the withdrawal of fluids from the coastal zone, primarily by the petroleum industry.

Huge quantities of fine sediment are transported by the river and deposited at its mouth, an average of up to 1.6 metric tons per day at the active lobe. This sediment is deposited so rapidly that it traps much water as it settles. The resulting mud, at places up to 90% water by volume, accumulates as thick sequences through the active delta region. As the weight of the new mud layers compresses the underlying ones, the water is squeezed out thereby compacting the sediment to a lesser volume, which causes the land surface to subside resulting in a relative rise in sea level.

While compaction of sediments has occurred for several thousand years, subsidence along a significant part of the northern coast of the Gulf of Mexico caused by the withdrawal of large volumes of fluid is a recent phenomenon caused by recent human activities. Nearly 100,000 wells have produced large quantities of oil and natural gas from the Mississippi Delta and also from the nearby coast and shelf of the Gulf of Mexico. Large volumes of ground water have also been taken for domestic and industrial use. This results in a dewatering effect similar to that of squeezing of the water from the delta muds, resulting in land subsiding and sea level rising. An example of this phenomenon is near Galveston, Texas where land has sunk nearly 2 m in this century (Figure 4.13). To reduce, or possibly eliminate, further subsidence, the major domestic and industrial uses are being shifted to surface water, and water is now being pumped back into the ground to replace that which had previously been withdrawn.

An example of the effects of the petroleum industry is in southeast Texas near the Neches River where an extensive wetland was present in 1956. After the development of an oil field and the drilling of several wells this wetland experienced significant subsidence and it became a lake, as shown in 1978 (Figure 4.14).

Figure 4.13 Photo of paved road that now descends into the sea that has risen near Houston, Texas. (*Source:* Courtesy of John Anderson).

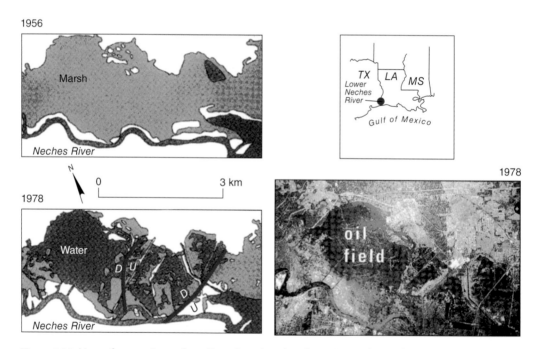

Figure 4.14 Maps of an area in southeast Texas based on data from 1956 and 1978 showing how an oil field changed a wetland into a lake. (*Source:* Adapted from White and Morton (1997). Reproduced with permission from the Coastal Education and Research Foundation, Inc.).

Another human impact on the coastal zone is not as easily remedied. Building large cities on thick accumulations of unstable sediments has been an environmental disaster. The water of the nearby Gulf has responded by flooding parts of the inland port cities of Houston and New Orleans. In fact, the city of New Orleans is a few meters below sea level and even more than that below the level of the Mississippi River.

If sea level in the lower Mississippi Delta continues to rise at a rate of $9–10\,mm\,year^{-1}$, and if this rate is sustained for 50 years, it would amount to 0.5 m of sea-level rise; enough to submerge large areas of coastal wetlands; destroying the ecosystem that is so important to the area.

Compaction is not only an important contributor to land subsidence in delta regions throughout the world, it also affects peat bogs, marshes, and other organic-rich sediment accumulations that hold large volumes of water. The influence of compaction on these environments is quite small, however, because the overall thickness of sediments in them is usually less than a few meters.

4.4 Isostasy

The Mississippi River delta is one example of regional subsidence, but broad regions of continental plates can also sink under a heavy load. Over a period of thousands of years, the mass of a huge ice sheet can cause the continental lithosphere to become depressed by about 100 m, the amount dependent on the thickness and density of both the ice mass and the underlying lithosphere. The vertical movements of the lithosphere are accommodated by the semi-plastic portion of the upper asthenosphere. As the ice melts, removing the overlying pressure of the ice from the land mass, the lithosphere rises, or rebounds. Such vertical adjustments of the lithospheric crust are called isostatic adjustments; isostasy is the condition of equilibrium of the earth's crust that takes place as the forces that tend to elevate the lithosphere are balanced by those that tend to depress it.

The depression and subsequent rebounding of the lithosphere as mass is added and removed is only one kind of isostasy. A change in density of the lithosphere will cause a similar isostatic crustal response and produce changes in sea level as a consequence. When the lithosphere is young and still hot, as it is when produced at a rift zone of the oceanic ridge system, it is relatively low in density. As it cools over several millions of years, it reduces in volume, becoming denser. This causes subsidence over the asthenosphere. Because most of this activity takes place in the ocean basin, the resulting rise in sea level over the subsiding sea floor is so small as to be imperceptible. However, there are a few places on the globe where coastlines are close to sites of plate divergence. At these places the local rise of the young lithosphere can produce an increase in sea level and slow inundation of land. The coasts of both the Red Sea and the narrow Gulf of California are on such diverging oceanic plates and are places where this type of local sea-level rise can be expected.

Another reason for isostatic adjustment of the earth's crust is related to the thick accumulation of sediments and volcanic material in the lithosphere. This currently takes place along thick, prograding coastal plains such as the Gulf and Atlantic coasts of the United States ("prograding" means building in a seaward direction). In both cases, thousands of meters of sediment accumulates over tens of millions of years causing crustal subsidence that produces a relative rise in sea level. The same phenomenon, but in a more rapid scenario, is associated with the thick accumulations of volcanic crust such as the islands in the Pacific Ocean. One of the best examples of this phenomenon is the Hawaiian Islands where about 5000 m of volcanic material is piled up on the thin oceanic crust. This enormous mass causes subsidence and a relative rise in sea level.

4.5 Changes in the Volume of the World Ocean

4.5.1 Advance and Retreat of Ice Sheets

During the Quaternary Period, the most recent period in the geologic timescale, from 2.6 million years ago to the present, eustatic sea level has changed very rapidly as compared to most times in the earth's history. It is likely that we have this opinion about rates of sea-level change simply because we do not have the detailed information about sea-level change throughout geologic time that we do about the last couple of million years. These geologically recent and rapid changes have been due to the advance and retreat of continental ice sheets and the polar ice caps that formed the extensive glaciers of the northern hemisphere and the Antarctic. The Pleistocene Epoch of the Quaternary Period, also called the "Ice Age," came to a close 18,000 years ago. At this time the last of the major ice sheets had begun to melt rapidly, except for Greenland and Antarctica. For the previous million or two years, there had been thick ice sheets in the northern hemisphere that repeatedly covered and withdrew from most of Europe, northern Asia and North America down to the Missouri and Ohio River valleys (Figure 4.15). Development of similar ice

Figure 4.15 Map of the northern hemisphere showing the distribution of glacial ice sheets during the peak condition of the Pleistocene. (*Source:* Flint and Skinner (1974)).

sheets in the Southern Hemisphere was limited because of the general absence of land in the high southern latitudes, except for Antarctica and the southern tip of South America.

The record of the Pleistocene Epoch that is preserved on the continents is generally considered to display four cycles, in each of which there was glaciation alternating with interglacial melting. Each of these four cycles has a name for both the glacial advance and the following melting portion of the cycle. For example the last portion of the cycle characterized by glacial advance is called the Wisconsinan in North America. It was preceded by the Sangamonian interglacial period when sea level was high and glaciers were smaller and covered a smaller area than they do at the present time. In Europe and other areas where similar glacial cycles took place there is a different terminology for each of the cycles.

For the first half of the twentieth century this interpretation of four glacial cycles was accepted and taught throughout not only North America, but the rest of the world. As we developed more capability for studying the ocean basins through sediment cores, the climatic history depicted showed that there were many more Pleistocene glacial cycles than the four shown in the land-based stratigraphic records (see Figure 4.10). The evidence for these numerous glacial cycles has been masked on land by deposits of the four larger and longer-lasting glacial cycles.

The discovery from the oceanic record of these numerous glacial cycles is primarily due to techniques for investigating ancient climatic conditions on the earth that did not become available until the late 1940s, just after World War II. Probably the most important of these is the use of oxygen isotopes to help interpret past climatic conditions. Isotopes are atoms of a given element with the same number of protons but different numbers of neutrons. The number of protons in an atomic nucleus determines the number of electrons, which determines an element's chemistry. Oxygen, for example,

typically has an atomic number of 16, but there is also a heavier isotope, oxygen-18, which has two extra neutrons. Both isotopes have very similar chemical properties and both are incorporated in the skeletons of organisms like corals and mollusks as part of the compound, calcium carbonate ($CaCO_3$). It was discovered at the University of Chicago in 1947 that the relative concentration of the two oxygen isotopes in the skeletons of animals was a function of the temperature of the ocean water in which they were living. This led researchers to realize that the temperature of the water at the time when an organism was living in it could be determined by analyzing the $^{16}O/^{18}O$ ratio of the calcium carbonate in the skeleton. This relationship was not applied to global climate interpretations in any significant fashion until 1955 when the skeletons of floating single-celled animals (foraminifers) from several deep-sea cores were analyzed using these techniques. These floating, single-celled animals are very common in the upper few hundred meters of the water column throughout the ocean. Their calcium carbonate skeletons settle to the bottom when the individuals expire, and the sediments in cores taken from the ocean floor contain many of these skeletons. Oxygen isotope ratios of samples taken from these cores demonstrated that there were numerous periods of significant temperature fluctuation in ocean waters in only 400,000 years (see Figure 4.11). These changes in ocean temperature were interpreted as being a consequence of climatic changes associated with glacial activity. These data demonstrated that the record preserved in the sediment of the ocean floor was much more complete and complex than that preserved on land.

Perhaps more importantly, the temperature cycles shown by the oxygen isotopic data in these sediment cores turned out to be in agreement with cycles that had been predicted many years earlier by Milutin Milankovitch, a Serbian astronomer. He developed a theory of climatic changes that was based on cycles of radiation received by the earth as it tilts relative to the sun. Three different astronomical conditions produced cycles (Milankovitch cycles) associated with variation in tilting and its effect on climate: variation in eccentricity of the earth's orbit around the sun, with a periodicity of 90,000 to 100,000 years; changes in the obliquity of the earth's plane of orbit and the angle it makes with the plane of the ecliptic, having a period of 41,000 years; and the precession, or wobbling, of the earth's axis, with a period of 21,000 years.

The coast as we see it now is the product of the most recent temperature cycle. The last of the great advances of glaciers, the Wisconsinan Ice Age, began about 120,000 years ago and lasted for more than 100,000 years. The formation of these glaciers took most of this time; they melted to their present size in only 18,000 years. Since the end of the Wisconsinan period we have been in what is called an interglacial period, characterized by global warming, glacial melting, and rapid rise in sea level. The Antarctic and Greenland ice sheets of today are remnants of the last Wisconsinan advance. The period of overall melting and related warming has been interrupted by a few "little ice ages," which are extended periods of abnormally cold weather, some of which have occurred during recorded history. The most prominent was chronicled in Europe between about 1450 and 1850. Sea level actually dropped during these short cold periods that occurred during the overall warming trend.

By calculating the difference in volume between the Wisconsinan ice sheets and the ones that remain today, it is possible to determine the volume of ice that has melted and the ways that this volume has changed global sea level. The next step is to extrapolate what would happen to sea level if the present rate of global warming continued until the remaining ice sheets melted. In order to make these determinations, it is necessary to estimate the area and thickness of the previously existing ice sheets. The surface area covered by the Antarctic ice sheet during the Wisconsinan advance was almost 14 million square kilometers; today it is about 12.5 million square kilometers, which is not a big difference. By way of contrast, the North American ice sheet once extended over more than 13 million square kilometers of land and now covers only 147,000 km^2, a loss of about 99%. There were also ice sheets covering Greenland, much of Europe and the northern part of Asia. about 12–15,000 years ago, these and other, smaller, ice sheets covered in total more than 44 million square kilometers, of which just under 15 million now remain, one-third of the former extent.

Whereas we have geologic and geographic information that enables us to determine the areal extent of these enormous ice sheets, it is much more difficult to determine the volume of the Wisconsinan ice sheets. A reasonable approach is to use the Greenland ice sheet as a model. This ice sheet is well known because of extensive petroleum exploration surveys as well as scientific and military studies. Glaciologists have estimated that the Greenland ice sheet holds about 2.5 million cubic kilometers of ice. By extrapolation of surface areas and by estimating average thicknesses, we interpret that 75 million cubic kilometers of ice were contained in the vast glaciers of the Wisconsinan ice age. Given the ratios mentioned above, then about 50 million cubic kilometers of ice have melted since the Wisconsinan Ice Age – roughly equivalent to 20 times the volume of the present Greenland ice cap. Bearing in mind that that there is a 10% decrease in volume when ice turns to water, then 45 million cubic kilometers of water were returned to the ocean. This obviously caused a major change in volume and therefore a dramatic rise in sea level. The mass of this volume of water was enormous and caused isostatic adjustment of the crust, both on the continents where mass was removed and on the oceanic areas where mass was added. This tremendous shift of mass from the continent to the ocean had, and is still having, a pronounced influence on sea level.

As the ice sheets were removed from the continents, the continental lithosphere began to rebound in an isostatic adjustment. If we assume that some areas were covered with 3000 m of ice at peak glacial conditions, the isostatic rebound after melting would have reached about 1000 m. This is determined by assuming that the 3000 m of ice had a density of 0.9 g cm^{-3} which is close to one-third of the density of the continental lithosphere. This would be equivalent to about 1000 m of lithospheric crust. The rate of the rebound has been slow, and in some places, such as Norway and Sweden, it is still ongoing (Figure 4.8); at the present time this area is rising a few millimeters each year.

While this was taking place on the continents, the ocean was experiencing the opposite condition. The added mass coming from the continents in the form of glacial meltwater caused a sinking of the ocean floor, which has a much younger and thinner crust than the continents. This combination of continental rebound and oceanic crustal depression with added volume of the ocean still produced an increase in sea level in most places despite the sinking floor of the ocean. There were, however, regional decreases in sea level due to rapid rates of continental rebound. Considering the combination of all of these phenomena, the overall eustatic increase in sea level over the past 18,000 years has been about 100 m, with local variations due to differences in continental rebound.

4.5.1.1 Continental Rebound

Because the Wisconsinan glacial ice melted slowly, the isostatic adjustment in the form of continental rebound was also slow. Most of the crustal rebound took place before the melting ice sheets shrank to near their present size—the mass of the remaining ice was no longer sufficient to cause significant subsidence. The rebound is still ongoing in the Hudson Bay area of Canada, and in Argentina and lower Chile as well as Scandinavia.

Along the coast of Sweden and Norway, the mean annual sea level is still dropping a few millimeters a year as a result of the rebound of continental crust. If the annual global sea level rise of 2–3 mm is added, the rate of uplift is up to 9 mm a year. If we assume that this has been going on for about 5000 years, the rebound has been a net uplift of about 30–35 m. It is even more pronounced in the Hudson Bay area of north-central Canada. Aerial photographs of these areas show parallel ridges marking the old shorelines and abandoned beaches as the land has risen, causing the ocean to recede from the land (see Figure 4.3).

The amount of isostatic adjustment diminishes toward the low latitudes because the ice formed later, it was thinner, and it melted more rapidly toward the warmer climate. This can be shown along the New England coast where the old shorelines slope downward from Maine to New York. Sea-level change over recorded history shows that there has been a drop of several millimeters a year in Maine and essentially no change in New York.

The Great Lakes originated from glacial activity, and display clear signs of a north-to-south tilt, as shown by the ancient shorelines. These shorelines are several meters higher on the northern part of the lakes as compared to the south sides. Lake Michigan is gradually getting shallower at the north end and deeper in the south. If the present rate of rebound continues, after about 3000 years the lake will drain into the Mississippi River system rather than through the other lakes into the St. Lawrence Seaway as it does now.

Wherever the high-latitude land mass is still responding to the melting of an ice sheet, coastal uplift leaves the shorelines well above present sea level (see Figure 4.4), abandoned by the receding water level. Eventually the rebound will slow and water level will rise again relative to the land mass.

Glaciation is discussed further in Chapter 17.

4.6 Post-Glacial Rise in Sea Level

The post-glacial rise in sea level began 18,000 years ago. It is characterized by a rapid and large rise in global sea level as a response to the melting of the Wisconsinan glaciers.

Geologists, oceanographers and climatologists who study sea-level fluctuations cannot agree among themselves on the details of the rise in sea level over this period, especially during the last few thousand years.

It is generally agreed that sea level was at its lowest position about 18,000 years ago, when the ice sheets of the Wisconsinan had reached their maximum extent. The lowest position of sea level, called the lowstand, is deduced by uncovering evidence of the oldest drowned shoreline now located beneath the waters of the continental shelf. This evidence might take the form of beach sand, marsh deposits, drowned wave-cut platforms, drowned river deltas or almost any sort of indication of an old shoreline. Other factors such as tectonic uplift and subsidence must also be taken into account before deciding the vertical position of the lowstand. Scientists know that sea level was more than 100 m below its present position; perhaps as much as 130 m. There is general agreement that sea level rose very rapidly for the first several thousand years and that it slowed about 6000–7000 years before present (BP) (see Figure 4.4). The annual rise during this period was near 10 mm year^{-1}, a rate fast enough for the sea to cover parts of many of the present coastal cities in a century.

During this period of rapid rise in sea level the shoreline moved so fast that the sand bars and barrier islands that protect so many of our coasts today had no time to build vertically. For these barriers to develop, the shoreline has to be either stable or very slowly moving to give waves and currents enough time to construct these long sand bodies. The lack of a stable shoreline for any length of time, coupled with moderate to high tidal ranges along the irregular coasts, produced tide-dominated coasts with widespread estuaries and tidal flats. Only when the rate of sea-level rise slowed did waves begin to mold the coast into a linear shape and barrier islands begin to form.

There is no general agreement about the position of sea level along the coast of North America during the past 3000 years or so,

when all of our coasts began to form barriers. There are three scenarios that have been proposed during this period: it has been stable at the present position; it has changed about a meter or so above and below its present position; or it has been gradually rising during the period—over only about 2–3 m. It is possible that each of these situations has prevailed at various places. Because of the resolution necessary to determine which of these three scenarios might apply at any given coastal location, we are still working on this question.

4.7   Current and Future Sea-Level Changes

The impact of the rapid increase in the population of the Earth has produced elevated levels of carbon dioxide. According to many authorities, this has led to global warming (see Box 4.1). The warming of the atmosphere over a long period of time causes an increase in the rate of melting of the ice sheets and also causes the ocean to warm up. Both of these phenomena result in an increase in the volume of water in the world ocean, thus an increase in sea level. Recent data from tide gauges around the world show that the rate of rise in eustatic sea level is increasing (Figure 4.16). Because tide gauges are available from all over the world this is one of the most prevalent types of sea-level data and must be studied very closely. The records from many of these stations have to be discarded because of instability of the location due to subsidence or compaction, or because of tectonic activity. For most locations, we have an average of about one hundred years of reliable data for the position of sea level.

Most of the data for the United States are collected through the efforts of the National Ocean Survey (NOS) which is part of the National Oceanographic and Atmospheric Administration (NOAA). Their personnel have studied and analyzed data from hundreds of tide stations. Similar efforts are

Box 4.1 Global Warming and the Future of Sea Level Rise

During the past decade or so, there has been much speculation about the future trends in sea level rise. Conservative estimates have a rise of about 50 cm over the twenty-first century whereas the most liberal predictions are in excess of a meter for that time period. Let us take a close look at the situation then make your own decision about "global warming."

It is well known that sea level rose tremendously at various rates during the past 18 000 years as extensive glaciers melted throughout many parts of the earth. This melting took place because of an increase in global temperatures: about 2–3 °C per year. That does not seem like much but it is enough to cause melting of snow and ice that had accumulated over many thousands of years. Obviously this melting added huge amounts to the volume of the oceans and sea level rose greatly; nearly 130 m in all.

The important aspect of this sea level rise is the rate at which it occurred. During the early part of the melting, sea level rose about a centimeter per year which is a catastrophic rate. Such a rate lasted for several thousand years and accounted for the vast majority of the rise. Although a centimeter a year (an inch each 2.5 years) seems like a pretty slow change, it causes great erosion of the coast and drowning of the wetlands (see Box Figure 4.1.1).

The addition of sediment to compensate for these processes is generally insufficient to prevent the sea from encroaching on the land. In a lifetime of this rate of rise, sea level will rise more than 2 ft (60 cm), enough to cause problems for many low-lying coastal communities. Recall that this rate of change occurred for thousands of years.

We can observe climate changes that have taken place over the past few decades. This comes from temperature records at various places around the world. The U.S. Geological Survey has been making observations on the ice cover in the Arctic where barrier islands are present on the north coast of Alaska. In 1980 the beaches on these islands were free of ice for only 80 days year^{-1}. At the present time there are 140 days in a year where there is no ice at the shore. These conditions scream "global warming." In addition, the adjacent Canadian Arctic is experiencing extensive melting of permafrost, another indication of climate change.

During most of the time that significant numbers of people have inhabited the coast, sea level has been rising very slowly; only a couple of millimeters or so per year. This goes unnoticed by almost everyone except those studying the position of sea level. Why did sea

Box Figure 4.1.1 Drowned wetlands.

level increase rapidly for an extended period and then slow down greatly for another extended period? The simple answer is that melting of the ice caps throughout the world slowed as a response to global temperature; it stopped warming up so fast.

Why this happened provides the more complicated and complete answer to the question. Part of it can be found in the various perturbations of the earth–sun system as recognized by Milankovich in the early twentieth century. He showed that there are various cycles to the way the earth moves in its orbit; the shape of the orbit, and the wobble and tilt of the earth on its axis. These cycles range from about 28,000 years to nearly 100,000 years. As a consequence of these fairly long cycles and their shorter-term variations the amount of insolation that is received on the earth as a whole, and in various parts of the earth, changes with time, resulting in fluctuations in the mean annual temperature (see Box Figure 4.1.2).

These are the well-documented natural changes in the earth's climate. What about the possible influence of humans on the global climate? Since the industrial revolution in the nineteenth century, we have been emitting large amounts of carbon dioxide into the atmosphere. This carbon dioxide absorbs heat radiated from the Earth (long-wavelength infra-red light) that would otherwise harmlessly radiate away into space; this absorption causes the temperature of the CO_2 to rise, thus atmospheric CO_2 is warming the atmosphere, and thus the term "greenhouse effect." Since the advent of the automobile this effect has been exacerbated. Much of the problem originates in North America and Europe, the so-called developed countries, but in recent years the industrializing countries of Asia have become major contributors.

During the past 100 years, about as long as we have had good records, it is evident that the rate of sea-level rise has increased. It is now about 3.0 mm year^{-1}. Is it coincidence that this increase in sea level parallels the great increase in CO_2 production, or is human activity ("anthropogenic") the cause of increased global temperatures and therefore the rise in sea level? Actually there is still some disagreement among scientists as to the answer. Most scientists favor anthropogenic factors as being very important, but many believe that the changes in sea level are a result of natural changes in the earth's climate. The case is not yet closed on global warming.

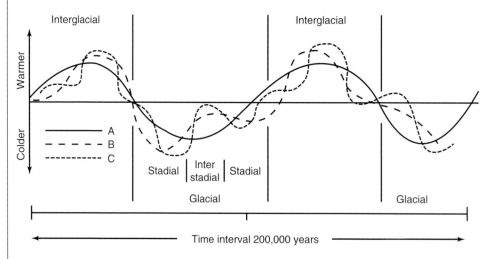

Box Figure 4.1.2 Diagram showing the Milankovich Cycles. Stadial stages are cold and interstadial are warm.

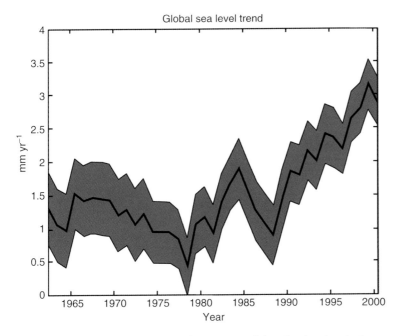

Figure 4.16 Plot of sea level rise over the past several decades showing the increase in the rate over the past several years. (*Source:* Merrifield et al., 2009. © American Meteorological Society. Used with permission).

being extended throughout many other countries to provide a decent global coverage of these data. Examples from studies of various parts of North America serve to illustrate the nature of these data and the trends that are present (Figure 4.17). The east coast of the United States shows a general increase in sea level with considerable short-term variation. In New England the same pattern is present but with a bit higher general rate of rise. The west coast of the United States shows more variation from one location to another but there is little overall increase in sea level for the period of record. This is due to the tectonic activity on this leading edge coast. The more stable east coast shows a rise that ranges from relatively slow in the north where isostatic rebound is still going on to more rapid rates of rise to the south where rebound is absent and some compaction is taking place.

The west coast, a crustal plate collision area, experiences great variety in sea-level conditions. In Alaska many locations show a decrease in sea level due to tectonic conditions of uplift; up to $14\,mm\,year^{-1}$ (see

Figure 4.17). The Gulf Coast of the U.S. shows great range in sea-level rise due to differences in the geologic setting. The Florida peninsula is a carbonate platform, one of the most stable and least compacting geologic provinces. Its rate of rise is only about $1.5\,mm\,year^{-1}$ which is probably a good reflection of the actual sea-level rise due to the increase in the volume of the ocean. By contrast, the Mississippi Delta region is experiencing an annual sea-level rise of $10\,mm$ (see Figure 4.12). This is due mostly to compaction of the thick sequences of mud there, and to the withdrawal of huge volumes of fluids by the petroleum industry.

On a global scale, the tide gauge records show a distinct but variable increase in sea level everywhere except for the Pacific Ocean (Figure 4.17). The coasts of the Pacific Ocean are mostly collision coasts where tectonic conditions override eustatic changes in sea level. We must remember that 100 years of data for sea-level changes represents an insignificant period of time in the cyclic systems of the Earth. It is inadequate to accurately predict long-term sea-level trends.

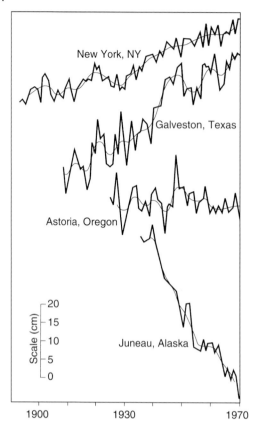

Figure 4.17 Sea level data from tide gauges over the past century for each of the coasts of the United States. (*Source:* Hicks, 1972. Reproduced with permission of ASBPA).

During the past several thousand years, large changes in sea level have occurred as climates displayed smaller cycles within the long-term warming and melting trends of several millenia. These short reversals on the long-term trend were sometimes a century or more long. One of the best examples of this took place a few hundred years ago in the aforementioned "Little Ice Age." During that time the existing ice sheets increased in size, and there was a slight reversal of the rise in sea level. This reversal of a long-term trend was best recorded in Europe and in China, and was documented in various historical records. We expect that similar short-term reversals are typical of most long-term trends in climate. Factors that can contribute to these climatic changes include variations in sun spots, shifting of oceanic currents and slight changes in earth–sun positions.

4.7.1 Impact of Increasing Rise in Sea Level on Modern Coastal Environments

The present coastal morphology is very young, even in the context of the Holocene sea-level rise. Virtually all of the coastal features we see have developed in less than 6500–7000 years. The interactions of coastal processes with sediments and rock cause rapid changes in erosion and deposition. When these conditions, including sea level, remain constant for a period of time an equilibrium situation develops. Changes in sea level can upset this equilibrium and affect virtually all of the environments in the coastal zone. A rising sea level causes the shoreline to move, which brings about changes in the coast.

Since the time of widespread human occupation of Earth, eustatic sea level has risen by only about a meter. Nevertheless, there have been important changes to various coastal environments over this period. Some examples include the development of the present lobe of the Mississippi Delta, formation of most of the barrier islands on the Florida Gulf Coast, and considerable erosion of bluffs on the west coast of the United States. The possibility of nearly doubling the eustatic rate of sea-level rise from about 1.5 to $3\,mm\,year^{-1}$ means that there will be about 30 cm of increase in a century. Such a rate over only a thousand years would produce an increase in sea level of about 3 m, enough to cause major changes in the nature of the shoreline and flood portions of many of the world's foremost cities like New York, London, Amsterdam and Los Angeles. This is a reasonable forecast. See Box 4.2.

Coastal management must be more closely regulated. Anthropogenic activity in some of the most fragile areas for sea-level rise has taken place. For example, there has been major industrial construction on the wetlands of the Mississippi Delta (Figure 4.18).

Box 4.2 Venice, Italy; A City that Is Drowning

A city located on the Adriatic Sea coast of northern Italy, Venice is a major historic treasure, and a very popular tourist stop for people from throughout the world. The city is commonly known for its canals, gondolas, and the absence of automobiles as well as for its excellent cuisine and music. It is located on a combination of more than 100 small islands within a large, shallow, backbarrier lagoon, and is offshore from the mainland. Venice began in the tenth century as a city-state with trading and fishing as its major industries. It became very famous and its merchants became very wealthy as the city developed into a major economic center of the eastern Mediterranean area.

The general setting is a few kilometers offshore where there are several low islands that have become developed as residential and commercial centers. These islands have essentially no natural relief and they rest on thick muddy sediments deposited by rivers that used to flow into the lagoon. The city is covered with buildings (Box Figure 4.2.1), most of which are several hundred years old and have great historic significance to Italy. Included are numerous churches, governmental buildings, plazas, old military installations, etc. The entire city was built within a meter or two of sea level

but with the protection of the barrier islands a few kilometers seaward of them.

As time passed, several things occurred to jeopardize the future of the city: the underlying mud began to compact, and the rate of global sea level rise increased. The combination of these phenomena resulted in a significant relative rise in sea level, essentially flooding the city (Box Figures 4.2.2 and 4.2.3). This has been exacerbated by the unfortunate decision to relocate the mouths of two rivers, the Sile River and the Brenta River, that naturally emptied into the Venice lagoon. It was feared that the discharge of sediment would fill up the lagoon. In order to prevent this from happening, the rivers' lower courses were changed so that both rivers now empty directly into the Adriatic Sea; one north of the Venice lagoon and one to the south. This has resulted in a lack of sediment to nourish the marshes which are vital to both the ecology of the lagoon and the protection of the developed islands from wave attack and erosion.

At the present time the city remains vital but the population, which peaked at more than 250,000 a couple of centuries ago, is now down to about 50,000. Much of this decrease is the result of the tremendous cost of renovating

Box Figure 4.2.1 Venice.

Box Figure 4.2.2 Common situation in the city of Venice, when high spring tide floods much of the city.

Box Figure 4.2.3 View of structure on the shoreline of the Venice Lagoon that is now uninhabitable due to sea level rise.

properties for residential or commercial use. Virtually all buildings fall within the historic preservation regulations and therefore must be restored in order to be occupied. Such expense is more than most people, or even corporations, can bear. Another deterrent to Venice residence is the lack of jobs.

The current rate of sea level rise is about $3\,mm\,year^{-1}$. For a city that is nearly at the level of spring high tide, the future is limited. The most popular location in Venice is San Marco Plaza, which is also one of the lowest parts in the city. A hundred years ago this area was flooded only a few times per year but with an increase in relative sea level of about 30 cm over that time, it is now flooded about 50 times each year, a couple of days during each spring tide plus times when winds blow for a sustained period. As sea-level rise increases and continues, the city will literally drown.

At the present time, construction is underway to build large floating gates that will prevent wind from blowing large amounts of water into the lagoon and raising its level beyond that which the city of Venice can withstand. These gates rest on the inlet channel floor during normal times; when strong winds begin to raise the water level in the lagoon, water is evacuated from them causing one end to float up above the surface acting as a dam against water being blown into the lagoon.

In the long term, there seems to be little that can be done to save this beautiful and historically significant city other than encasing it in a large dike. It would be a very expensive and time-consuming project, but it might be the only way.

Figure 4.18 Large industrial installation for the petroleum industry on the Mississippi Delta.

Such construction is destined to be destroyed by sea-level rise in the near future, as are the native wetlands.

4.8 Summary

Maybe more than any other coastal process, sea-level change influences the entire world. Global and slow changes are always taking place, and regional or local changes may be rapid and catastrophic. Sea-level change is, in reality, everywhere! The current concern about global warming and its influence on sea-level change is a front page story in many parts of the world. Rates of rise have increased over the past century but the long-term future is still a matter of speculation.

Ranging from changes on the order of meters in less than a day associated with earthquakes to only a millimeter or so in a year, these changes can impact coastal management in many ways. Location of development, types of construction, density of occupation and other factors must all be done while considering the local sea level situation. We are still wrestling with the causes for these increases in global sea-level rise; are they natural or anthropogenic?

References

Bloom, A. (1978). *Geomorphology: A Systematic Analysis of Late Cenozoic Landforms*. Englewood Cliffs, NY: Prentice-Hall.

Flint, R.F. and Skinner, B.J. (1974). *Physical Geology*. New York: Wiley.

Hicks, S.D. (1972). On the classification and trends of long term sea-level series. *Shore Beach* 40: 20–23.

Komar, P.D. and Enfield, D.B. (1987). Short-term sea-level changes and shoreline erosion. *SEPM Spec. Publ.* 43: 17–27.

Merrifield, M.S., Merrifield, S.T., and Mitchum, S.T. (2009). An anomalous recent acceleration of sea-level rise. *J. Clim.* 22: 5772–2781.

Shackleton, N.J., Berger, A., and Peltier, W.R. (1990). An alternative astronomical calibration of the lower Pleistocene timescale base on ODP site 677. *Trans. R. Soc. Edinburgh Earth Sci.* 81: 251–261.

White, W.A. and Morton, R.A. (1997). Wetland losses related to fault movement and hydrocarbon production, southeastern Texas. *J. Coastal Res.* 13: 1305–1320.

Suggested Reading

Carlile, B. (2009). *After Ike*. College Station, TX: Texas A & M Press.

Church, J.A., Woodworth, P.L., Aarup, T., and Wilson, W.S. (2010). *Understanding Sea Level Rise and Variability*. New York: John Wiley and Sons.

Davis, R.A. (2011). *Sea-level Change in the Gulf of Mexico*. College Station: TX, A & M Press.

Englander, J. (2012). *The Rising Tide on Main Street: Rising Sea Level and the Coming Coastal Crisis*. Boca Raton, FL: The Science Bookshelf.

Jamin, H. and Mandia, S.A. (2012). *Rising Sea Level: An Introduction to Cause and Impact*. Jefferson, NC: McFarland and Company.

Murray-Wallace, C. and Woodrofe, C.D. (2014). *Quaternary Sea Level Change: A Global Perspective*. Cambridge: Cambridge University Press.

Titus, J.G. and Narayanan, V.K. (1995). *The Probability of Sea-level Rise*. U. S. Environmental Protection Agency: Washington, DC.

Williams, M., Dunkerley, D., DeDeckker, P. et al. (1998). *Quaternary Environments* (chapters 5 and 6), 2e. London: Arnold Publishers.

5

Weather Systems, Extratropical Storms, and Hurricanes

5.1 Introduction

Storms are a frightening but fascinating manifestation of Earth's surface energy (Figure 5.1). They have a dramatic influence on coastlines, both in terms of sediment movement along the coast, and damage to homes and other structures. Storm waves and storm-induced currents are the primary agents responsible for removing sand from beaches and causing long-term shoreline recession. In a matter of hours storms can produce dramatic changes in coastal morphology, as evidenced by the breaching of barriers, relocation of tidal channels, formation of dune scarps and the deposition of extensive aprons of sand along the landward side of barriers.

Property damage resulting from major hurricanes can be enormous. In 1988 Hurricane Gilbert cut a wide swath through the Caribbean Islands resulting in 6.5 billion dollars to damage in Jamaica alone. In the wake of Hurricane Hugo that hit Charleston, South Carolina in 1989, 3 billion dollars were spent in reparations just to the city. Three years later Hurricane Iniki (1992) swept through the Hawaiian Islands causing a billion dollars in damage. During the same year Hurricane Andrew produced a staggering 25 billion dollars in property damage as it crossed over southern Florida into the Gulf of Mexico. Eventually it made a second landfall in western Louisiana where it wreaked devastation totaling another 1.6 billion dollars. Two more recent storms have sent the price tag of damage much higher. Hurricane Katrina impacting the central Gulf of Mexico coast in 2005, and Hurricane Sandy wreaking destruction along New Jersey, New York and New England in 2012, resulted in $108 billion and $75 billion of damage respectively.

As recorded in history and well chronicled during the last decade, the passage of major hurricanes not only results in substantial property loss, they have also proved to be "killers" to the unprepared coastal population. The greatest loss of life in the United States due to natural causes is attributed to the 1900 Hurricane that killed at least 8000 people on Galveston Island along the Texas Coast. Most of these deaths were due to drowning resulting from a 6-m high storm surge that covered the island, most of which was less than 2.5 m in height. A storm surge is the super-elevation of the water surface produced by a hurricane's strong onshore winds and low atmospheric pressure (Figure 5.2). In response to the Galveston disaster a substantial seawall was constructed along the ocean shoreline and, at the same time, much of the island was raised some 1.5 m using sediment dredged from a nearby site. The storm-related deaths that occurred in Galveston pale in comparison to the loss of life that has taken place in Bangladesh, located in the apex of the Bay of Bengal. Like the State of Louisiana the southern part this country has been built through deltaic sedimentation. Much of the lower portion of the Ganges–Brahmaputra Delta (Bangladesh) is highly populated. Most of these people live on land

Beaches and Coasts, Second Edition. Richard A. Davis, Jr. and Duncan M. FitzGerald.
© 2020 John Wiley & Sons Ltd. Published 2020 by John Wiley & Sons Ltd.

(a) (b)

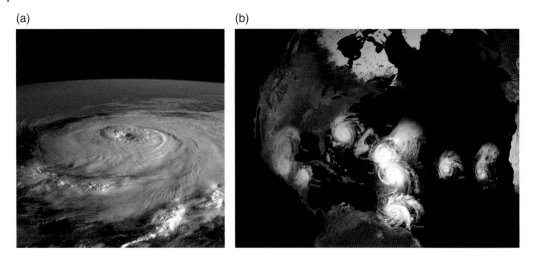

Figure 5.1 (a) Photograph of Hurricane Elena taken in 1985 from Space Shuttle Discovery. (*Source:* NASA-Johnson Space Center, https://upload.wikimedia.org/wikipedia/commons/b/b3/Hurricane_Elena.jpg). (b) Hurricane season begins in late spring and extends through mid fall. In late August of 1995 five major storms were present simultaneously in the North Atlantic Ocean while another storm, tropical storm Gil, was present in the northeast Pacific Ocean. Most Atlantic storms are born off the coast of northwest Africa and are steered westward by the Easterlies or Trade Winds. (*Source:* NASA, https://upload.wikimedia.org/wikipedia/commons/3/3b/2010_Hurricane_Season_Composite.jpg).

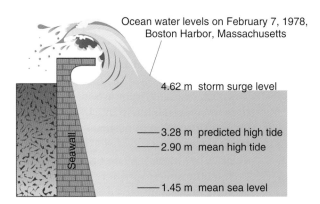

Ocean water levels on February 7, 1978, Boston Harbor, Massachusetts

4.62 m storm surge level

3.28 m predicted high tide
2.90 m mean high tide

1.45 m mean sea level

Seawall

Figure 5.2 A storm surge is a super-elevation of the water surface caused by the combined effects of wind stresses pushing water onshore and atmospheric low pressure. During the Blizzard of 1978, the storm of record for much of northern New England, large astronomic tidal conditions combined with record storm surges caused significant lowland flooding and extensive coastal damage. The high tide levels allowed large storm waves to break over seawalls, against fore dune ridges, and across barrier systems. The amount of beach erosion and damage to coastal structures and dwelling is closely related to storm surge levels.

that is less than 4 m above sea level. An intense cyclone in 1970 moved through the northern Bay of Bengal and pushed a dome of water onshore that reached 10 m above normal water levels. An estimated half-million people lost their lives as a result of this storm. A view of the area after floodwaters subsided, revealed a land almost completely stripped of human occupation. A second tropical cyclone in 1991 (Cyclone Gorky) took another 200,000 lives. What makes barriers and other depositional coasts so vulnerable to intense storms is the combination of storm-induced elevated water levels and the inherent low elevation of many coastal regions.

This chapter describes the fundamentals of atmospheric circulation, because global wind patterns largely control the pathways of storms and major weather fronts. A basic discussion of air flow associated with high- and low-pressure systems provides a foundation for understanding the two primary types of coastal storms; those that form above the tropics, called extra-tropical cyclones, and those that are generated in the tropics, called tropical cyclones. Typically, these storms have different strengths and modes of formation. Similar factors such as storm path, speed and magnitude, as well as the configuration of a coast, govern the extent and type of damage resulting from these storms; these points can be illustrated with case histories of various storms. The chapter presents classifications of hurricanes and extratropical storms are presented so that the effects of various magnitude storms can be compared. The discussion will emphasize hurricanes that impact the Gulf of Mexico and eastern Atlantic Ocean, although similar storms elsewhere in the world will be considered. Likewise, primary attention is given to the northeast storms that affect the northeast coast of the United States.

5.2 Basic Atmospheric Circulation and Weather Patterns

5.2.1 Wind

Wind is defined as the horizontal movement of air. It is caused by differences in atmospheric pressure, which in turn are produced by differential heating or cooling of air masses. On weather maps the distribution of atmospheric pressure is represented by isobars, lines connecting points of equal pressure (Figure 5.3). As seen on weather maps, wind flows at a slight angle to the isobars towards the central low pressure area. Wind speed increases with the steepness of

the pressure gradient, the steeper a pressure gradient the more closely spaced are the isobars. The pressure gradient can be thought of as the slope of a hill; a ball rolling down the hill will go faster the steeper the slope. Likewise, the greater the contrast in pressure (pressure gradient) the higher the wind speed.

5.2.2 Atmospheric Circulation

The movement of air masses over the surface of Earth is one of the processes whereby temperature differences between the equator and poles are accommodated. The Sun's preferential heating of the equatorial region compared to polar areas causes a strong temperature gradient between the low and high latitudes. A simple heat budget of Earth reveals that between approximately 35° North and 40° South, more energy is absorbed from incoming solar radiation than is radiated back into space (Figure 5.4). Above these latitudes the curvature of Earth produces a deficit of heat caused by a combination of less incoming radiation per unit area and greater reflectance of solar radiation by the ice-covered poles. While the uneven heating of Earth has always existed, the equator does not grow warmer with time, nor do the poles grow colder. Offsetting the heat imbalance are mechanisms of heat transfer involving both the atmosphere and oceans which move heat poleward and cold toward the equator. The California Current, which transports cold water southward along the Pacific Coast, and the Gulf Stream, which moves warm water northward along the Atlantic Coast, are examples of this heat transfer. The north–south movement of air masses accomplishes the same task of equilibrating energy over Earth's surface, as seen when polar air invades southern Canada and the continental United States during winter or when hurricanes move into the north Atlantic during late summer and early fall.

The global wind patterns are also a product of differential heating. The Sun's concentrated incoming radiation in the equatorial

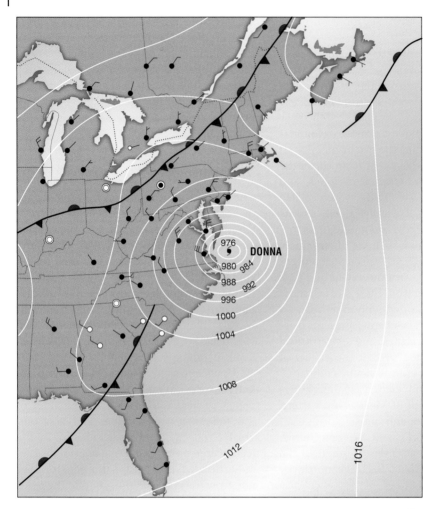

Figure 5.3 Atmospheric pressure is measured in millibars (1 millibar is equal to 100 N m^{-2}). Standard sea-level pressure is 1013.25 millibars. Isobars are lines on a weather map connecting points of equal pressure. Commonly, weather maps show concentric isobars around low and high-pressure systems. In the northern hemisphere, wind flows at a slight angle to the left of isobars, causing a counterclockwise windflow in low-pressure systems and a clockwise flow outward from high-pressure systems. Wind velocity increases as the pressure gradient steepens, or, as seen in weather maps, as isobars become more closely spaced. (*Source:* NOAA.)

latitudes heats the surrounding air and evaporates water from the oceans. As this warm moist air rises, it expands and cools, resulting in water condensation and rainfall in equatorial regions. In the highly simplified case of a non-rotating Earth one would expect that the dry air would flow outward from the equator toward the poles, resulting in further cooling of the air. At the poles the now cold, dry, dense air would sink and then flow back along the Earth's surface toward the equator. In this ideal model two convection cells would operate, one in each hemisphere (Figure 5.5).

However, due to the Coriolis effect, the actual global air circulation is much more complex. The Coriolis effect is produced by the Earth's rotation and causes all moving objects to be deflected to the right in the Northern Hemisphere and to the left in the Southern Hemisphere. The Coriolis effect applies to ocean currents, moving air masses, and even jet planes flying across the globe.

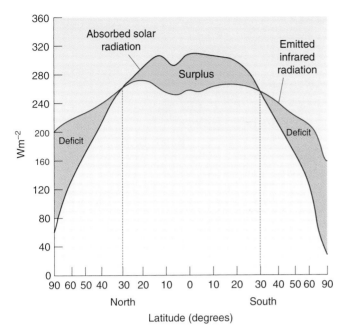

Figure 5.4 Over the Earth's surface there is an uneven distribution of incoming solar radiation and outgoing infrared radiation. As the graph illustrates, this condition leads to a surplus of heat in lower latitudes and a deficit of heat in higher latitudes. Ocean currents and weather systems are responsible for equalizing the geographic heat imbalance. The radiation budget of Earth balances because total heat losses equal total heat gains.

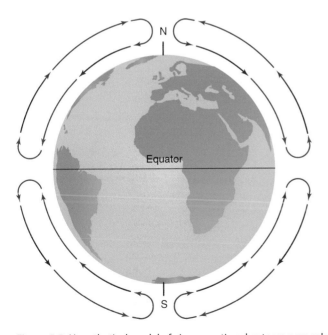

Figure 5.5 Hypothetical model of air convection due to uneven solar heating on a non-rotating Earth. Heated air in equatorial regions rises and moves poleward ultimately being replaced by cooled air descending from the poles and flowing toward the equator.

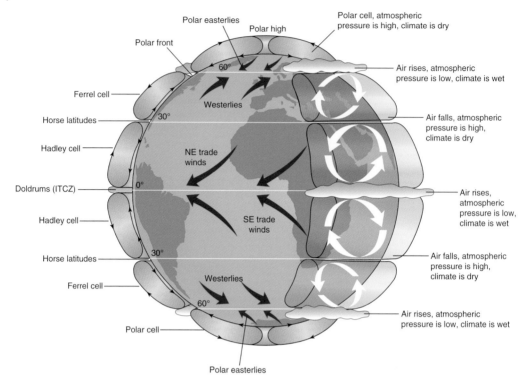

Figure 5.6 Simplified air circulation model for a rotating Earth containing the Hadley, Ferrel and Polar convection cells north and south of the equator.

Differential heating of the atmosphere coupled with the Coriolis effect produces a global air circulation model with six convection cells; three each in the Northern and Southern Hemisphere (Figure 5.6). As described in the simplified model, preferential heating causes warm moist air to rise in the equatorial region. The rising air is replaced by surface air flowing toward the equator from the northern and southern latitudes. The rising air expands and cools, producing precipitation. Thus, the equatorial region is noted for hot temperatures, low pressure, water-laden clouds, and rain. High above the equator the air mass flows towards the pole, all the while continuing to cool and lose moisture. At about 30° latitude the cooler, dryer, denser air mass descends, thus completing an atmospheric circulation cell known as the Hadley Cell. As the air sinks at 30° it compresses, producing high pressure, dry air, and variable winds. The

regions around latitude 30° are known as the horse latitudes. Folklore tells us they are so-named because ships sailing into these latitudes became becalmed. As the ship expended its animal feed and began running out of water, the crew was forced to throw horses and other farm stock overboard.

A second convection cell exists between 30 and 60° latitude, which is called the Ferrel Cell. This cell is formed because some of the air that sinks at 30° latitude flows toward the pole upon reaching Earth's surface. At the same time a secondary low pressure system stationed at about 60° latitude coincides with a rising air mass and precipitation. Aloft, this air mass cools and moves toward lower latitudes.

The Polar Cell is the third circulation cell and occurs between 60 and 90°. It results from upper air masses moving toward the pole and descending at the poles while at the same time surface air flows south.

5.2.3   Prevailing Winds

Now that the atmospheric circulation has been established, we can use air flow along Earth's surface to understand the global prevailing wind patterns (Figure 5.6). In the Northern Hemisphere circulation in the Hadley convection cell would seem to indicate that air should flow from north to south between 30 and 0° latitude. Remembering, however, that the Coriolis effect causes all moving masses to be deflected to the right (in the Northern Hemisphere), the flow of air from high pressure at 30° N to low pressure at the equator is actually from the northeasterly quadrant toward the southwest. These winds are called the Easterlies because the winds blow from the east. They are also referred to as the "Trade Winds." The early merchants who sailed from Europe bound for the New World gave them this name because the word "trade" was used by the British of that day to mean constant and steady, and this is how they described the winds that helped them sail across the Atlantic. In the southern hemisphere the pattern is mirrored, the Coriolis effect causing moving air masses to be deflected to the left, resulting in the convergence of the northeast and southeast Trade Winds at the equator. This region is called the intertropical convergence zone (ITCZ), and scientists have shown that it profoundly affects ocean currents and weather patterns in the equatorial area.

In the northern mid-latitudes, surface winds associated with the Ferral circulation cell are deflected to the east by the Coriolis effect. These winds are known as the Westerlies and affect most of the continental United States. The West Coast is known as a windward coast because the westerlies blow onshore, augmenting the wave energy in this region. Conversely, the East Coast is a leeward coast because the westerlies blow offshore. The prevailing winds diminish wave energy, partially explaining why average shallow water wave heights are more than twice as large on the West Coast as they are along the East Coast. The westerlies are also responsible for steering the mid-latitude weather systems, including hurricanes.

In the Northern Hemisphere, air flowing southward from the pole forms the Polar Easterlies. At about 50–60° latitude the polar easterlies meet the westerlies establishing the Polar Front. This convergent zone produces a near-permanent boundary separating the polar cold dense air from the warm tropical air mass. The variable weather that characterizes much of the United States reflects a latitudinal wandering of the Polar Front.

5.2.4   Cyclonic and Anticyclonic Systems

Although exceptions do exist, low atmospheric pressure is most often associated with rainy or stormy weather whereas high pressure is a sign of fair weather. A well-developed low pressure system is characterized by a gyre of air that rotates in a counterclockwise direction (in the Northern Hemisphere, clockwise in the Southern Hemisphere) around a central low pressure cell (Figure 5.7). These are called cyclonic systems and may be hundreds of kilometers is diameter. The counterclockwise movement of air is produced by the convergence of surface currents. Low pressure systems contain ascending air masses, which many people have observed in film footage of tornadoes. As air streams in to replace the rising air, the currents are deflected to the right by the Coriolis effect, producing a counterclockwise circulation. Hurricanes, extratropical storms, and tornadoes are all types of cyclonic systems, although tornadoes are considerably smaller systems.

High-pressure systems occur where air masses are sinking. As the air mass descends toward the ground, it flows outward from a central high pressure cell. In the Northern Hemisphere the outward-flowing currents are deflected to the right due to the Coriolis effect, resulting in a rotating air mass which circulates in a clockwise direction. These are called anticyclones. The Bermuda High is an

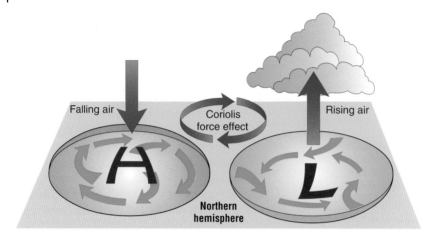

Figure 5.7 Gyres of moving air are created around low and high-pressure systems due to the Coriolis effect. Air masses flow in a counter-clockwise direction in low-pressure systems (cyclonic circulation) and in clockwise direction in high-pressure systems (anticyclonic circulation).

example of an anticyclonic system. This high-pressure system stabilizes over Bermuda during mid-summer and is responsible for transporting the uncomfortable hot, hazy, humid air from the Gulf of Mexico to the northeastern Atlantic seaboard.

5.2.5 Land-breezes and Sea-breezes

Anyone who has spent the summer along the seashore or has sailed along the coast is familiar with the systematic breezes that characterize the coastal zone. Onshore and offshore winds are a result of differential heating of the land surface versus that of the ocean. Under fair-weather conditions when the sun has just risen the air is usually calm because the air over the ocean and land have similar temperatures. However, as the sun ascends in the sky, the land surface preferentially warms compared to that of the ocean. In turn, the hot land surface warms the overlying air, causing it rise. Over the ocean the air remains cool and dense. Thus, a pressure gradient develops between the relatively low pressure over the land and the higher pressure over the ocean. This results in an onshore flow of air that is referred to as a sea-breeze (Figure 5.8). The sea breeze strengthens during mid-day and reaches a maximum

during mid to late afternoon before diminishing by late evening. During the night, the land surface radiates heat back to the atmosphere at a much higher rate than the ocean. This reverses the temperature and pressure gradient along the shore, resulting in a flow of air from the land toward the ocean, called a land-breeze. The extent and magnitude of the land and sea-breezes are a function of ocean and daytime temperatures, coastal morphology, vegetation, and other factors.

5.3 Mid-latitude Storms

Whereas tropical storms and hurricanes dominate coasts of low latitudes, extratropical storms and weather fronts are the major weather systems impacting mid-latitude coasts. A transition zone exists between these two regions where both weather systems are common. As the name implies, extratropical cyclones form north of the Tropic of Cancer or in the case of the Southern Hemisphere, south of the Tropic of Capricorn. These storms are associated with low-pressure systems and affect the Pacific, Gulf, and Atlantic Coasts of the United States and other mid-latitude coastlines of the world. Like all cyclonic weather systems, extratropical storms are air masses that

Figure 5.8 Land and sea breezes are a common diurnal pattern of winds along many coasts. They are formed due to temperature differences between the land and adjacent seawater. During the day ascending air over the land produces onshore sea breezes, whereas at night rising air over the sea causes offshore land breezes.

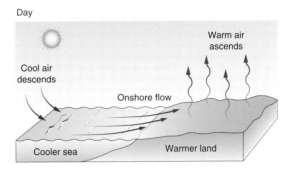

Day

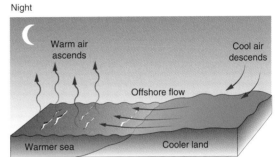

Night

rotate around a central low pressure having a counterclockwise circulation (Figure 5.7). Another important type of mid-latitude weather system is the front. The passage of frontal systems along the Gulf Coast strongly influences the coastal processes of this region. Frontal systems are discussed first, as they are the precursors to extratropical cyclone development.

5.3.1 Frontal Weather

A front is defined as a narrow transition zone (25–250 km wide) between two air masses having different densities (Figure 5.9). Fronts may extend for more than 1000 km. One air mass is usually warmer and more humid than the adjacent air mass. The boundary between the two is usually inclined, with the warmer lighter air rising over the colder, denser air. A cold front is one in which the cold air mass advances thereby displacing the position of the warm air. The opposite occurs during the passage of a warm front. Generally, the two air masses travel with nearly the same speed and in about the same direction. Cold fronts usually advance at a slightly more rapid rate

(35 km h^{-1}) as compared to warm fronts (25 km h^{-1}). In addition, cold fronts tend to be accompanied by more energetic weather and often contain concentrated precipitation and severe wind.

In North America, cold fronts commonly are initiated by cold air (a polar air mass) sweeping down from Canada meeting warm air from the south (Figure 5.10). Fronts travel west to east across the country and may extend all the way to the Gulf of Mexico, where they can produce strong winds and surf. The northern Gulf region experiences more than 20 cold fronts each year lasting from 12 to 24 hours depending on the speed of the storm and whether it becomes stalled or not. Due to the overall low tidal range of most of the Gulf Coast (<1 m) cold fronts can be effective agents in substantially augmenting or diminishing tidal elevations. Along the Louisiana coast cold fronts create westerly to southwesterly winds, which cause higher tide levels than would be expected due to astronomic forcing alone. This condition enables storm waves to break higher along the beach, leading to the over-washing of low barriers. At the same time

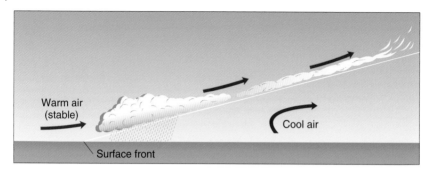

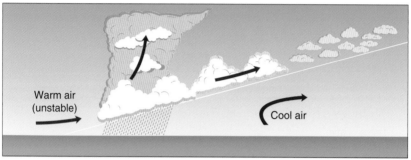

Figure 5.9 Weather fronts are formed at the boundary between two air masses of contrasting densities, which is usually due to differences in temperature. Fronts mark a change in the weather and are generally associated with moderate to intense precipitation.

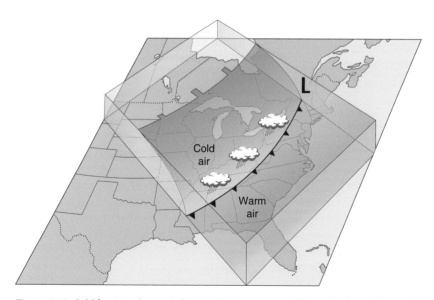

Figure 5.10 Cold fronts in the central United States are caused by polar air moving south from Canada displacing a warm air mass. The front moves eastward across the country and may stretch from the Gulf of Mexico to the Great Lakes.

elevated tides increase the flow of ocean water into the bays and marsh systems that back the barrier islands. This generates strong landward currents through tidal inlets during the passage of the front, and strong seaward currents when the floodwaters exit the inlets. As the front passes, there is a dramatic shift from southerly to northerly winds that occurs within a few hours or less. Strong northerly winds suppress the waves in the nearshore of the Gulf-facing beaches, while at the same time generating substantial waves in the larger bays behind the barriers. In fact, scientists have shown that these bay waves, which can approach 3 m in height, are responsible for chronic shoreline erosion on the backside of barriers along long stretches of the northern Gulf of Mexico. Thus, cold fronts are important agents in modifying beaches and tidal inlets in the Gulf Coast region.

5.3.2 Cyclogenesis

The process of extra-tropical cyclone formation, cyclogenesis, was first described by Norwegian scientists during World War I, and eventually the theory was published by J. Bjerknes in 1918. Even though it was devised using limited ground observations, the basic tenets of the model are still deemed acceptable today despite major advances in the collection and analysis of weather data. Cyclones develop along advancing frontal systems in which the two air masses have a slight component of differential movement. At the surface of the front this condition is manifested by the two air masses moving in opposite directions to one another (Stage 1, Figure 5.11). The next stage of cyclogenesis coincides with a disturbance along the front that is produced by topographic irregularities, such as mountain ranges, temperature differences, or other factors. The end result of the disturbance is a

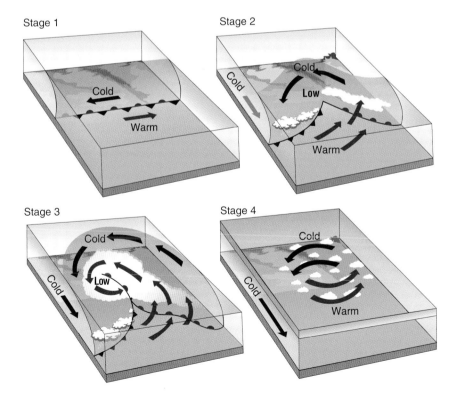

Figure 5.11 Model of cyclogenesis as proposed by Norwegian scientists and published by J. Bjerknes.

wave-like form, similar to a breaking ocean wave, in which the warm low-density air penetrates into the cold air mass (Stage 2, Figure 5.11). Extra-tropical cyclone formation also appears to be strongly linked with upper air circulation where currents flow west to east in long meanders. This pattern serves to initiate or reinforce the counter-clockwise rotation of air around a central low pressure (Stage 3, Figure 5.11). The demise of a cyclone occurs when the cold air supplants the rising warm air, the sloping gradient between the two air masses ceases, and the storm runs out of energy (Stage 4, Figure 5.11).

5.3.3 Extratropical Storms

5.3.3.1 Occurrence and Storm Tracts

Although extratropical cyclones can occur in the mid-latitudes at any time of the year, they happen most frequently in the hemisphere that is going through winter because this is when the temperature contrasts between the polar air masses of high latitudes and the warmer air of the lower latitudes reach their maximum. In the United States, extratropical storms are most common from late fall through early spring. The weather conditions that lead to extratropical cyclone formation often originate in the Pacific Ocean. Full development of the storm is associated with several locations throughout North America due to the complexity of the factors governing the continent's weather. These sites include the continental southwest, the Midwest, and the southeast United States, southern Alberta, Canada, and in the Gulf of Mexico and the Florida–Bahamas regions (Figure 5.12). The continental storms travel on an eastward path across the country eventually moving northeast into the northern Atlantic where they dissipate. Those that pass through the Gulf states and the Bahamas region move northward along the eastern seaboard. Other storms track through the Great Lakes and into the Gulf of Saint Lawrence. Because most extratropical storms track along the east coast, they are major storm producers of this region, particularly north of North Carolina. Hurricanes have a greater frequency and influence along the coastal states south of Virginia.

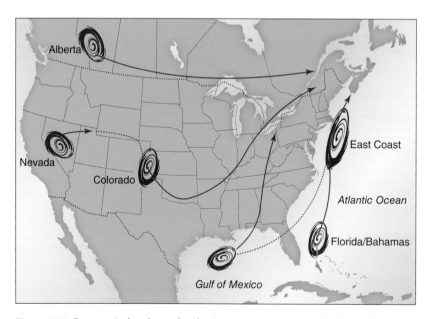

Figure 5.12 Extratropical cyclones develop in many regions across the United States. Regardless of their origin, the pathway of these storms is toward northeast Canada and the northern Atlantic Ocean.

5.3.3.2 Northeasters

"Northeaster" or "Nor'easter" is the name given to the extratropical cyclones that pound the northeast coast of the United States and Canada with driving rains, strong winds, elevated tide levels, and storm waves. During the winter, northeasters can produce blizzard conditions blanketing the northeast with more than 2 ft of snow. They are called northeasters because the winds associated with these storms come from the northeast (Figure 5.13). (Remember that winds are named for the direction from which they blow.) In fact, it is the path of the extratropical cyclone that determines the wind direction as well as the type and severity of the storm. Northeast storms occur when the eye of the storm tracks in a northeasterly path offshore of the coast, eventually moving east of Cape Cod and Nova Scotia. Under these conditions the counter-clockwise air circulation associated with the cyclone generates onshore winds that blow out of the easterly quadrant. The eastern seaboard of the United States is particularly susceptible to these storms because most of this coast faces eastward and is directly exposed to northeast winds and waves.

The Ash Wednesday Storm of 1962 was one of the most powerful and damaging northeasters ever to strike the east coast of the United States during historic times. The storm lasted for more than five tidal cycles and impacted a 1000 km of mid-Atlantic coastline, stretching from North Carolina to Long Island, New York. Wind gusts exceeded hurricane force, producing deepwater wave heights greater than 10 m. The storm hit

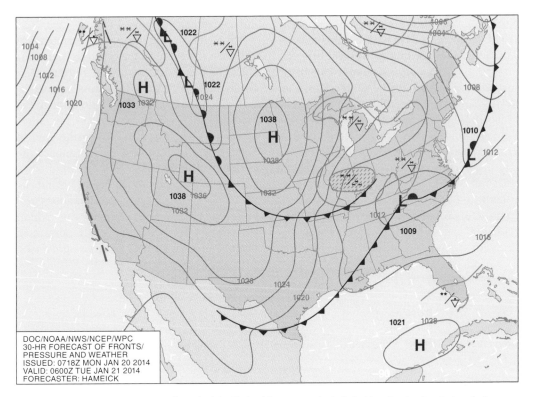

Figure 5.13 Along the eastern seaboard of the United States, particularly in New England, extratropical cyclones travel offshore of the coast and generate strong winds and storm waves from the northeast. Hence, these storms are called "northeasters." The weather map illustrates the conditions that existed during a northeast storm on January 25, 1979. In the Boston, Massachusetts area, winds reached 60 km h⁻¹ (40 mph) and the region was blanketed with over 2 ft of snow. (*Source:* NOAA.)

during perigean spring tides, which contributed to the elevated water levels. The large storm surge and erosive waves removed beaches and carved deeply into adjacent dunes. Along many shorelines storm waves carried beach sands across barriers forming extensive overwash deposits. The storm breached numerous barriers creating more than a dozen ephemeral tidal inlets. Along populated sections of shoreline, rows of houses and other buildings were destroyed. Ultimately the "storm of the century" caused more than $300 million in damage and accounted for numerous deaths.

A classification of northeasters has been devised by scientists Robert Dolan and Robert Davis, who have studied 1347 storms affecting the North Carolina coast over a 42-year period. Their scheme is based on the power of the northeaster, which is calculated by multiplying the storm's duration times the square of the maximum significant wave height (Table 5.1). They have divided storms into five classes, with the weakest storms, Class #1 and #2, comprising about 75% of all northeasters impacting the Outer Banks of North Carolina. As seen in Table 5.1, Class #4 northeasters have a frequency of 2.4% with a significant deep-water wave height of 5 m and an average duration of 63 hours. The extreme northeaster, Class #5, has a recurrence interval of 67 years. These storms have a significant wave height of 7 m, an average duration of 96 hours, and are agents of permanent change along the coast, as was the case with the Ash Wednesday Storm of 1962.

5.3.3.3 Shoreline Vulnerability

Many of the same factors that control the severity of hurricanes also apply to the impact of northeasters. The storm's size and intensity are of paramount importance as these elements control wave height, magnitude of storm surge, and, to some extent, the storm's duration. Astronomic tidal conditions can significantly augment or reduce the effects of the storm surge thereby affecting the elevation where waves break along a shoreline. For example, the Blizzard of 1978, which is the storm of record along much of the New England coast, occurred during extraordinarily high tidal elevations in which the predicted mean high tide was 60 cm higher than mean high-water levels. The coincidence of high astronomic tides with the blizzard was a major factor contributing to coastal flooding, beach erosion, overwashing of barriers, inlet formation, and wave-induced damage (Figure 5.14).

Although the majority of extratropical cyclones travel in a northeasterly path offshore of the coast, producing northeasterly winds and waves, some low-pressure systems take an inland track, for example through the Hudson or Connecticut Valleys. In some instances, storms move through the Gulf of Saint Lawrence. In these circumstances the cyclonic wind regime produces wind from the southerly quadrant. In New England the shorelines have a variety of orientations and exposures to wave energy. In this region the storm tract is particularly important in determining the impact of storm processes.

Table 5.1 Dolan and Davis classification.

| Storm class | Frequency of storms | | Significant wave height (m) | Duration (h) | Power (m² h) | |
	Number	Percentage			Mean	Range
1 Weak	670	49.7	2.0	8	32	≤71.63
2 Moderate	340	25.2	2.5	18	107	>71.63 to 163.51
3 Significant	298	22.1	3.3	34	353	>163.51 to 929.03
4 Severe	32	2.4	5.0	63	1455	>929.03 to 2322.58
5 Extreme	7	0.1	7.0	96	4548	>2322.58

Figure 5.14 View of houses destroyed in Scituate, Massachusetts during the Blizzard of 1978. Storm waves and the onshore movement of gravel toppled foundations, leading to the collapse of numerous houses along this section of coast. Many of these same houses received extensive damage resulting from the Halloween Eve northeast storm of 1991.

The passage of the common northeaster primarily affects the eastward-facing shorelines of outer Cape Cod and the north and south shores of Massachusetts Bay as well as the New Hampshire and southern Maine coasts. Storms that travel west of New England generate southerly wind and waves that impact the southward shores of Rhode Island, Cape Cod, and the central peninsula coast of Maine.

Another factor that influences the vulnerability of beaches to storm erosion and damage to adjacent structures is the interval between storms. The width and elevation of the berm strongly affects erosion of the abutting dune during storms. Beaches with wide accretionary berms can withstand the onslaught of an intense northeaster because the berm constitutes a large quantity of sacrificial sand that can be eroded before the dune or adjacent dwelling is destroyed. If a previous storm has removed the sand buffer, then the impact of the next storm will be much greater than the first. One of the reasons that the Halloween Eve Storm of 1991 caused significant damage to New England was that Hurricane Bob, which passed through the region a month and half earlier, had left many of the shorelines with little sand on the beaches.

5.4 Hurricanes and Tropical Storms

5.4.1 Low Latitude Storms

Hurricanes and their forerunner, tropical storms, are the major storms of the tropics. Like other weather systems they may move beyond their typical latitudinal range given the right set of circumstances. Most of us are familiar with tropical storms and hurricanes due to the wide media coverage they receive, particularly when a major coastal region in the United States is impacted.

Tropical storms and hurricanes are known by different names in other parts of the world. In the western Pacific they are called typhoons, in the Indian Ocean they are known as cyclones, and in Australia they are given the name "willi-willis." A tropical storm is a low-latitude cyclonic system that may intensify to be reclassified as a hurricane if wind velocities surpass 119 km h^{-1}. During an average year, approximately 20 tropical storms form in the equatorial Atlantic and of these 8–10 reach hurricane strength. Most hurricanes that make landfall in the United States do so in the Gulf of Mexico or along the Florida and North Carolina coasts. In a study of hurricane frequency for the southeast United States coast, Robert Muller and Gregory Stone of Louisiana State University showed that Morgan City, Louisiana and the Florida Keys have the shortest recurrence of major hurricanes (wind speed > 179 km h^{-1}, 111 mph) (Table 5.2). The Florida Keys (including Key West and Key Largo) have survived over 32 hurricanes during the past century, of which seven were major hurricanes.

5.4.2 Origin and Movement of Hurricanes

5.4.2.1 Formation

Tropical weather is generally considered to be that occurring between the Tropic of Cancer and the Tropic of Capricorn (23.5 N–23.5 S lat.). Here there is little change in day-length, seasons are subtle to non-existent, year-round temperature is warm to hot, and major changes in weather patterns are linked to the dry and wet seasons. In these latitudes the winds typically blow from the southeast, east, or northeast depending on the latitude. Weather systems of the Tropics, such as hurricanes, are steered by the trade winds, in contrast to those of the mid-latitudes that move west to east (the Westerlies).

The first indication of the potential development of a storm in low latitudes is the presence of a tropical wave. This feature is identified on weather charts as a bending of the streamlines, which show pathways of airflow within the wind system. Tropical waves form over western Africa and move westward into the Atlantic Ocean where they gain strength over the warm water of the tropical latitudes (Figure 5.15). The warm water reduces pressure along the wave, transforming it into a trough. In a typical year, about 60 of these develop during the hurricane season – about one every three to four days. They have a wave length of about 2500 km and travel westward with speeds of 10–40 km h^{-1}. Some of these troughs intensify and develop into tropical disturbances, which is the infancy stage of a tropical cyclone. Tropical disturbances are characterized by a line of thunderstorms that maintain their identity for a day or so. These weather systems have a rotary circulation, counterclockwise in the northern hemisphere and clockwise in the southern hemisphere. Further strengthening of these storms produces a tropical depression, which is a weather system having maximum cyclonic wind velocities up to 61 km h^{-1} (38 mph). Storms with wind velocities greater than 61 km h^{-1} but less than 119 km h^{-1} are classified as tropical storms.

In about 10 % of cases, the developing tropical storm receives sufficient energy from the warm ocean waters to reach hurricane strength. Hurricanes are one of Earth's largest weather systems with wind velocities of at least 119 km h^{-1} (74 mph) but some exceeding 250 km h^{-1} (155 mph). The stages in the genesis of a hurricane are listed in (Table 5.3). The conditions necessary to produce a hurricane include:

- Warm ocean temperatures (>26 °C), which occur from the beginning of June until the end of November in the Northern Hemisphere and during the opposite time of year in the Southern Hemisphere. The conditions extend late in the season because ocean waters cool slowly in the fall. Most hurricanes form in August and September when ocean waters are at their warmest.
- Vertical movement of warm moist air rising within the storm from the ocean surface upward to a height of 10–20 km. As the air rises it cools. Eventually the water vapor contained in the air condenses releasing huge quantities of energy in the form of heat.
- The Coriolis effect produces the spinning of the hurricane. Air that flows toward the

Table 5.2 Hurricane frequency in southeast United States.

	Hurricane strikes	Major hurricane strikes	Hurricane return period (years)	Major hurricane return period (years)
South Padre Island, TX	8	3	12	33
Port Aransas, TX	9	2	11	50
Port O'Connor, TX	13	2	8	50
Galveston, TX	13	3	8	33
Cameron, LA	5	2	20	50
Morgan City, LA	10	7	10	14
Boothville, LA	16	5	6	20
Gulfport, MS	10	2	10	50
Dauphin Island AL	15	4	14	25
Pensacola Beach, FL	14	2	7	50
Destin, FL	10	3	10	33
Panama City Beach, FL	10	2	10	50
Apalachicola, FL	11	0	9	100
Cedar Key, FL	3	1	33	100
St Petersburg, FL	5	1	20	100
Sanibel Island, FL	8	1	12	100
Marco Island, FL	12	3	8	33
Key West, FL	17	2	6	50
Key Largo, FL	15	5	7	20
Miami Beach, FL	15	2	7	50
Palm Beach, FL	13	4	8	25
Vero Beach, FL	10	1	10	100
Cocoa Beach, FL	4	0	25	100
Daytona Beach, FL	3	0	33	100
Jacksonville Beach, FL	2	0	50	100
St Simons Island, GA	1	0	100	100
Tybee Island, GA	5	0	20	100
Folly Beach, SC	3	1	33	100
Myrtle Beach, SC	7	2	14	50
Wrightsville Beach, NC	11	2	9	50
Atlantic Beach, NC	10	3	10	33
Cape Hatteras, NC	15	3	7	33

Source: Study of hurricane frequency of the southeastern USA by Robert Muller and Gregory Stone of Louisiana State University.
TX, Texas; LA, Louisiana; MS, Mississippi; AL, Alabama; FL, Florida; GA, Georgia; SC, South Carolina; NC, North Carolina.

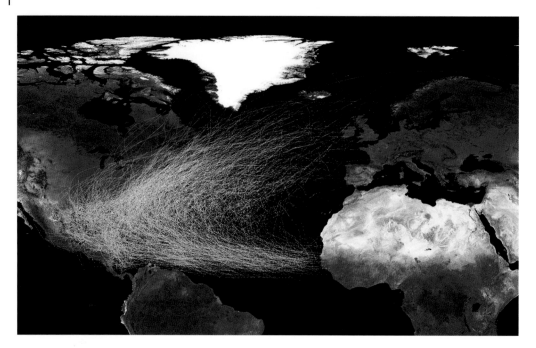

Figure 5.15 Pathway of hurricanes in the central Atlantic during the past approximately 100 years. Note that tropical storms and hurricanes are born off the west coast of northern Africa and travel westward toward the Caribbean and Gulf of Mexico. These storms eventually turn north and eastward before expiring in the northeast Atlantic. (*Source:* NASA, https://upload.wikimedia.org/wikipedia/commons/3/31/Atlantic_hurricane_tracks.jpg).

Table 5.3 Storm genesis.

Tropical disturbance	Prevalence of thunderstorms, infancy stage of tropical cyclone
Tropical depression	Cyclonic wind pattern, wind speeds up to 61 km h^{-1} (38 mph)
Tropical storm	Highly organized ocean storm visible in satellite imagery, wind speeds from 61–119 km h^{-1} (38–73 mph)
Hurricane	Spiraling cloud pattern around a central eye, wind speeds exceed 119 km h^{-1} (73 mph)

center of the low-pressure system to replace the air that is rising in the hurricane is deflected (to the right in the Northern Hemisphere, to the left in the Southern Hemisphere). This causes the air mass to rotate. Hurricanes generally do not form within 5° of the equator because the Coriolis effect is very weak in this region.

5.4.2.2 Hurricane Pathways

Once formed, hurricanes and developing tropical storms move in a variety of pathways, all having a general westerly direction (Figure 5.16). South Atlantic hurricanes usually travel across the Caribbean Sea and then either enter the Gulf of Mexico or move northward up the western margin of the Atlantic. Storms that move into the Gulf of Mexico continue west or northwest, making a landfall encompassing the shoreline between the Yucatan Peninsula and the western Florida panhandle. It is rare that these storms swing northeast and cross the western Gulf Coast of Florida. The last hurricane to do so was Donna in 1960.

Atlantic tropical storms that do not enter the Gulf generally move northward with a slight westerly component. Although they do not ordinarily cross the Florida peninsula, they may travel close enough to impact coastal communities with high surf, strong

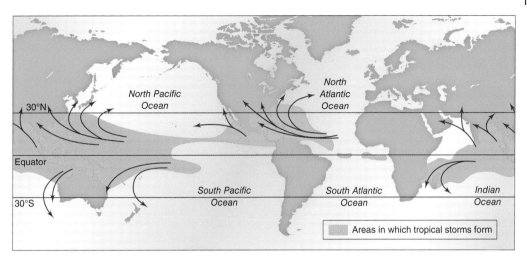

Figure 5.16 Hurricanes and tropical storms form in the tropics above 5° north and south of the equator. Trade winds steer these storms westward, where they eventually degrade over continental areas. In other cases, after traveling westward the storms move east and poleward into the colder waters of the mid-latitudes, where they die.

winds, and possible lowland flooding. These storms commonly make a landfall in the Carolinas or in rare instances, as far north as New England. Recent examples are Hurricane Hugo, which hit South Carolina in 1989, and Hurricane Fran that came across the Outer Banks of North Carolina in 1996.

Hurricanes and tropical storms in other ocean basins of the Northern Hemisphere have pathways that are similar to the Atlantic systems (Figure 5.16). In the western Pacific storms travel westward toward the Philippines and Southeast Asia as well as swinging northward where they impact the coasts of China, Japan and Korea. In the northern Indian Ocean cyclones move northwestward, making landfalls along Bangladesh, India and Pakistan, and along the Arabian Gulf. The mirror image of this pattern takes place in the southern hemisphere, where tropical cyclones move westward curving to the south.

5.4.3 Anatomy of a Hurricane

Conditions that lead to the formation of a full-blown hurricane start with what is called a seedling. In addition to warm water, there must be weather conditions that enhance the upward spiraling of winds, the "vorticity" of

the storm. High humidity, lack of vertical wind shear, and wind surge are also conducive to hurricane formation. Wind surge adds bursts of high-speed flow to the center of the disturbance, causing upward circulation and intensification of the storm.

Although hurricanes vary greatly in size, intensity, speed and path, they have many common characteristics. A satellite view of a hurricane reveals spiraling bands of thunderstorm-like clouds (Figure 5.17). Some of these cloud systems contain abundant moisture and some do not, which explains why the intensity of rainfall is so variable during the passage of a hurricane. Hurricanes may have a single cloud band or more than seven, each extending from the center of the storm outward to a distance of about 80 km. The storm itself can range in diameter from about 125 km to more than 800 km; the average is 150–200 km.

Most people know that the center of the hurricane is called the **eye** and that within this zone winds are weak to perfectly calm. The eye is 5–60 km in diameter, averaging about 20 km. This part of the hurricane is cloudless – many people comment about seeing the sun or stars as the eye passes over them. It is also common for birds to be

Figure 5.17 Satellite view of Hurricane Andrew on the 23rd, 24th, and 25th of August 1992 as it moved westward across Florida toward Louisiana. (*Source:* NASA. https://www.google.com/search?biw=1300&bih=74 3&tbm=isch&sa=1&ei=nY4CXaTuPLL45gKxvIbQCw&q=NASA+hurricane+Andrew+photos+in+Atlantic&oq=N ASA+hurricane+Andrew+photos+in+Atlantic&gs_l=img.3...69831.72211..74044...0.0..0.70.411.7......0....1.. gws-wiz-img.......35i39.GcQZp545poY#imgrc=7CZErm9B-6k9YM:)

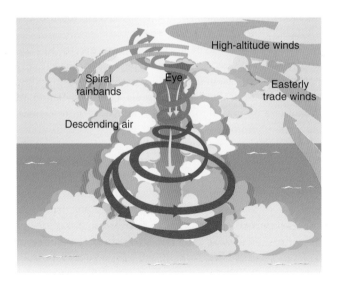

Figure 5.18 Internal structure of a hurricane illustrating the overall counter-clockwise flow of air at the base of the storm. Toward the center of the hurricane the air mass spirals upward and eventually flows outward at the top of storm. The vertical dimensions of the hurricane in this diagram are greatly exaggerated.

trapped within the eye; the strong winds beyond the storm's center make escape impossible (Figure 5.18).

The fact that the strongest wind occurs near the center of the storm is due to conservation of angular momentum. The Law of Conservation of Momentum states that the product of an object's velocity around its center and its distance from the center squared is constant. The law can be stated mathematically as:

$$MVD^2 = constant$$

where M is the mass of the object, V is its velocity around the center, and D is its distance from the center. This principle can be illustrated by thinking of a figure skater, who is doing a spin on the ice. As a skater pulls her arms in toward her body, she decreases the diameter of her rotation. To maintain the same angular momentum the skater's rotational velocity increases drastically. By doing the opposite and extending her arms straight outward, the skater slows. Likewise, the winds in a hurricane gradually diminish in velocity toward the periphery of the storm.

Adjacent to the eye of the hurricane is a wall of clouds, which may reach nearly 20,000 m in height. The eye wall, as it is called, contains abundant water vapor. This vapor moves upward and eventually condenses, releasing vast amounts of energy in the form of heat, which strengthens the storm. During a single day, a hurricane releases enough energy to supply the United States' electrical power needs for an entire year.

Because of the numerous conditions affecting the development of a hurricane, their ultimate size and intensity vary considerably from one storm to another. The primary factors in determining the intensity of a given storm are wind velocity and barometric pressures. Storm surge, which is largely responsible for the damage resulting from a hurricane, is difficult to predict because of the variations in speed of the storm, the diversity in bathymetry of the inner continental shelf and the configuration of a coast.

Most people are aware that hurricanes are named. This practice was initiated because of the confusion caused when several hurricanes are active at one time. Prior to naming the storms they were identified by their location; latitude and longitude. Names were first given during World War II and were allocated in alphabetical order, such as Able, Baker, Charlie; the commonly used designations for the alphabet by the military. This practice continued until 1953 when female names began to be used; also in alphabetical order. This style of designating

storms lasted for 25 years, then, in 1978, the policy was changed to include both male and female names. A set of names is chosen years in advance with names alternating between male and female continuing through the alphabet. Names of hurricanes that have severely impacted the United States are permanently retired from the list. Hurricanes that have achieved this status include Camille (1969), Hugo (1989), and Andrew (1992). Different names are used for north Atlantic and eastern Pacific storms. The name is applied from the time a storm achieves tropical storm level until it has completely dissipated.

Hurricanes lose their power when they move over cool ocean waters or onto land. Generally by 40° latitude the waters are too cold to supply the large amounts of moisture needed to fuel the storm.

Likewise, when a hurricane moves over land sources of water vapor are greatly reduced. The ability of the storm to take up moisture is further lessened by the cooling effect of the land. Finally, the friction imparted by the land surface rapidly diminishes the low-level storm winds. These factors contribute to an increase in barometric pressure and a spreading out of the storm, leading to its general unraveling and loss of identity. Hurricanes typically last about a week to ten days, though some have been known to last as long as a month.

5.4.4 Hurricanes at the Coast

Many coastal regions around the world, including numerous sites along the Gulf coast and the eastern seaboard of the United States, are low-lying and moderately to densely populated, with numerous dwellings, buildings, and other infrastructure. This combination is highly vulnerable to a major storm. Devastation to natural environments, destruction of property, and injury or even death to people, are all typical hurricane impacts to the coastal zone. This section will consider what happens when a hurricane approaches and passes over a coastal area.

5.4.4.1 Factors Affecting their Severity

The strength, speed, and size of the storm are major elements in determining how the hurricane will affect the coast. In addition, the gradient and width of the inner continental shelf are important considerations. In some regions, such as the New York Bight, the configuration of the coast is also a critical factor. All of these variables contribute to the size of waves, magnitude of the storm surge and overall impact of the hurricane. Remember that storm surge, or storm tide as meteorologists frequently call it, is the super-elevation of the ocean water surface above the predicted tide level. It is the storm surge that allows the high-energy storm waves to break high against a dune ridge, across a barrier, or over a seawall. We can better assess the impact and behavior of different hurricanes and compare those of increasingly higher category, if first we understand the factors that define and influence them:

- *Magnitude* – encompasses both hurricane size and intensity. Hurricane size governs the length of coast that is affected by the storm as well as the duration of high-velocity winds and high-energy waves. The greater the intensity of the hurricane, the higher the wind velocity and the larger the wave heights and storm surge.
- *Speed of Storm* – determines amount of time that storm winds can transfer their energy to the water surface waves and pile water onshore. Generally, slower-moving storms produce higher waves and larger storm surges than faster moving storms of equal magnitude.
- *Path of Storm* – determines the landfall of a hurricane and areas along the coast of greatest storm impact. In the northern hemisphere when a hurricane moves onshore, areas to the right of its landfall will experience the strongest winds and greatest storm damage.
- *Coastline Configuration* – is an important factor in large deeply embayed coastlines. In this setting certain hurricane tracks can significantly amplify storm surge levels as water is forced into a funnel-shaped embayment.

The circulation of wind within a hurricane and the storm's speed and pathway toward land dictate the relative intensity of the storm and the amount of damage the hurricane will inflict along a given stretch of coast. In the northern hemisphere the wind orbiting the central low pressure of the hurricane travels in a counter-clockwise direction. At the same time that this wind is blowing in circular fashion, the hurricane is also moving over the ocean surface with a speed that ranges from 5 to 40 km h^{-1}. This forward movement of the hurricane has an additive effect on wind blowing in the same direction as path of the storm and a subtractive effect where the storm wind is blowing in the direction opposite to the storm's forward motion. To illustrate this point, consider a moderate-sized hurricane in the Gulf of Mexico having average wind speed of 180 km h^{-1} (Figure 5.19). The storm is moving northward with a speed of 30 km h^{-1}. East of the storm's eye hurricane wind blows in a northerly direction with a velocity of 210 km h^{-1} (180 + 30 km h^{-1}). At the same time, wind west of the eye blows southward with an effective velocity of only 150 km h^{-1} (210–30 km h^{-1}). Not only is the effective wind velocity different on either side of the storm, but also as the storm passes the coast the shoreline east and west of the eye experiences very different wind and wave conditions. East of the eye there is a wind "set-up" where wind continues to blow onshore building the storm surge. Conversely, west of the eye there is a "set down" producing lower water level and waves due to the wind blowing offshore. Damage from a hurricane is always greater to the right of its landfall than to the left.

Hurricane Frederick was a moderate-sized storm that struck the Mississippi coast in 1979. Meteorological and oceanographic data collected during the storm illustrate the relationship among the storm's low pressure, its wind velocity, and storm surge pattern (Figure 5.20). The hurricane moved onshore from the Gulf of Mexico having an average forward speed of 15 km h^{-1}. The eye of the storm made landfall at Dauphin Island,

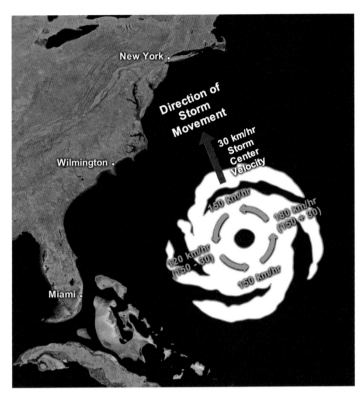

Figure 5.19 The forward speed of a hurricane increases wind velocities on the advancing right side of the storm while diminishing wind velocities on the advancing left side of the storm.

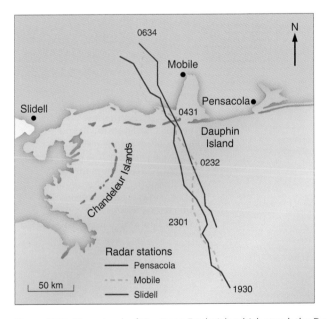

Figure 5.20 Storm track of Hurricane Frederick, which struck the Dauphin Island along the Mississippi coast on 13 September 1979. (*Source:* Courtesy of Shea Penland, University of New Orleans.)

which is located along the western flank of Mobile Bay. As seen in Figure 5.21, the highest wind velocity of 126 km h^{-1} that was recorded at the coast coincided with the period of lowest pressure. Generally the lowest pressure is found near the eye of the hurricane, and it gradually increases toward the perimeter of the storm. The strongest wind corresponds with the steepest pressure gradient, which occurs just beyond the eye of the storm. During the passage of Hurricane Frederick maximum tidal elevations recorded along the coast ranged from 1.0 to 3.8 m above mean sea level. As illustrated in Figure 5.22, water elevations were much greater east of Dauphin Island than in the coastal regions to the west. Note that the greatest storm surge took place approximately 30 km east of the eye. This pattern of higher water-level, flooding, and damage that occurs east of the storm center is due to the hurricane's counter-clockwise wind circulation. When the hurricane was centered over Dauphin Island wind was still blowing onshore east of the island. At the same time, wind was already blowing offshore along the coasts of eastern Louisiana and western Mississippi. The longer time hurricane winds blow onshore, the greater the storm surge and resulting damage.

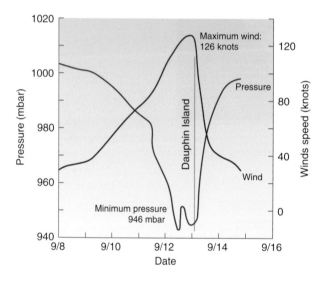

Figure 5.21 Graph of atmospheric conditions at Dauphin Island, Mississippi during the passage of Hurricane Frederick. Maximum wind velocities coincided with lowest atmospheric pressures. (*Source:* Courtesy of Shea Penland, University of New Orleans.)

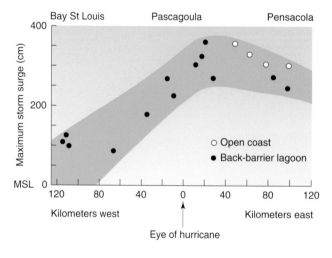

Figure 5.22 Storm surge values along the Gulf Coast from Mississippi to the Panhandle of Florida resulting from Hurricane Frederick. The greatest storm surges occurred east of the landfall due to the counterclockwise wind circulation associated with hurricanes and cyclonic storms. (*Source:* Courtesy of Shea Penland, University of New Orleans.)

Box 5.1 The Fury of Hurricane Camille

Hurricane Camille slammed into the Mississippi coast late on August 17, 1969. Despite hurricane warnings and evacuations, 143 people lost their lives along the Gulf Coast. Another 113 people drowned in Virginia floods caused by intense rainfalls spawned by Camille.

This destructive storm was born in early August when a tropical wave traveled westward off the coast of Africa. By August 14, it had reached tropical storm status and later that day intensified to a hurricane 100 km southeast of Cuba. When it passed over the western tip of Cuba on August 15, atmospheric pressure had dropped to 964 millibars and wind velocity reached 185 km h^{-1} (115 mph). Once in the Gulf of Mexico, the hurricane traveled northwestward at 23 km h^{-1} and intensified dramatically (Box Figure 5.1.1). By the afternoon of August 16 an Air Force plane measured a low pressure of 905 millibars and wind velocities of 260 km h^{-1} (160 mph). The last flight into the storm was made on the afternoon of August 17; by that time minimum pressure had dropped to 901 millibars (26.61 inches of mercury, 30 inches is normal), and surface winds had increased to more than 322 km h^{-1} (200 mph). This was the second-lowest pressure ever recorded in the United States and was only surpassed by the great Labor Day Hurricane of 1935 (892 millibars; 26.35 in.) that swept through the Florida Keys killing 408 people.

Early on August 17, police and civil defense officials used television and radio messages to call for an immediate evacuation of coastal regions along the Mississippi shore, knowing that a category 5 hurricane was located just 400 km south of Mobile Bay and its landfall was imminent. That night at 10:30 p.m. one of the strongest storms ever witnessed by mankind came onshore. Maximum wind velocities are unknown because all instruments stopped working before the storm reached its greatest intensity. However, estimates of winds velocity based on surface pressures and previous flights into the hurricane were calculated at 324 km h^{-1} (202 mph). While wind of that magnitude can rip a structure apart, it is the accompanying wall of water known as the "storm surge" that accounts for most of the destruction and loss of human life during a storm. The extreme low pressure and highvelocity winds

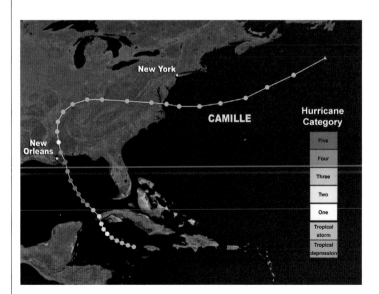

Box Figure 5.1.1 Storm track of Hurricane Camille which slammed into the Mississippi coast late on August 17, 1969, killing 143 in the Gulf Coast region. (*Source:* NOAA.)

of Camille produced a wall of water that measured 24.6 ft above mean sea level at Pass Christian, located 20 km east of the landfall. Not only does a storm surge of this magnitude flood areas far inland from the coast, but it also allows monstrous waves to break against and over anything along the coast, including dune ridges, sea walls, buildings and other structures (Box Figure 5.1.2). On the offshore barriers of Ship and Cat Island, debris marks indicated that hurricane waves broke across the tops of trees covering the island.

Hurricane Camille ranks as one of the deadliest and costliest storms in United States history. Certainly the amount of damage caused by the storm is attributable to its category 5 status, but the loss of life was also due to the fact that some of the coastal residents failed to respect the danger of an intense hurricane and did not respond to repeated warnings.

Stories chronicled by the National Hurricane Center help to demonstrate the magnitude and destructive force of Hurricane Camille. Perhaps the fate of the Richelieu Apartments

Box Figure 5.1.2 Before and after photographs of Richelieu Motel, destroyed by Hurricane Camille. (*Source:* From U.S. National Weather Service.)

and its occupants best illustrate the storm's immense power. The Richelieu was a three-story, brick-front building located about 100 m from the ocean. Standing between the building and the beach was a 2.4 m-high seawall and a four-lane highway. Civil Defense personnel pleaded with the occupants of the Richelieu to evacuate, but 25 people planned a "hurricane party" instead. The entire structure was destroyed by the storm, and only two people survived (Box Figure 5.1.3). One of the survivors was Mary Ann Gerlach who remained in the Richelieu with her husband because they had lived through previous hurricanes in Florida and thought that this one would be no more challenging. When the storm surge and waves began dismantling the Richelieu, Mary Ann managed to jump out a second story window. Miraculously, she washed ashore 7.2 km (4.5 miles) from the apartment after being in the water for almost 12 hours. A small boy from the Richelieu also survived the hurricane. He was saved by the father of the family, who lived next to the apartment building. After the man's house was broken apart by gigantic waves he swam onshore and found temporary safety by clinging to the top of a tree. He succeeded in saving his son by grabbing the 10-year old boy

Box Figure 5.1.3 Before and after photographs of the Trinity Church, destroyed by Hurricane Camille. (*Source:* From U.S. National Weather Service.)

as he floated from the Richelieu. His wife and two daughters perished. Another group of people went to the Trinity Episcopal Church seeking shelter; of the 21 at the church only 13 survived the storm (Box Figure 5.1.3). These stories, and experiences with other major storms, have contributed to a growing public awareness that intense hurricanes can strike with tremendous force, inflicting death and injury ono those who don't heed their fury.

The importance of coastline configuration has been clearly demonstrated in a study of New York Bight by Nicolas Coch of Queens College (New York). Eastern Long Island and northern New Jersey meet at near right angles to one another, forming a funnel-shaped embayment. Coastlines with this type of configuration are susceptible to amplification of the storm surge as waters are constricted by the adjoining landmasses. Dr. Coch showed that if a moderate-sized hurricane tracked in a northwesterly direction across northern New Jersey, the storm surge would significantly heighten from the inner shelf to the mouth of the Hudson River. Hurricane Sandy proved him to be correct, as illustrated in Figure 5.23. Note the significant increase in storm tide level toward New York Harbor and into the embayment. The same trend occurred at the west end of Long Island Sound. The large storm surges associated with extensive loss of life in Bangladesh have been caused in part by the same funneling effect that is produced in upper Bay of Bengal.

5.4.4.2 Hurricane Categories

The strength of a hurricane is formally classified using the Saffir–Simpson Scale and is based on the maximum wind velocity, barometric pressure, storm surge level, and expected damage (Table 5.4). The scale has five categories and each higher level represents a substantial increase in hurricane intensity (Table 5.5). A list of the most intense hurricanes in the United States in terms of loss of life and property damage is provided in Table 5.6. It is evident from this table that death tolls have decreased in recent times because meteorological forecasts and evacuation procedures have improved (see Box 5.2).

Most hurricanes fall within categories 1–3. One of the difficulties of accurately determining the maximum wind in severe storms such as a category 4 or 5 hurricane is the inability of instruments to withstand the storm. As an example, there has been considerable disagreement about the maximum wind velocity of Hurricane Andrew as it struck south of Miami in August of 1992. Instrumentation at the National Hurricane Center in Coral Gables sustained measured sustained wind of 210 km h^{-1} (125 mph) with gusts of about 275 km hr^{-1} (165 mph) at the time that it became disabled. Calculations indicate that peak wind velocities were probably greater than 235 km h^{-1} (140 mph). Inland from the shoreline high-velocity wind exerts the greatest influence on vegetation and manmade structures. The path of a hurricane is often evidenced by broken and uprooted trees, destroyed homes, and disarray of public infrastructure.

Damage to structures due to hurricanes is associated with lowland flooding, wave attack, and strong wind. Because of the current stringent rules and regulations regarding construction in the coastal zone, there are usually few problems with wind damage. In hurricane-prone areas, most damage is now confined to roofs, where shingles are commonly torn from roofs. Under extreme conditions such as during a category 4 or 5 hurricane, damage can be much greater and the building itself can be destroyed. A good example of how recent construction codes have protected dwellings

(a)

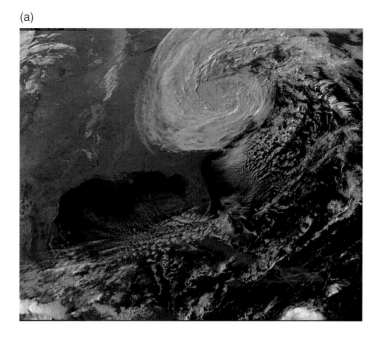

(b)

Hurricane Sandy, Maximum Water Levels

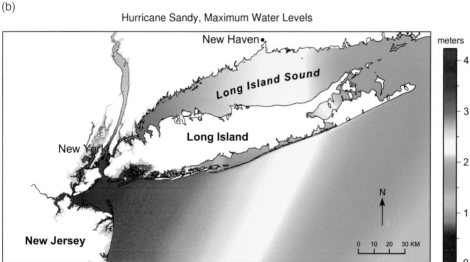

Figure 5.23 (a) Westward track of Hurricane Sandy across the New Jersey coast. (*Source:* NASA, https://en.wikipedia.org/wiki/Hurricane_Sandy). (b) Computer-simulated maximum storm tide conditions due to Hurricane Sandy. (*Source:* Courtesy of Joe Holmes, RPI.)

during passage of moderate hurricanes is the impact of Hurricane Opal in 1995. This severe storm made landfall in the Panhandle of Florida just east of Pensacola. Of the nearly 1500 homes subjected to the storm's fury approximately half were constructed prior to the new regulations and the other half were built in compliance with new building codes. There were hundreds of homes severely damaged or destroyed during the storm, but all of

Table 5.4 Saffir–Simpson hurricane intensity scale.

Scale number (category)	Central pressure (millibars)	(inches)	Wind speed (mph)	(km h^{-1})	Storm surge (feet)	(meters)	Damage
1	≥980	≥28.94	74–95	119–154	4–5	1–2	No real damage to building structures. Damage primarily to unanchored mobile homes, shrubbery, and trees. Also some coastal road flooding and minor pier damage.
2	965–979	28.50–28.91	96–110	155–178	6–8	2–3	Some damage to roofing material, and door and window damage to buildings. Considerable damage to vegetation, mobile homes, and piers. Coastal and low-lying escape routes flood 2–4 hours before arrival of hurricane eye. Small craft in unprotected anchorages break moorings.
3	945–964	27.91–28.47	111–130	179–210	9–12	3–4	Some structural damage to small residences and utility buildings with a minor amount of curtainwall failures. Mobile homes are destroyed. Flooding near the coast destroys smaller structures, with larger structures damaged by floating debris. Terrain continuously lower than 1.5 m (5 ft) above sea level may be flooded inland as far as 9.6 km (6 miles).
4	920–944	27.17–27.88	131–155	211–250	13–18	4–6	More extensive curtainwall failures with erosion of beach areas. Terrain continuously below 3 m (10 feet) above sea level may be flooded, requiring massive evacuation of residential areas inland as far as 9.6 km (6 miles).
5	<920	<27.17	>155	>250	>18	>6	Complete roof failure on many residences and industrial buildings. Some complete building failures, with small utility buildings blown over or away. Major damage to lower floors of all structures located less than 4.5 m (15 feet) above sea level and within 457 m (500 yards) of the shoreline. Massive evacuation of low areas on low ground within 8–16 km (5–10 miles) of the shoreline may be required.

Table 5.5 Damage comparison of different category hurricanes.

Category	Relative hurricane destruction potential
1	1 (reference level)
2	4 times the damage of a category 1 hurricane
3	40 times the damage
4	120 times the damage
5	240 times the damage

Source: Based on e.mpirical analysis over the past 42 years by Dr. William M. Gray, Colorado State University meteorologist.

those built under present guidelines suffered only minor damage.

Storm surges and the accompanying large waves can, however, cause major damage, including the destruction of structures. For this reason, current zoning in Florida requires that the first level of occupancy must be above the 100-year storm surge level. This level is based on existing data and predicted frequency of storm surge. For example the Gulf Coast of central and southern Florida is predicted to experience a 4 m (13-ft) storm surge on a 100-year return interval. This is based in part on historical data of previous storms and by computer modeling of storm conditions using variables such as shelf gradient, wind velocity, wave size and others. Qualification for federally supported insurance in the coastal zone requires compliance with these regulations.

5.5 Summary

Hurricanes and extratropical cyclones are a dramatic expression of Earth's weather system. Through recorded history severe storms have accounted for permanent changes to the coastal landscape, billions of dollars of damage, and the unfortunate loss of many lives. Hurricanes are tropical storms having wind velocities exceeding 119 km h^{-1}; the largest have storm surges greater than 7 m and winds attaining 220 km h^{-1}. They form in the tropics and are steered by the prevailing global wind patterns. Hurricanes affecting the United States are initiated off the west coast of Africa. They intensify over the warm Atlantic Ocean waters and travel in a westerly direction, making landfall along the Gulf of Mexico or along the east coast, most commonly from Florida north to the Outer Banks of North Carolina. Rarely do hurricanes strike the West Coast of the United States. Hurricanes degenerate after moving over land or cold water.

Extratropical cyclones form above the tropics and are commonly associated with cold fronts. These storms are usually weaker than hurricanes, but occur more frequently. They develop over the continental United States, southwestern Canada, in the Gulf of Mexico, and off the east coast of Florida. These storms generally move eastward and northward, eventually traveling offshore of New England and passing east of Nova Scotia. They generate northeasterly winds and waves and therefore are called "Northeasters." Northeasters last for one to two days and are accompanied by wind velocities of 40–65 km h^{-1} and storm surges ranging from 0.2 to 1.2 m.

Both hurricanes and extratropical storms are low-pressure systems with counterclockwise wind patterns. The strength, speed and size of these cyclonic systems control the severity of the storm, including the amount of erosion and structural damage. Storm track, gradient of the shelf, configuration of the shoreline, and astronomic tidal conditions are other important factors governing storm processes. See Box 5.1.

Table 5.6 Compilation of the deadliest and costliest hurricanes in the United States history.

Costliest Hurricanes

Name	Damage (Billions USD)	Year	Storm classification at peak intensity	Areas affected
Harvey	$198.63	2017	Category 4 hurricane	Texas, Louisiana, South and Central America, Caribbean, Yucatan Peninsula
Katrina	$108.0	2005	Category 5 hurricane	Bahamas, Florida, U.S. Gulf coast, Ohio Valley, northeast U.S., eastern Canada
Sandy	$75.0	2012	Category 3 hurricane	Caribbean, U.S. east coast, Eastern Canada
Ike	$37.5	2008	Category 4 hurricane	Greater Antilles, Texas, Louisiana, midwestern U.S., eastern Canada, Iceland
Wilma	$29.4	2005	Category 5 hurricane	Greater Antilles, Central America, Florida
Andrew	$26.5	1992	Category 5 hurricane	Bahamas, Florida, U.S. Gulf coast
Ivan	$23.3	2004	Category 5 hurricane	Caribbean, Venezuela, U.S. Gulf coast
Irene	$16.6	2011	Category 3 hurricane	Caribbean, U.S. east coast, eastern Canada
Charley	$16.3	2004	Category 4 hurricane	Jamaica, Cayman Islands, Cuba, Florida, The Carolinas
Matthew	$15.09	2016	Category 5 hurricane	Colombia, Venezuela, Caribbean, U.S. east coast
Rita	$12.0	2005	Category 5 hurricane	Cuba, U.S. Gulf coast
Hugo	$10.0	1989	Category 5 hurricane	Caribbean, U.S. east coast
Frances	$9.8	2004	Category 4 hurricane	Caribbean, eastern U.S., Ontario
Georges	$9.72	1998	Category 4 hurricane	Caribbean, U.S. Gulf coast

Deadliest Hurricanes

Name	Deaths	Year	Storm classification at peak intensity
Great Galveston Hurricane (TX)	8000	1900	Category 4 hurricane
FL (Lake Okeechobee)	2500	1928	Category 4 hurricane
Katrina (LA/MS/FL/GA/AL)	1200	2005	Category 3 hurricane
Chenière Caminanda (LA)	1100–1400	1893	Category 4 hurricane
Sea Islands (SC/GA)	1000–2000	1893	Category 3 hurricane
GA/SC	700	1881	Category 2 hurricane
Audrey (SW LA/N TX)	416	1957	Category 4 hurricane
Great Labor Day Hurricane (FL Keys)	408	1935	Category 5 hurricane
Last Island (LA)	400	1856	Category 4 hurricane
Miami Hurricane (FL/MS/AL/ Pensacola)	372	1926	Category 4 hurricane
LA (Grand Isle)	350	1909	Category 4 hurricane
FL (Keys)/S. TX	287	1919	Category 4 hurricane
LA (New Orleans)	275	1915	Category 4 hurricane
TX (Galveston)	275	1915	Category 4 hurricane
Camille (MS/LA)	256	1969	Category 5 hurricane
New England	256	1938	Category 3* hurricane
Diane (NE U.S.)	184	1955	Category 1 hurricane
GA, SC, NC	179	1898	Category 4 hurricane
TX	176	1875	Category 3 hurricane
Southeast Florida	164	1906	Category 2 hurricane

Box 5.2 Deadly and Costly Hurricanes of the Twenty-first Century

Hurricane Katrina in 2005 and Hurricane Sandy in 2012 impacted metropolitan areas, causing catastrophic damage and loss of human life. These hurricanes were also the costliest storms in US history (105.8 and 68 billion dollars in damage, respectively; Table 5.6). Despite these similarities, they were very different hurricanes in terms of their origin, track, size and intensity, and type of impact. The notoriety of Katrina is due to its flooding of downtown New Orleans, extensive loss of life, and widespread damage ranging across three states. Sandy was the largest Atlantic hurricane on record, simultaneously causing damage to structures along the Lake Michigan shoreline and collapsing power lines in Nova Scotia, 1200 miles away. A significant storm surge accompanying Sandy flooded large areas of New York City, including its subway system and most of the tunnels entering Manhattan.

Hurricane Sandy developed from a tropical wave in the western Caribbean Sea, which quickly strengthened to Tropical Storm Sandy within hours. The storm moved northward and gradually intensified, reaching its maximum category 3 status offshore of Cuba. After diminishing in size, Sandy re-intensified to a category 2 hurricane 300 miles east of Delaware Bay (Box Figure 5.2.1). During this period, Sandy had grown to a huge hurricane producing waves >3 m high along the entire coast from the Outer Banks of North Carolina to Cape Cod in Massachusetts. At landfall, Sandy's tropical storm-force winds spanned 943 miles or almost three times the size of Hurricane Katrina (Box Figure 5.2.2). Its size was a product of several factors, including an extra-tropical system to the west that pumped upper cold air into the hurricane, transforming the nature of the storm into more of a "north-easter" having lower wind velocity but a much greater breadth than most hurricanes.

Hurricane Katrina formed as tropical depression over the southeastern Bahamas and moved northwest, reaching hurricane force winds two day later just off the southeast coast of Florida. After passing across Florida it looped northward toward the Mississippi River

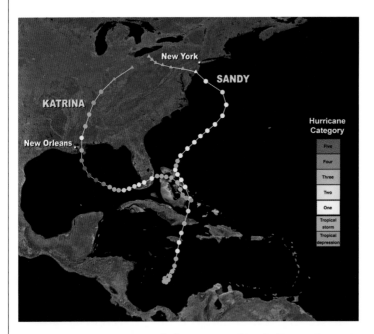

Box Figure 5.2.1 Hurricane tracks for Katrina and Sandy. (*Source:* From U.S. National Weather Service.)

Box Figure 5.2.2 Radar images of Hurricanes Katrina and Sandy. *Source:* From Indian Space Research Organization. OceanSat-2 missions and NASA's Jet Propulsion Laboratory's QuikSCAT.

delta, greatly intensifying due the unusually warm waters in the Gulf of Mexico (Box Figure 5.2.1). It strengthened to a category 5 hurricane 200 miles south-southeast of the delta with maximum sustained wind velocity of 175 mph. Before Katrina made landfall at the delta, it had diminished to a category 3 hurricane. Katrina continued moving northward making a second landfall along the Mississippi coast, where it gradually moved inland and decreased to a tropical storm. Compared to Sandy, Katrina was only one-third the size, but had much stronger winds that produced a much higher storm surge (Box Figure 5.2.2).

Both hurricanes had similar impacts along barrier shorelines. Elevated water levels facilitated wave-overtopping of dune ridges producing overwash fans and, in some cases, the formation of new tidal inlets (Box Figure 5.2.3). In residential areas on the Gulf Coast and in coastal areas along New Jersey, New York, and Connecticut, houses were flooded and damaged, coastal defense structures were dismantled, and infrastructure was destroyed. Flooding of the New York City metropolitan region was due to several factors, but most importantly related to the westward track of the storm across central New Jersey and the

May 21, 2009

November 5, 2012 ≋USGS

Box Figure 5.2.3 Photographs illustrating barrier breaching and inlet formation along Fire Island. (*Source:* USGS.)

funnel-shaped shorelines formed: (i) by northern New Jersey (particularly Sandy Hook) and Long Island, and (ii) at the western end of Long Island Sound (Box Figure 5.2.1). The westerly track of the hurricane meant that surge waters were forced into narrowing embayments causing an amplification of the storm surge. In addition, Sandy's impact coincided with high tide, a worst-case scenario. Many areas within the embayment are low-lying or defended by low seawalls that are particularly susceptible to wave overtopping.

The devastating impact of Katrina was a result of insufficient levee protection and the extreme storm surge that varied from 9 ft at

the birdfoot section of the Mississippi delta to 27.8 ft along the Mississippi coast at Pass Christian, a record level for the United States. The huge storm surge was caused by a number of factors, including the magnitude and track of the hurricane and configuration of the continental shelf. Even though Katrina was much smaller in size than Sandy, it was still a large, intense hurricane reach category 5 status 200 miles south of the delta, with gusting winds greater than 120 mph. While the storm traveled due north, its southerly wind regime continuously added to the height of the storm surge. In addition, the shallow sloping continental shelf rimming the Gulf of Mexico

Box Figure 5.2.4 Photographs of the Ninth Ward in New Orleans when it was still flooded and after the area drained. Large red barge is present in both photos. Note the empty lots where houses once existed. *Source:* Top photo from https://media.npr.org/assets/img/2015/08/14/gettyimages-97657258_wide-f00d13e5a774570c4d92838cfcd36320fc638e5d.jpg?s=1400. Bottom photo by FitzGerald (author).

constricted and amplified the height of a surge. As the hurricane swept across the delta plain, surge waters traveling northward toward New Orleans were heightened due to the presence of large, dredged access channels. Catastrophically, flood waters moving through these channels coupled with large waves over-topped and compromised levees that pro-tected residential areas within Saint Bernard Parish and the Ninth Ward section of New

Orleans, resulting in more than 1000 lives being lost, mostly due to drowning (Box Figure 5.2.4). Levees also failed in other regions of New Orleans producing extensive flooding in large portions of the city.

We are certain that impacts of major hurricanes to coastal regions will continue into the future, and due to global warming, these storms are expected to occur more frequently and with greater ferocity. The ultimate defense of our vulnerable coastal cities will depend on massive and highly expensive engineering projects.

Suggested Reading

Coch, N.L. (1995). *Geohazards; Natural and Human*. Englewood Cliffs, NJ: Prentice-Hall.

Elsner, J. and Kara, A.B. (1999). *Hurricanes of the North Atlantic: Climate and Society*. New York: Oxford University Press.

Forbes, C., Rhome, J., Mattocks, C., and Taylor, A. (2014). Predicting the storm surge threat of hurricane sandy with the National Weather Service SLOSH model. *J. Mar. Sci. Eng.* 2: 437–476. https://doi.org/10.3390/jmse2020437.

Henry, J.A., Portier, K.M., and Coyne, J. (1994). *The Climate and Weather of Florida*. Sarasota, FL: Pineapple Press.

Robertson, C., and Fausset, R. (2015). Ten years after Katrina, *New York Times*, August 26, 2015. https://www.nytimes.com/interactive/2015/08/26/us/ten-years-after-katrina.html Accessed August 23, 2018.

Pielke, R.A. Jr. and Pielke, J.A. Sr. (1997). *Hurricanes, their Nature and Impacts on Society*. New York: Wiley.

Sallenger, Abby (2009). *Island in a Storm: A Rising Sea, a Vanishing Coast, and a Nineteenth-Century Disaster that Warns of a Warmer World*. New York, Public Affairs.

Simpson, R.H. and Riehl, H. (1981). *The Hurricane and its Impact*. Baton Rouge: Louisiana State University Press.

Williams, J.M. and Duedall, I.W. (1997). *Florida Hurricanes and Tropical Storms*, Revised Edition. Gainesville, FL: University Press of Florida.

6

Waves and the Coast

Waves are a surface disturbance of a fluid (gas or liquid) in which energy is transferred from one place to another. In the case of the coast, we are concerned with the interface between the ocean and the atmosphere, but waves also occur between different liquid masses and between different gaseous masses. For example, waves occur between water masses of different densities caused by temperature and/or salinity contrasts; these are termed internal waves. Waves also occur within the atmosphere such as between a warm, light air mass and one that is cold and heavy.

When the surface of the fluid is disturbed, the perturbation is transferred from one location to another although the medium itself, water in case of the coast, does not move with the propagating disturbance.

Water waves occur in a wide range of sizes (Figure 6.1) and may be caused by various phenomena. The primary type of wave that influences the coast is what is called a progressive surface wave, produced by wind. In this type of wave, energy travels across and through the water in the direction of the propagation of the wave form. These water surface waves are called gravity waves because gravity is the restoring force. The movement is due to restoring forces that cause an oscillatory or circular motion that is basically sinusoidal in its form (Figure 6.2), That is, it is shaped like a sine curve; perfectly symmetrical and uniform. Both gravity and surface tension are important restoring forces that maintain waves as they propagate. Surface tension is important in very small waves called *capillary waves*—waves that are less than 1.7 cm long. Larger waves are called *gravity waves* because the primary restoring force is gravity. These are the typical wave we see at the coast or on any water surface.

A wave has several components that are important in describing it and its motion (Figure 6.2). The *wave length* (L or λ) is the horizontal distance between two like locations on the wave form; crest to crest, trough to trough, etc. The *wave height* (H) is the vertical distance between the base of the trough and the crest; it is twice the amplitude. The steepness of the wave is the ratio of the height to the length (H/L). Another important characteristic of a gravity wave is its *period* (T), the time, measured in seconds, required for a complete wavelength to pass a reference point. Because the wave length is so difficult to measure due to the constant movement of the wave, the period is typically used as a proxy for the length. The relationship between the two is generally given as

$$C = \frac{L}{T} \tag{6.1}$$

Where C is the celerity or velocity of wave propagation. Another way of measuring wave propagation is by the *frequency* (f), that is the number of wave lengths passing per second. Therefore, a 10-second wave would have a frequency of 10^{-1}.

Beaches and Coasts, Second Edition. Richard A. Davis, Jr. and Duncan M. FitzGerald.
© 2020 John Wiley & Sons Ltd. Published 2020 by John Wiley & Sons Ltd.

QUALITATIVE WAVE POWER SPECTRUM

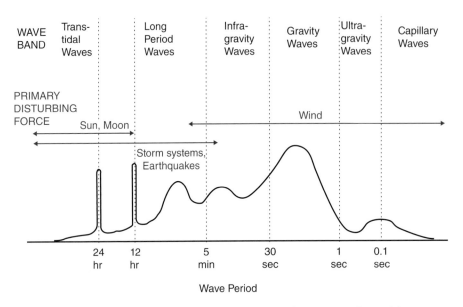

Figure 6.1 Graph of showing the frequencies various types of water waves from tidal waves to capillary waves. (*Source:* Courtesy of NOAA, https://sos.noaa.gov/copyright-information).

Transverse Wave

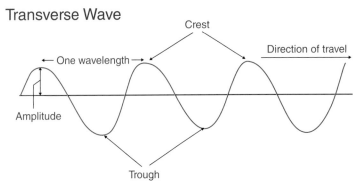

Figure 6.2 Typical waveform showing the various elements of a gravity wave.

The actual transfer of energy from the wind to the water surface is complicated and not completely understood. It is well-known, however, that the size of waves depends upon three primary factors: how fast the wind is blowing in a particular direction (velocity); the length of time that the wind blows (duration); and the distance over the water that the wind blows ("fetch"). Any one or a combination of these factors can be limiting on the wave size. In many situations the theoretical limitation of wave size is caused by the fetch; these are typically termed fetch-limited basins. Any location can have any wind velocity and the wind can blow for any length of time. What cannot change, however, is the size of the water body that is being subjected to the wind and therefore the fetch. Good examples are Lake Michigan and Gulf of Mexico, both large bodies of water to be sure,

Figure 6.3 Photograph of complicated sea conditions.

Figure 6.4 Plot of a wave spectrum showing the range of frequencies recorded at one location in a brief time series.

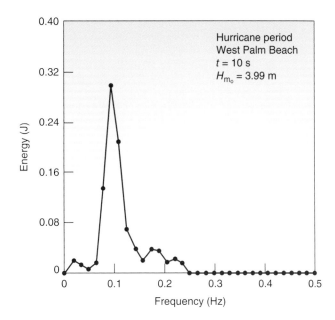

but ones that are not nearly as big as an ocean. Huge waves cannot form in these water bodies even during hurricanes or other severe storms.

Wave form and its components are discussed here in relatively simplified terms; however, the actual wave conditions in nature are extremely complicated. Typically there are several families of waves of different sizes (Figure 6.3) moving in different directions all superimposed at any location in the sea. These waves combine to produce a wave field that can be recorded and analyzed. The data

are in the form of a wave spectrum which can be separated into its component wave forms, each with its own period or frequency and height (Figure 6.4). One of the most important aspects of the spectral analysis of a wave field is the determination of the *significant wave height*, the most commonly used wave-measuring category (Figure 6.5). This value (H_s) is the average height of the highest one-third of the waves occurring during the time period being analyzed. It is the significant wave height that is commonly used as an index of wave energy.

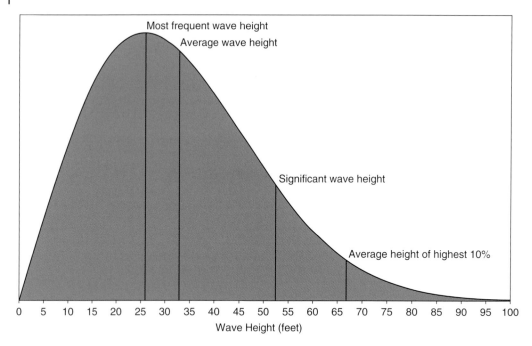

Figure 6.5 Graph showing significant wave height and other important wave statistics. (*Source:* Courtesy of NOAA, https://sos.noaa.gov/copyright-information).

6.1 Water Motion and Wave Propagation

Recall from the previous section that only the wave form is propagated in gravity waves, not the water itself. That being the case, we need to consider how the water actually moves within gravity waves: the water moves in an orbital path with the circulation in each orbit being in the direction of wave propagation. As the wave moves toward the coast, the surface water is moving landward on the crest of the wave and seaward in the wave trough. In doing so, the wave form moves toward the shore but the water itself moves only in circles. This orbital motion in the wave extends well below the water surface. The diameter of the orbital motion at the water surface is equal to the wave height and this diameter decreases with depth. At a depth of approximately one-half the wave length of a surface wave, the orbital motion is very slow, and the orbits are very small; sediment on the bottom is not moved (Figure 6.6).

Anyone who has done SCUBA diving knows that even if it is quite rough on the surface, there is a depth below which you do not feel wave motion. Because this orbital motion is forward on the crest, backward on the trough with vertical motion half way between, a fishing float or a ball appears to actually move up and down as the waves pass but without progressing significantly itself. Where wind is present, close inspection of the actual path of water particles shows that there is slight net movement of water in the direction of propagation due to friction between the wind and the water surface. At the coast, this wind may push the water landward and produce what is called *setup*, a temporary elevation of water level. Setup is a phenomenon that is the major factor in *undertow*, a process that is discussed in the chapter on beaches (Chapter 13).

When the depth of water is more than half of the wave length of the surface wave, the orbital motion of water within the wave is not influenced by the ocean floor as the wave propagates. As the wave moves

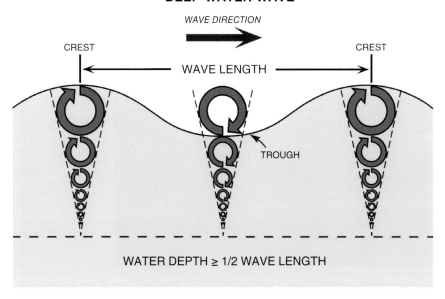

DEEP-WATER WAVE

WAVE DIRECTION

CREST — WAVE LENGTH — CREST

TROUGH

WATER DEPTH ≥ 1/2 WAVE LENGTH

Figure 6.6 Diagram of water motion in waves showing the decrease in the diameter of the orbital paths with depth. (*Source:* Courtesy of Joe Holmes, RPI).

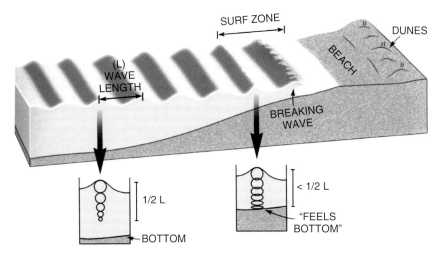

SURF ZONE

DUNES

BEACH

(L) WAVE LENGTH

BREAKING WAVE

1/2 L

BOTTOM

< 1/2 L

"FEELS BOTTOM"

Figure 6.7 Changes in the orbital paths under wave surfaces as they move toward the shore. (*Source:* Courtesy of Joe Holmes, RPI).

into increasingly shallow water during its approach to the coast, these orbital motions begin to interact with, or "feel" the bottom at a depth about equal to half the wave length (Figure 6.7). This is called *wave base* and this interference causes the orbits of water particles to become deformed and to slow down. At first the circular motion becomes squashed into a football shape and eventually becomes simply back and forth motion. This condition progresses as the wave moves into increasingly shallow water. The slowing of the wave causes it to steepen—it is being compressed somewhat like an accordion. At the same time that this is happening, bottom friction causes the wave to slow down at the

Figure 6.8 Photograph showing how waves can break over each of multiple sand bars as well as at the shoreline.

bottom, but not at the surface. These conditions cause the wave to eventually become so steep that it is no longer stable and its shape collapses. This is the breaking of waves that we see in the *surf* along the coast. This instability due to excessive steepness takes place when the inclusive angle of the wave form seeks to be less than 12° or the steepness exceeds 1 : 7. After breaking, restoring forces cause the wave to reform and continue its progression, perhaps to break again before reaching the shoreline.

Multiple breaking of waves in shallow water at the coast is produced by a bar and trough topography of the seabed (Figure 6.8). As the wave moves into shallow water and begins to feel bottom, slow its forward speed and steepen, it will eventually break. This

first break is commonly over the crest of a sand bar that may rise a meter or so above the gently sloping nearshore gradient. Landward of this bar is a trough of deeper water and it is over this trough that the wave reforms after initial breaking. The water shallows again and breaking takes place again, over a second sand bar. It will reform again and finally break at the shore. This situation, with two sand bars and waves breaking over them (see Figure 6.8) is typical of many of the gently sloping nearshore areas of the world. It is also possible for a coast to have just one bar or three bars. Under conditions of small waves, breaking will take place only over the shallow bar and not the deeper one. The fairly narrow coastal area where wave breaking occurs is called the surf zone.

Box 6.1 The Duck Pier

In 1977, the U.S. Army Corps of Engineers established their Field Research Facility (FRF) at Duck, North Carolina near the northern end of the Outer Banks. The central element of this facility is a research pier 590 ft (180 m) long

that extends to a depth of 7 m (see Box Figure 6.1.1). This pier was constructed to enable monitoring and the conducting of research on a variety of coastal processes across the inner shoreface and surf zone. Other elements of

Box Figure 6.1.1 Oblique aerial photo of the Duck Pier. (*Source:* Courtesy of Andrew Morang).

the facility include research laboratories, sophisticated computer capabilities, and a conference room. An observation tower rises 13 m above the adjacent dunes, and various specialized vehicles for taking measurements in the rigorous conditions of storms are available. The staff of 16 includes coastal scientists and engineers, computer specialists and technicians.

The specialized vehicles and other equipment permit the FRF to make observations and take measurements that are otherwise not possible. A specially constructed motorized vehicle called a CRAB (Coastal Research Amphibious Buggy) can move across the surf zone to survey with centimeter accuracy to a depth of 9 m. It can also deploy instruments and provide a stable platform for other activities such as vibracoring, side-scan sonar surveying and sediment sampling. A Sensor Insertion System (SIS) is a large crane that can extend out up to 24 m from the pier and can carry wave gauges, current meters, suspended sediment sampling devices and other instruments. The SIS can be moved on a track along the entire length of the pier.

One of the main tasks of the FRF is to continuously monitor coastal processes and change in order to provide a large data base.

Wave height, period and direction are monitored by pressure transducer arrays beyond the pier and wave height and period are measured at three locations along the pier. There are s wave-rider buoys at various locations offshore. The water current profile is measured, as are various meteorological parameters, and water parameters such as temperature, salinity and light penetration are also included in this data base. The changes in bathymetry and shoreline position are measured and related to the processes. All of these data are compiled into monthly and annual reports and are available to the public. Some of the process data are available online in near-real time.

In addition to the regular collection of data by the FRF staff, the facility also hosts visiting researchers from other Corps of Engineers locations and from universities. There have been huge experiments at the FRF that have involved more than 100 researchers each. Such events as "Super Duck" and "Sandy Duck" were held several years ago and lasted at least two weeks each. Investigators brought their own instruments and personnel to interact with the total group. In order to be invited to participate in these large and complex experiments, a principal investigator had to submit a proposal of research and relate it to the overall objectives. These events have provided a tremendous amount of comprehensive data for a selected time period.

One criticism of the FRF is that there has been too much time, effort, and money expended at this single location instead of having multiple locations where a wealth of data are collected. Funding is always a limiting factor so such shortcomings are tolerable given the budget of the supporting agencies. At the present time FRF personnel are conducting projects at several locations away from Duck, most at other sites in North Carolina but also at Bethany Beach, Delaware and Ocean City, Maryland.

6.2 Wind Wave Types

Although the theoretical and simple wave form is a sinusoidal curve, that form is not common in nature. Wave shape depends on the conditions of wind, water depth and the progression of the wave itself. Although wind is responsible for the production of most gravity waves, it is common for the waves to travel well beyond the area where the wind blows or for waves to continue long after the wind stops blowing. Waves that are directly under the influence of wind are called *sea waves* (Figure 6.3). They typically have relatively peaked crests and broad troughs. In nature, sea waves tend to be complicated by multiple sets of various sized waves superimposed. Whitecaps occur when the wind blows off the tops of sea waves.

A common wave type is the *swell wave*, which actually has a waveform that approaches a sinusoidal shape (Figure 6.9). Swell develops after the wind stops or when the wave travels beyond the area where the wind is blowing. Swell waves commonly have a long wave length and small wave height, thus having a very low steepness value, much lower than sea waves.

The breaking of waves as they enter shallow water takes on different characteristics depending upon the type of wave that is present in deep water, and the conditions of breaking as the waves move into the surf zone. For example, swell waves are long and low, therefore they begin to feel the bottom in relatively deep water. They gradually slow and steepen until they break as *plunging breakers* (Figure 6.10). This type of breaking

Figure 6.9 Photo of swell waves showing low amplitude and long period as they approach the shore. (*Source:* R. A. Davis).

Figure 6.10 Waves that steepen in the nearshore and break almost instantaneously are called plunging waves.

Figure 6.11 Waves that steepen in the nearshore and break over time and distance are called spilling breakers. (*Source:* R. A. Davis).

wave is typified by a large curling motion with an instantaneous crashing of the wave characterized by a sudden loss of energy. It is typical of gently sloping nearshore zones.

The other common type of breaking wave is the *spilling breaker* (Figure 6.11). This type is most common as sea waves with their shorter wave length and higher steepness enter shallow water. As these waves break they do so slowly over several seconds and some distance. The wave looks like water spilling out of a container. Surging breakers

Figure 6.12 Waves break as surging breakers in very shallow water.

(Figure 6.12) and collapsing breakers are other types of breaking waves; they look very much like each other, having an appearance that is similar to spilling breakers. Surging breakers typically develop on or near the beach as the wave runs up to the shoreline. The wave steepens, and just as it begins to break, it surges up the beach.

6.3 Distribution and Transfer of Wave Energy

As waves move into shallow water and are influenced by the bottom and by various natural features or structures made by humans, they may experience changes in their energy distribution and/or direction. This occurs in three primary ways: refraction, diffraction and reflection—waves act much like light. Waves typically approach the shoreline at some angle; the crest may have an orientation ranging from nearly perpendicular to parallel to the shoreline. As the wave enters shallow water and begins to be slowed by interference with the bottom, this slowing of the rate of movement of the wave takes place at different times and places along the crest of the wave. The result is a bending or

refraction of the wave as it passes through shallow water on its way to the shoreline (Figure 6.13). As the waves bend or refract, they cause a vector of energy to move along the shoreline in the form of a longshore current (discussed in Chapter 7), and they cause energy to be distributed according to the relationship between the bending wave and the shoreline. The distribution of wave energy can be represented by orthogonals, lines constructed perpendicular to wave crests. This construction enables us to see where wave energy is concentrated and where it is dispersed, as shown by converging or diverging orthogonals respectively.

If we have a uniformly sloping nearshore with waves approaching a straight shoreline at an acute angle, we would expect the wave energy to be uniformly distributed along the shoreline. If the shoreline and/or the nearshore topography is irregular, then the refraction of the waves will be affected by these irregularities and the distribution of wave energy will be complicated. As a result of refraction, wave energy is concentrated at headland areas along the coast and is dissipated in embayments (Figure 6.14). This combination of conditions causes headlands to erode and embayments to accumulate sediment.

(a)

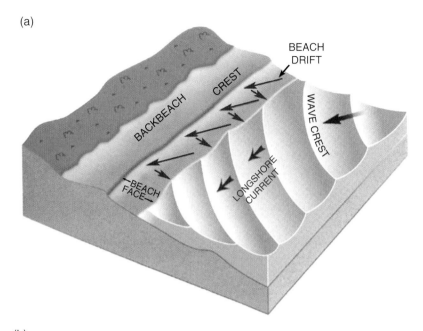

(b)

Figure 6.13 (a) Schematic diagram of waves refracting across the surf zone (*Source:* Courtesy of Joe Holmes, RPI), and (b) photograph of the same phenomenon.

As waves pass an impermeable obstacle such as a jetty, breakwater or other type of structure, the wave energy is spread along the crest behind the obstacle, it is *diffracted*. Part of the wave crest is stopped by the structure and the rest of it passes by (Figure 6.15). As the wave passes the structure, energy is transferred laterally along the wave crest to the sheltered areas behind the obstacle. In this way waves gradually progress behind the obstruction although they will have smaller heights than the wave that is

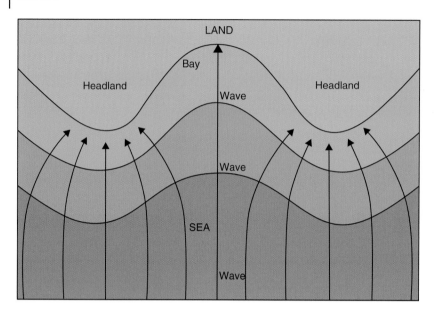

Figure 6.14 Diagram showing how refraction concentrates wave energy on the headland and disperses it in the embayments.

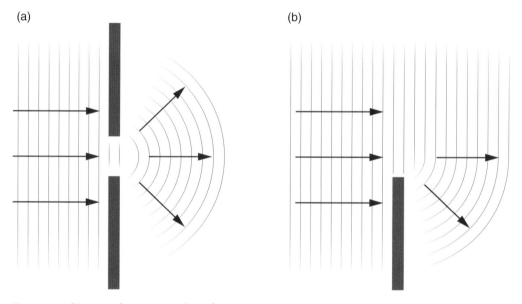

Figure 6.15 Diagrams showing spreading of wave energy as (a) waves pass through an opening and (b) as they pass by an obstacle. (*Source:* Courtesy of http://physics.taskermilward.org.uk).

unaffected by the structure. This is the reason that boats anchored behind breakwaters or other shelters still feel waves. It should also be noted that the obstacle also influences wave energy beyond the structure itself. The same phenomenon occurs when waves pass through a constricted passageway to open water (Figure 6.16).

Waves can also have some or all of their energy reflected as the wave meets the beach

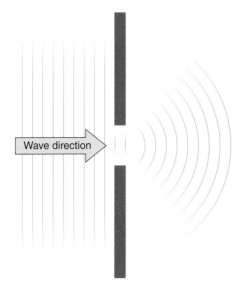

Figure 6.16 Vertical aerial photo showing wave diffraction as the waves pass through a narrow bedrock opening into a bay on the Spanish coast. (*Source:* Courtesy of Barcelona Field Studies Center).

(Figure 6.17), a structure or rock exposures. The amount of reflected energy and the direction in which the reflected energy is directed is dependent upon the amount of energy absorbed by the coast or structure, and the angle of approach. If a wave approaches and hits a vertical and imperme-

able structure without any interference by the bottom, 100% of its energy will be reflected. This would be the situation if a seawall or breakwater presented a solid vertical surface, such as one of poured concrete. A vertical rock outcrop would present the same situation. A structure that is sloping and/or has an irregular surface will absorb some of the wave energy and will reflect only part of it. The same is true of a steep beach: some of the wave energy is reflected and some of it dissipated or absorbed. Typically, only a small percentage of incident wave energy will be reflected at a beach.

Structures like seawalls are problematic because the impact of wave energy causes scour at the base of the wall that may cause it to fail, and the scour also causes considerable beach erosion. In many situations riprap rock is placed in front of the seawall to dissipate and absorb wave energy and thereby diminish erosion (Figure 6.18).

The direction along which this reflected energy proceeds is in accordance with the law of *reflection*; the angle of reflection is equal to the angle of incidence. In other words waves approaching a vertical sea wall act just like light rays on a mirror. A wave that approaches a sea wall at an angle of 45° will be reflected

Figure 6.17 A wave reflected from the steep beach face on the island of Curacao.

Figure 6.18 Vertical concrete seawall on the Florida Gulf Coast with boulder riprap in front to dissipate wave energy.

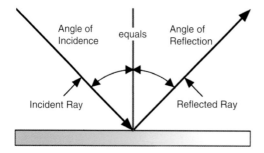

Angle of Incidence   equals   Angle of Reflection

Incident Ray   Reflected Ray

Figure 6.19 Diagram showing how incoming waves are reflected from a seawall at the same angle as the incoming wave. (*Source:* Courtesy of DOSITS, Discovery of Sound in the Sea).

at the same angle (Figure 6.19). Incident waves on a rocky coast may experience just this sort of condition.

6.4   Other Types of Waves

Tsunamis

A *tsunami* is a large, rapidly moving wave that can be very destructive when it reaches the coast. The word tsunami is Japanese for "very long harbor wave." This name has been applied to these waves because it is in harbors that much of the damage occurs.

The name is typically applied to waves produced by seismic or other events that cause major movements on, or of, the crust of the ocean floor. Such events cause a disturbance of the sea surface, a tsunami; commonly referred to as a seismic sea wave. Earthquakes, volcanic eruptions, landslides and possibly even strong, deep-water currents may trigger tsunamis. A good way to envision the generation of a tsunami is that the water container (the ocean floor) is disturbed, which in turn causes a disturbance of the water surface.

The wave length of a tsunami is generally up to hundreds of kilometers, wave height may be about a meter in deep water and the wave travels at hundreds of kilometers per hour. Because of its great wave length, a tsunami will begin to be affected by the sea floor at great depth. This causes a very marked slowing of the velocity accompanied by steepening of the wave, resulting in a wave height of several meters at the coast. Many catastrophic tsunamis have seriously affected coastal areas of the world, for example in Japan and many of the small islands of the Pacific, especially Hawaii. The west coast of North America has also experienced large

Figure 6.20 Huge wave generated by the 2011 tsunami as it came ashore. (*Source:* U.S. Marine Corps photo by Lance Cpl. Ethan Johnson).

tsunami waves. Not only do these huge waves destroy property (Figure 6.20) and erode the beaches, several have caused a high loss of life. Among the recorded disasters of this type were the tsunamis of 1692 in Jamaica, 1755 in Portugal, 1896 in Japan, and 1946 in Hawaii. The eruption of Krakatoa in the southwest Pacific in 1883 produced a tsunami that carried a large ship 3 km inland to an elevation of 9 m above sea level. The 1964 earthquake in Alaska produced a tsunami that caused severe damage to Crescent City, California. Loss of life has been greatly reduced since the establishment of a network of seismic monitoring stations that covers the entire Pacific Ocean. This network was constructed in the 1950s and 1960s, and now makes it possible to predict the development, movement, landfall and size of tsunamis. This does not make them less dangerous, nor does it do anything to prevent damage or loss of property. It only provides warnings of several minutes or a few hours to permit rapid evacuation of coastal areas. Most public beach access sites on the west coast of the United States now have a sign warning of tsunamis.

Standing Waves

All of the previously discussed wave types are progressive gravity waves. The waveform moves forward and gravity is the restoring force. There are special conditions under which waves do not propagate but can influence the coast. Such waves are called *standing waves*—waves trapped in a container or a restricted body of water such as an embayment, a harbor or a lake. The wave length of a standing wave is equal to the diameter or length of the water body in which the wave develops. There is a node in the middle about which there is no motion and an alternating up and down motion at each end of the water body (Figure 6.21). The best example of a small-scale standing wave is the liquid motion that takes place when you walk with a cup of coffee or a pan of water; the surface sloshes back and forth.

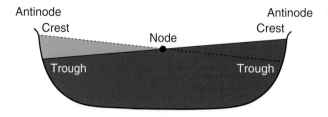

Antinode
Crest
Node
Antinode
Crest
Trough
Trough

Figure 6.21 Standing wave in a vessel shows a node in the middle above which the water surface moves.

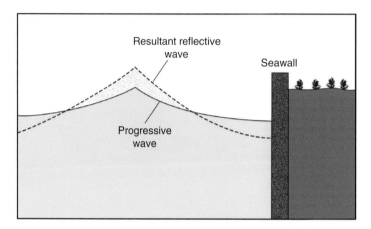

Resultant reflective wave

Seawall

Progressive wave

Figure 6.22 Reflected waves. (*Source:* Courtesy of Joe Holmes, RPI).

Most common and most noticeable among the standing waves in the natural environment is the *seiche* [pronounced "saysh"]. Seiches are usually a result of weather conditions that cause rapid changes in barometric pressure and/or wind conditions. Water is piled up on one side of the basin and then is released as the wind stops quickly or when the barometric pressure rises quickly. As a consequence, there is an alternating extremely high and extremely low water level at the shore as the standing wave sloshes back and forth across the water body. The Great Lakes tend to experience the most dramatic seiches, with changes in lake level of over a meter with the passage of a strong frontal system. Seiche waves have swept people from the beach and drowned them in Lake Michigan. Lake Erie is the most susceptible due to its shallow depth and its orientation along the dominant wind direction as storms pass through this area. It has experienced seiches of over two meters.

Wave reflection can also create a type of standing wave that is instantaneous in nature.

As waves approach a sea wall, steep beach or other obstacle with the wave crests parallel to the obstruction, the crest is reflected back and causes an instantaneous increase in the height of the next incoming wave (Figure 6.22).

6.5 Wave-Generated Currents

There are three types of coastal currents that are produced by waves: longshore currents, rip currents and undertow. All three of types of current develop as the result of shoreward progression of waves, and all three can play a role transporting sediment within the surf zone.

Longshore Currents

There is a slow, landward transport of water as gravity waves move landward and break in the surf zone. The rate at which this occurs is related to the refraction pattern of the waves as they move through the nearshore and surf zone. This refraction produces a current that flows essentially parallel to the shoreline; the *longshore current* or sometimes called the littoral current (see Figure 6.13).

As the wave refracts during steepening and breaking, there is both a shoreward and a shore-parallel vector to its direction. This condition produces the longshore current with a velocity that is related to the size of the breaking wave and to its angle of approach to the shoreline. The longshore current is essentially confined to the surf zone and has an effective seaward boundary at the outermost breaker line and a landward boundary at the shoreline. It acts much like a river channel with the greatest velocity near the middle and has been called the "river of sand." Under some conditions, wind blowing along shore will enhance the speed of the longshore current. These currents typically move at a few tens of centimeters per second but even moderate storm conditions generate speeds of more than a meter per second.

Longshore currents work together with the waves to transport large volumes of sediment along the shoreline. As the waves interfere with the bottom when the waves steepen and break, large amounts of sediment are temporarily put into suspension. The sediment is most concentrated at the bottom of the water column but is present throughout. This suspended sediment and bed-load sediment is then transported along the shoreline by longshore currents.

Because waves may approach the coast in wide range of directions, the longshore current may flow in either direction along the shoreline, depending on the angle of wave approach. As a result of this back and forth transport of sediment there might be a large amount of sediment flux over a designated period of time, but the net littoral transport will be in one direction and may be only a small portion of the total. We typically speak of *littoral transport* in terms of the annual amount in a particular length of coast. For example it may be 100,000 m^3 year^{-1} from north to south. Remember, this is the net transport, the total or gross amount may be many times that in each direction. Longshore currents and littoral sediment transport occurs in any coastal environment where waves are refracted as they move into shallow water and approach the shoreline. Whereas large volumes of littoral drift take place on open ocean beaches, longshore currents and resulting sediment transport are also common on bays, estuaries and lakes.

Rip Currents

The previously mentioned landward transport of water as waves move to the shoreline produces "setup," an increase in water level at the shore. This is largely the result of friction between the wind and the water surface as the progressive waves move through the surf zone. Setup is basically similar to storm surge but it is smaller in scale and is limited to a narrow zone along the shore. This elevation of the water is typically a few centimeters. The setup water is essentially piled up against the shoreline in an unstable condition because of the inclination of the water level. That situation cannot persist.

If this unstable condition exists along a barred coast or along some of the steeper coasts, the setup produces seaward flowing currents that are rather narrow and that create circulation cells (Figure 6.23) within the surf zone. These narrow currents are called *rip currents*. They commonly flow at speeds of a few tens of centimeters per second and they can transport sediment. Some are strong enough that they excavate shallow channels that are essentially normal to the orientation of the coast (Figure 6.24). Rip currents are connected with feeder currents that are part of the longshore current system. Rip current is the phenomenon that is commonly and mistakenly called a "rip tide," even though it is not related to tides. Rip currents can be a danger to swimmers and they cause several drownings each year at locations where they are common. In actuality they should not be a problem to swimmers because they are narrow and the swimmer simply needs to move a short distance along shore to escape their seaward path.

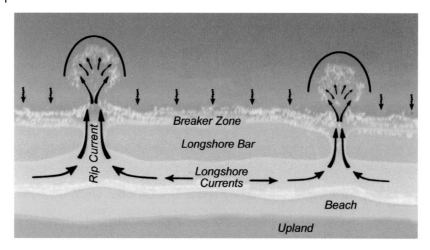

Figure 6.23 Diagram showing water circulation in a rip current system. (*Source:* Courtesy of Joe Holmes, RPI).

The size, orientation and spacing of rip currents are related to wave conditions and nearshore bathymetry. In the absence of longshore sand bars and along a smooth, embayed shoreline, the rip current spacing is typically regular and fairly persistent. If incoming waves are essentially parallel to the shoreline they will produce a nearly symmetrical circulation cell with feeder currents that are symmetrical in each direction. Rip currents that form when waves approach at an angle generally develop asymmetrical cells and related feeder currents.

Rip currents are not major transporters of sediment although they may carry small amounts of fine sediment (Figure 6.25).

Undertow

Another nearshore circulation phenomenon that is caused by shoreward movement of water and setup is called *undertow*. Here, the water that piles up at the shoreline is returned along the bottom to alleviate the unstable condition produced by the setup. This is a type of circulation that is commonly confused with rip currents and that may also cause problems for swimmers. Water that is transported toward the shoreline by forward motion of water as waves progress landward and from friction between wind and water is returned seaward in a strong and essentially continuous current. Flow is commonly up to 50 cm s^{-1} during strong onshore waves, the only time when undertow is noticeable.

Undertow is not confined to rather high-energy wave conditions in the surf zone; it is also common in places where longshore bars are absent or far offshore and relatively deep. In some areas rip currents will persist during low- to moderate-energy wave conditions and then lose their definition during high-energy wave conditions; undertow will then become the primary mode of seaward return of water from unstable conditions of setup.

6.6 Summary

Waves are probably the most important factor in coastal development. They move sediment directly or through the generation of wave-driven currents. Wave activity can cause erosion and can also transport sediment to the coast. Although there are still many aspects of wave–sediment interaction that we do not understand, the basic characteristics of wave mechanics are well known. Part of the difficulty is that

(a)

(b)

Figure 6.24 Rip current channels on the coasts of (a) Florida and (b) Australia.

we are dealing with the relationships between fluid motion and a solid substrate. Another difficulty is that much of the interaction between waves and the coast takes place during severe conditions of storms. Making observations under such conditions is extremely difficult, although great progress is being made.

Waves are primarily at work on the open coast and have only minor influence on the more protected environments such as estuaries, tidal flats and wetlands.

Figure 6.25 Rip current carrying sediment offshore. (*Source:* R. A. Davis).

Suggested Reading

Bascom, W. (1980). *Waves and Beaches.* Garden City, NY: Anchor Books.

CERC (1984). *Shore Protection Manual,* 2 volumes. Vicksburg, MS: Coastal Engineering Research Center, Waterway Experiment Station, Corps of Engineers.

Davidson-Arnott, R. (2009). *Introduction to Coastal Processes and Geomorphology.* Cambridge: Cambridge University Press.

Dean, R.G. and Dalrymple, R.A. (1998). *Water Wave Mechanics for Engineers and Scientists,* 2e. Englewood Cliffs, NJ: Prentice-Hall.

Holthuijsen, L.H. (2010). *Waves in Oceanic and Coastal Waters.* Cambridge: University of Cambridge Press.

Hudspeth, R.T. (2005). *Waves and Wave Forces on Coastal and Ocean Structures.* Singapore: World Scientific Publishing Company.

U.S. Army, COE (2002). *Shore Protection Manual,* vol. 1. Washington, DC: U.S. Army, Corps of Engineers.

7

Tides of the Ocean

7.1 Introduction

Anyone who has spent time along the seashore knows that the ocean level changes on an hourly basis. Fishermen plan their activities around high and low tide, for instance those who dig clams when tidal flats are uncovered or gather mussels at low tide. Boaters are aware that some channels are only navigable at high tide when waters are deep enough for their boats to pass. Large vessels sailing to port normally enter at slack water or during an ebbing tide. Moving against the tidal flow provides greater steerage for ships than traveling with the currents. The rise and fall of the tides is one of the major rhythms of Planet Earth. It has a dramatic effect on shoreline processes and coastal landforms. Were it not for the tides there would be no tidal inlets and few natural harbors along East and Gulf Coast of the United States as well as along many other barrier coasts of the world. Those familiar with the coast recognize that times of high and low tide occur approximately an hour later each day. More astute observers know that daily tides gradually change in magnitude over the course of a month and that these variations are closely related to phases of the Moon.

Tides are a manifestation of the Moon and Sun's gravitational force acting on the Earth's hydrosphere. With wave lengths measuring in thousands of kilometers, tides are actually shallow-water waves affecting the world's oceans from top to bottom. The surface expression of tides is most dramatic in funnel-shaped embayments where vertical excursions of the water surface can reach more than 10 m in such areas as the Gulf of Saint-Malo, France (Figure 7.1) or the Gulf of San Matias, Argentina and even as high as 15 m in the Bay of Fundy, Canada. Pytheas, a Greek navigator, recorded the Moon's control of the tides in the fourth century BCE. However, it was not until Sir Isaac Newton (1642–1727) published *his Pliloshophiae naturalis principia mathematica* (*Philosophy of natural mathematical principles*) in 1686, that we finally had a scientific basis for understanding the tides.

In this chapter we discuss the origin of the tides and how their magnitude is governed by the relative position of Earth, the Moon and the Sun. It will be shown that the complexity of the tides is a function of many factors including the elliptical orbits of Earth and the Moon, the angles of inclination of the orbits of Earth and Moon, and the presence of continents that partition the oceans into numerous large and small basins. The phenomena of tidal currents and tidal bores are also discussed.

7.2 Tide-Generating Forces

7.2.1 Gravitational Force

As a basis for understanding the Earth's tides, we begin with Newton's universal **Law of Gravitation,** which states that every particle

Beaches and Coasts, Second Edition. Richard A. Davis, Jr. and Duncan M. FitzGerald.
© 2020 John Wiley & Sons Ltd. Published 2020 by John Wiley & Sons Ltd.

Figure 7.1 Mont Saint Michel in the Gulf of St.-Malo is surrounded by water at high tide due to a 12-m tide that inundates the tidal flats and even floods some of the parking lots.

of mass in the universe is attracted to every other particle of mass. This force of attraction is directly related to the masses of the two bodies and inversely proportional to the square of the distance between them. The law can be stated mathematically as:

$$F = G \frac{M_1 M_2}{R^2}$$

where F is the force of gravity, G is the gravitational constant, M_1 and M_2 are the masses of the two objects and R is the distance between the two masses. From the equation it is seen that the force of gravity increases as the mass of the objects increases and as the objects are closer together. Distance is particularly critical because this factor is squared (R^2) in the equation. Thus, the two celestial bodies producing the Earth's tides are the Moon, owing to its proximity, and the Sun, because of its tremendous mass. The other planets in the Solar System have essentially no effect on the tides due to their relatively small mass as compared to the Sun and their great distance from Earth as compared to the Moon. Although the attractive forces of the Moon and Sun produce slight tides within the solid Earth and large oscillations in the atmosphere, it is the easily deformed liquid component of the Earth (the hydrosphere) where tides are most clearly visible.

As illustrated in Figure 7.2, the Sun is 27 million times more massive than the Moon but 390 times further away. After substituting the respective mass and distance values for the Moon and Sun into Newton's Gravitation equation, it is seen that the attractive force of the Sun is approximately 180 times greater than that of the Moon. However, we still know that the Moon has a greater influence on Earth's tides than does the Sun.

7.2.2 Centrifugal Force

To understand how gravitational force actually produces the tides it is necessary to learn more about orbiting celestial bodies including the

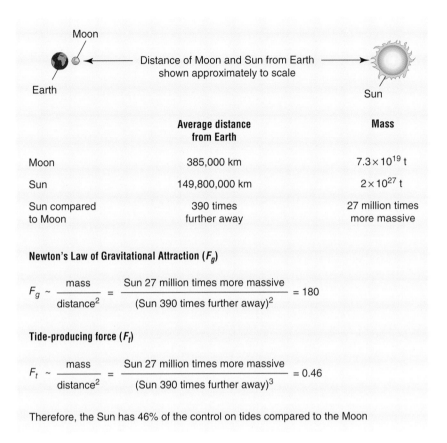

Moon

Distance of Moon and Sun from Earth shown approximately to scale

Earth

Sun

	Average distance from Earth	Mass
Moon	385,000 km	7.3×10^{19} t
Sun	149,800,000 km	2×10^{27} t
Sun compared to Moon	390 times further away	27 million times more massive

Newton's Law of Gravitational Attraction (F_g)

$$F_g \sim \frac{\text{mass}}{\text{distance}^2} = \frac{\text{Sun 27 million times more massive}}{(\text{Sun 390 times further away})^2} = 180$$

Tide-producing force (F_t)

$$F_t \sim \frac{\text{mass}}{\text{distance}^2} = \frac{\text{Sun 27 million times more massive}}{(\text{Sun 390 times further away})^3} = 0.46$$

Therefore, the Sun has 46% of the control on tides compared to the Moon

Figure 7.2 Earth's tides are primarily controlled by the Moon because tidal force is inversely proportional to distance between the masses *cubed*. Thus, even though the Sun is much more massive than the Moon, because it is also much further away from Earth than the Moon its influence is a little less than one-half that if the Moon.

Moon–Earth system and the Earth–Sun system. First, it is important to recognize that a centrifugal force is counteracting the gravitational attraction between the Moon and Earth (Figure 7.3). Centrifugal force is a force that is exerted on all objects moving in curved paths, such as a car moving through a sharp right bend in the road. The centrifugal force is directed outward and can be felt by a car's driver as he or she is pressed against the car's left door through the turn. If the Moon were stopped in its orbit, centrifugal force would disappear and gravitational force would cause Earth and the Moon to collide. Conversely, if the gravitational force ceased between the two bodies, the Moon would career into space.

Thus far, we have been careful not to say that the Moon orbits Earth. In fact, Earth and Moon form a single system in which the two bodies revolve around a single center of mass. Because Earth is approximately 81.5 times more massive than the Moon, the center of mass of the system, called the **barycenter**, must be 81.5 times closer to Earth's center than the Moon. The barycenter can be determined by knowing that the average distance between the center of Earth and the center of the Moon is 385,000 km; by dividing 385,000 by 81.5 we calculate that the center of mass is 4724 km from Earth's center. Earth's radius is 6380 km, and therefore the center of mass of the Earth–Moon system is located 1656 km (6380 − 4724) beneath the

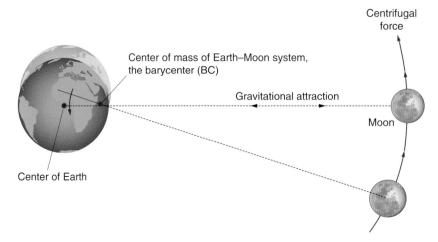

Figure 7.3 The Earth–Moon system rotates around a common center of mass called the barycenter, which is within Earth. The gravitational attraction between Earth and the Moon is balanced by the centrifugal force due to the Moon's motion around the barycenter.

surface of Earth. The Earth–Moon system can be visualized by considering a dumbbell with a much larger ball at one end (81.5 times greater) than the other. If this dumbbell were thrown end over end, it would appear as though the large ball (Earth) wobbled and the small ball (Moon) orbited the large ball.

It should be understood that because the entire Earth is revolving around the center of mass of the Earth–Moon system, every unit mass on the surface of Earth is moving through an orbit with the same dimensions. The average radius of each orbit is 4724 km (NOAA n.d.). (The movement of Earth around the Earth–Moon center of mass should not be confused with Earth spinning on its axis, which is a separate phenomenon and plays no part in establishing the differential tide-producing forces.) Thus, if every unit mass on the surface of Earth has the same size orbit, then it follows that the centrifugal force on the unit masses must also be equal.

7.2.3 Tide-Producing Force

Ocean tides exist because gravitational and centrifugal forces are unequal on Earth's surface (hydrosphere) (Figure 7.4). In fact, the gravitational attraction between the Moon

and Earth only equals the centrifugal force on Earth at their common center of mass (the barycenter), which is 1656 km inside the Earth's surface. Thus, if we consider a unit mass at the surface of Earth at a site facing the Moon, this mass experiences a force of attraction by the Moon that is greater than the centrifugal force due to its rotation about the barycenter. The larger gravitational force is explained by the fact that at this location the distance to the Moon is less than the distance between the Earth's center of mass and the Moon's center of mass at which the two forces are equal. (Remember in Newton's gravitation equation that as distance decreases, the force of gravity increases.) Conversely, for a unit mass on the opposite side of Earth the centrifugal force due to its rotation about the barycenter exceeds the gravitational attraction exerted by the Moon because this site is farther away from the Moon than the point at which the forces balance. Thus, the unequal forces on either side of Earth cause the hydrosphere to be drawn toward the Moon on the near side of Earth and to be directed away from the Moon on the opposite side. This produces two tidal bulges of equal size that are oriented toward and away from the Moon. These forces also result in depressions in the hydrosphere that

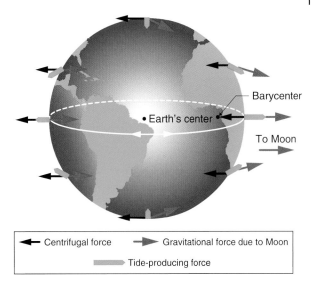

Figure 7.4 Tides are generated in Earth's hydrosphere due to differences between the centrifugal force and the Moon's attractive force. The centrifugal forces that result from the rotation of the center of the Earth around the barycenter, are essentially equal over Earth's surface and directed away from the Moon. This somewhat simple treatment accounts for two tidal bulges being produced on opposite sides of Earth.

are located halfway between the two bulges on either side of Earth. If we disregard the curvature of Earth, the tides can be thought as a long wave with the crest being the bulge and the depression being the trough. This waveform is called the **tidal wave** and should not be confused with a tsunami, which sometimes is inappropriately referred to as a tidal wave.

The above description reveals that forces generating Earth's tides are very sensitive to distance. The tide-generating force is derived by calculating the difference between the gravitational force and the centrifugal force. A simplified form of the relationship is given by:

$$F \approx \frac{M_1 M_2}{R^3}$$

The tide-generating force F is proportional to the masses M_1 and M_2 and inversely related to the cube of the distance between the bodies R^3. When these computations are performed for unit masses over the surface of Earth, it is seen that the resulting vectors are oriented toward and away from the Moon (Figure 7.4). Note also that *distance is cubed* in the equation, which explains why the Moon exerts a greater control on Earth's tides than does the Sun. As illustrated in Figure 7.2, after substituting the respective mass and distance values into the above equation, it is calculated that the tide-generating force of the Sun is only 46 % of the Moon's.

7.3 Equilibrium Tide

The **equilibrium tide** is a simplified model of how tides behave over the surface of Earth given the following assumptions:

1) Earth's surface is completely enveloped with water with no intervening continents of other landmasses.
2) The oceans are extremely deep and uniform in depth such that the seafloor offers no frictional resistance to movement of the overlying ocean water.
3) There are two tidal bulges that remain fixed toward and away from the Moon.

7.3.1 Tidal Cycle

In our initial discussion of the equilibrium tide model, we will neglect the effects of the Sun. If we consider a stationary Moon, then Earth passes under the two tidal bulges each time it completes a rotation around its axis (Figure 7.5). In this idealized case the wave length of the tidal wave, which is the distance between the tidal bulges, would be half the

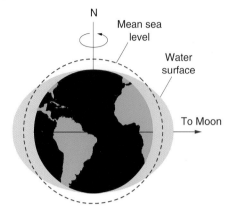

Figure 7.5 Under idealized equilibrium tide conditions Earth passes beneath two equal and opposite tidal bulges every 24 hours. This situation assumes an absence of continents, a uniform depth of the ocean, and the Moon aligned with Earth's equator.

circumference of Earth. High tide coincides with Earth's position under the bulges and low tide corresponds to the troughs located midway between the bulges. The tidal period would be 12 hours (interval between successive tidal bulges). The rhythm of tidal changes referred to as the tidal cycle can be better conceptualized if we choose a position on the

equator directly facing the Moon and record how the water level fluctuates at this site through time as Earth spins on its axis. At 12:00 midnight, the Moon is overhead and the ocean is at high tide because Earth is directly under the maximum extent of the tidal bulge. After high tide the water level gradually drops reaching low tide six hours later at 6:00 a.m. During the next six hours the tide rises, attaining a second high tide at 12:00 noon. The cycle repeats itself over the next 12 hours.

7.3.2 Orbiting Moon

In the real world the Moon is not stationary, rather it completes an orbit around the center of the Earth–Moon system in a period of 27.3 days. The Moon moves in the same direction as Earth spins on its axis and therefore after Earth completes a full 24-hour rotation the Moon has traveled 13.2° of its orbit. For Earth to "catch up" to the Moon, it must continue to rotate for an additional 50 minutes. Thus the Moon makes successive transits above a given location on Earth in a period of 24 hours and 50 minutes, which is called the lunar day (Figure 7.6). Because the Moon is

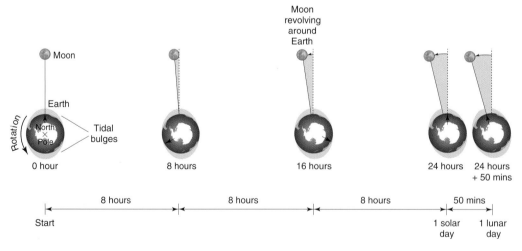

Figure 7.6 Cartoon illustrating why times of high and low tide occur 50 minutes later each successive day. The arrow represents a stationary position on Earth. Tidal bulges are oriented directly toward and away from the Moon and time zero coincides with a high tide. In 24 hours the arrow is back to the origin, however the moon is no longer directly overhead. As Earth completes one rotation on its axis, the Moon travels 1 day in its 27.3-day orbit around Earth. Thus, Earth must spin on its axis an extra 50 minutes before the arrow is aligned with the moon and the ocean reaches high tide.

Figure 7.7 In its orbit around Earth, the Moon moves above and below the equator. When the Moon is not aligned with the equator, successive tidal bulges, and the corresponding tidal ranges, are unequal. This phenomenon is called a semi-diurnal inequality.

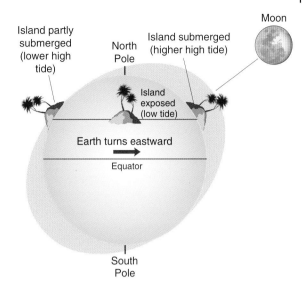

Island partly submerged (lower high tide)

North Pole

Island submerged (higher high tide)

Moon

Island exposed (low tide)

Earth turns eastward

Equator

South Pole

moving, high and low tides do not take place every 12 hours as discussed in the simple model above, rather they occur every 12 hour and 25 minutes. The time interval between high and low tide is about 6 hours and 13 minutes. When there are two cycles in a day (actually 24 hours and 50 minutes) they are called semi-diurnal tides.

7.3.3   Inclination of Moon's Orbit

So far in our discussion of the tides, we have simplified matters by envisioning a Moon that is always directly overhead of the equator. However, the Moon's orbit is actually inclined to the plane containing the equator. Over a period of a month the Moon migrates from a maximum position 28.5° north of the equator to a position 28.5° south of the equator and back again. When the Moon is directly overhead of the tropics the tides are called *tropic tides* and when it is over the equator they are called *equatorial tides*. Because the Moon is the dominant control of the tides it follows that when the Moon is positioned far north or south of the equator the tidal bulges will also be centered in the tropics. This arrangement of the tidal bulges leads to a **semi-diurnal inequality** meaning that successive tides have very different tidal ranges (tidal range is the vertical difference

in elevation between low and high tide). For example, if you are along the east coast of Florida at tropic tide conditions, during one tidal cycle the tide will come up very high and then go out very far generating a relatively large tidal range. During the next tidal cycle, a low high tide is followed by high low tide, producing a small tidal range. As viewed in Figure 7.7, the diurnal inequality is explained by the fact that when the tidal bulges are asymmetrically distributed about the equator, Earth will rotate under very different sized tidal bulges. This translates to unequal successive high and low tides. It should be noted that during equatorial tide conditions there is little to no inequality of the semidiurnal tides, whereas they reach a maximum during tropic tides.

7.4   Interaction of Sun and Moon

In an earlier section it was shown that the Sun's tide-generating force is a little less than half that of the Moon (46%). It is important to note that, just like the Moon, the Sun also produces bulges and depressions in Earth's hydrosphere. These are called solar tides and they have a period of 12 hours unlike the 12-hrs and 25 minutes period of the lunar tides.

The period is 12 hours because Earth passes through two solar bulges every day (24 hours). One way of explaining the interaction of the Moon and Sun is to show how the Sun enhances or retards the Moon's tide generating force. In order to do this we must first understand how the phases of the Moon correlate with the position of Earth, the Moon, and Sun.

New Moons and full Moons result when Earth, the Moon, and Sun are aligned, a condition referred to as *syzygy* (a great Scrabble word worth many points). A new Moon occurs when the Moon is positioned between Earth and the Sun, whereas a full Moon results when the Moon and Sun are on the opposite sides of Earth. When the Moon forms a right angle with Earth and the Sun (*quadratic* position), only half of the Moon's hemisphere is illuminated. This phase occurs during the Moon's first and third quarter.

The Moon cycles through these different phases over a period of 29.5 days.

During new and full Moons, when Earth, the Moon, and Sun are all aligned, the tide-generating forces of the Moon and Sun act in the same direction and the forces are additive (Figure 7.8). Conceptually, one can envision the Sun's bulge sitting on top of the Moon's bulge. Between the bulges the Sun's trough further depresses the Moon's trough. The double bulges and double troughs lead to very high high-tides and very low low-tides. This condition is called a **spring tide** and is characterized by maximum tidal range (Figure 7.9).

During quadratic conditions the Sun's effects are subtractive from the Moon's tide-generating force. Because the Moon is positioned 90° to Earth and the Sun, its bulge coincides with the Sun's trough and its trough is positioned at the Sun's bulge. The superposition of bulges

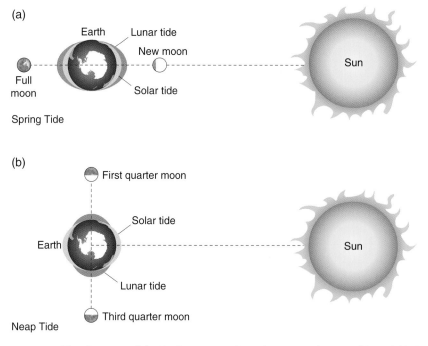

Figure 7.8 The alignment of the Earth, Moon, and Sun determines the size of the tidal bulges and the magnitude of the tidal range. (a) During periods of full and new moons when the Earth, Moon, and Sun are aligned the lunar and solar tidal forces are additive and their tidal bulges are additive ("spring tides"). These conditions produce relatively large tidal ranges. (b) When the Moon, Earth, and Sun are at right angles during half-moon conditions the tidal forces are subtractive and tidal ranges are relatively small ("neap tides").

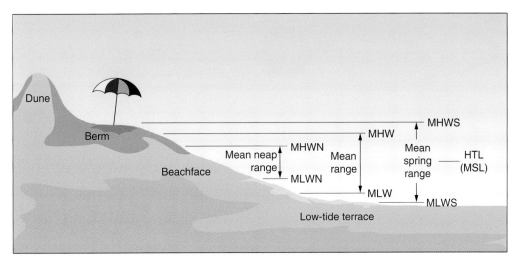

Figure 7.9 Tidal ranges and the ensuing elevation of high and low tides are a function of the position of the Earth, Moon, and Sun. During syzygy spring tides produce very high high-tides and very low low-tides. Conversely, during quadratic conditions neap tides create low high-tides and high low-tides. Mean tidal ranges occur during the periods in between syzygy and quadratic alignments.

and troughs causes destructive interference and produces low high tides and high low tides. This condition is called a **neap tide** and is characterized by minimum tidal ranges (Figure 7.9). When Earth, Moon, and Sun are arranged in positions between syzygy and quadratic we experience **mean tides** with average tidal ranges. Spring and neap tides occur approximately every 14 days, whereas mean tides occur every 7 days.

7.5 Effects of Orbital Geometry

Remembering that the tide-producing force is particularly sensitive to distance, it is understandable that the geometry of both Earth and the Moon's orbits affects the tides. Earth revolves around the Sun in an elliptical orbit and the Sun is situated at one of the foci of the ellipse. In early January Earth is nearest to the Sun at a position called **perihelion** (Figure 7.10). Six months later (July) Earth is at **aphelion** furthest from the Sun. The difference in distances is approximately 4%. The Moon's orbit around the center of the Earth–Moon system is also elliptical. When the Moon is close to Earth it is referred

to as **perigee** and when it is most distant it is called **apogee**. There is a 13% difference between perigee and apogee.

If we consider all the various factors that influence the magnitude of the tides, we begin to understand why tidal ranges and high and low tidal elevations change on a daily basis. Tide levels are especially important during storms. In early February 1978 a major northeast storm, the Blizzard of 1978, wreaked havoc in New England dropping over two feet of snow and completely immobilizing the residents for several days. The Blizzard of 1978 was a particularly menacing storm causing widespread beach erosion, destruction of hundreds of coastal dwellings, and hundreds of millions of dollars worth of damage to roadways and other infrastructure. One of the reasons why this storm was so severe was due to the astronomic conditions at the time of the storm. Earth, the Moon and the Sun were in syzygy, so the storm hit during spring tide conditions. At the same time the Moon was at perigee and Earth and Sun were close to a perihelion position. Syzygy, perigee, and perihelion combined to raise high tide levels 0.55 m above normal. The extreme astronomic tides

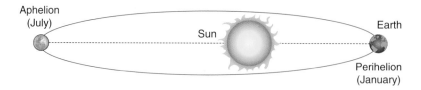

Figure 7.10 The elliptical orbit of Earth affects its distance from the Sun and therefore the magnitude of the tide-producing force during the year. Earth is closest to the Sun during perihelion (January) and furthest away during aphelion. Likewise the elliptical path of the Moon around Earth produces periods when the Moon is relatively close to Earth (perigee) and periods when it is relatively far away (apogee).

coupled with the 1.4-m storm surge caused extensive flooding. Storm waves elevated by high water levels broke directly against foredune ridges, across barriers and over seawalls. Had the storm hit during quadratic, apogean, and aphelion conditions, high tide waters would have been 1.1 m below the February Blizzard levels and damage would have been an order of magnitude less (Figure 7.11).

7.6 Effects of Partitioning Oceans

We have been treating tides as if Earth were completely enveloped by a uniformly deep ocean. However, we know that continents and island archipelagoes have partitioned the hydrosphere into several interconnected large and small ocean basins whose margins are generally irregular and quite shallow. Because oceans do not cover the surface of Earth, the tidal bulges do not behave as simply as they have been presented thus far. In addition to the complexities imparted by the presence of landmasses, the equilibrium tide concept is further complicated by the fact that Earth spins faster in lower latitudes and slower in higher latitudes than the tidal

wave. Thus, the oceans do not have time to establish a true equilibrium tide. Finally, the ocean tides are affected by the **Coriolis effect**, which is generated by the Earth's rotation. This causes moving objects, including water masses, to be deflected to the right in the northern hemisphere and to left in the southern hemisphere. For example, the Gulf Stream that flows northward along the margin of North America is deflected northeastward toward Europe due to Coriolis.

In the dynamic model of ocean tides we no longer envision static ocean bulges that remain fixed toward the Moon and under which Earth spins; rather, the tidal bulges rotate around numerous centers throughout the world's oceans (Figure 7.12). An individual cell is called an **amphidromic** system and the center of the cell around which the tidal wave rotates is known as the **amphidromic point** or nodal point. The rotation of the tidal wave is due to the Coriolis effect and is counterclockwise in the Northern Hemisphere and clockwise in the Southern Hemisphere.

To understand the behavior of an amphidromic system let us first begin with a hypothetical square-shaped ocean basin that responds to the Moon's tide-generating forces (Figure 7.13). The tidal wave that develops

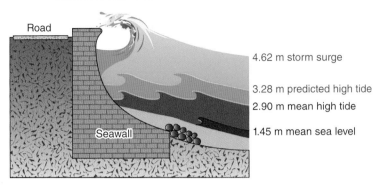

Figure 7.11 Cartoon of seawall along Winthrop Beach, Massachusetts that was overtopped by storm waves during the Blizzard of 1978. Note that if the storm had occurred during low astronomic tidal range conditions, fewer waves would have broken over the seawall and the overall damage to the New England shoreline would have been far less severe.

under these conditions exhibits elements of both a standing wave and a progressive wave. As Earth spins and the Moon travels from east to west over the hypothetical basin, the tidal bulge sloshes against the western side of the basin in an attempt to keep abreast of the passing Moon. As Earth continues to rotate, the bulge of water begins to flow eastward back toward the low center of the basin. However, the Coriolis effect deflects this water mass to the southern margin of the basin and water piles up there. This in turn creates a water surface that slopes northward and the process is repeated. The end result is a tidal wave that rotates in a counter-clockwise direction around the basin with a period of 12 hours and 25 minutes. High tide is coincident with the tidal bulge and low tide occurs when the bulge is along the opposite side of the basin.

When a line is drawn along the crest of the tidal wave every hour for a complete rotation, the resulting diagram looks like a wheel with spokes. It depicts how the tidal wave rotates within the hypothetical basin and its center is the amphidromic point. The spokes are called **co-tidal lines** and they define points within the basin where high tide (and low tide) occurs at the same time (Figure 7.12). If points of equal tidal range are contoured within the basin a series of semi-concentric circles are formed around the amphidromic

point. These contours having equal tidal range are referred to as **co-range lines**. Ideally, tidal range is zero at the amphidromic point and gradually reaches a maximum toward the edge of the basin. Due to the land barriers and other factors the world's oceans are divided into approximately 15 amphidromic systems. This does not include smaller seas that have their own amphidromic cells such as the Gulf of Mexico (1 system), Gulf of St. Lawrence (1 system), and the North Sea (3 systems).

7.7 Tidal Signatures

In the ideal case, we expect two tidal cycles daily (actually 24 hours and 50 minutes.). However, highly variable basinal geometries of the world's ocean and modifications of the tidal wave as it shoals across the continental shelf as well as other factors have combined to produce a variety of tidal signatures throughout the world's coastlines. There are three major types of tides (Figure 7.14):

Diurnal Tides

Coasts with diurnal tides experience one tidal cycle daily with a single high and low tide. They have a period of 24 hours and 50 minutes. This type of tide is rare and

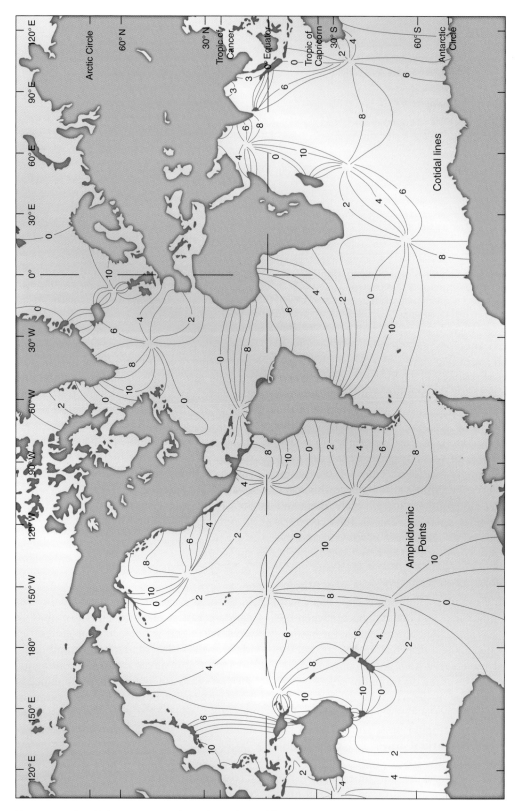

Figure 7.12 Amphidromic systems throughout the world's oceans. The tidal wave rotates in a counterclockwise direction around amphidromic points in the northern hemisphere and in a clockwise direction in the southern hemisphere. The lines radiating from the amphidromic points are co-tidal lines. They indicate hypothetical times in which the crest of the tidal wave passes through the ocean basins.

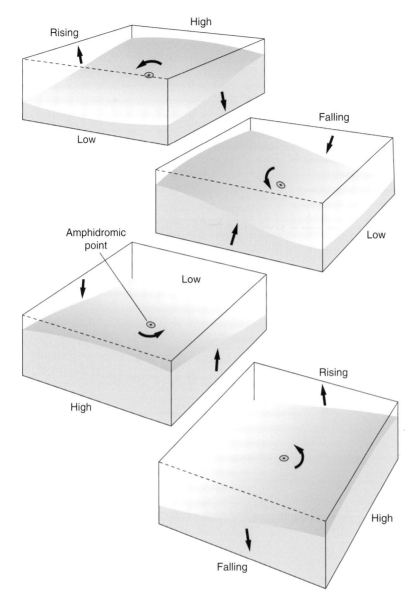

Figure 7.13 Hypothetical tidal wave rotating around in a square-shaped basin. Note that water elevation changes (tidal range) increase outward from the amphidromic point. The Coriolis Effect causes rotation of the tidal wave.

commonly associated with restricted ocean basins including certain areas within the Gulf of Mexico, the Gulf of Tonkin along Southeast Asia, and the Bering Sea. It is in these areas where distortions of the tidal wave produce a natural tidal oscillation coinciding with 24 hours and 50 minutes. The open southwest coast of Australia is an exception to this general trend.

Semidiurnal Tides

This is the most common type of tide along the world's coast. It is characterized by two tidal cycles daily with a period of 12 hours and 25 minutes. Seldom, however, are the two tides of the same magnitude except when the Moon is over the equator or at locations near the equator. As discussed earlier, when the Moon is over the tropics successive tidal

(a)

(i) Diurnal

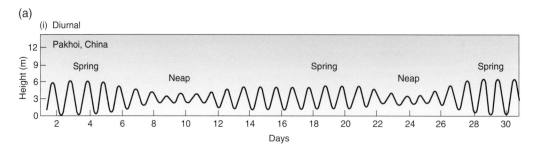

(ii) Equal semi-diurnal

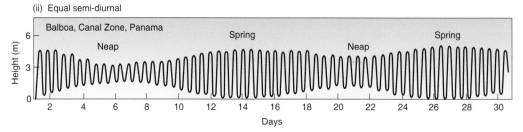

(iii) Unequal semi-diurnal

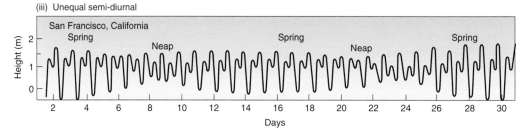

(iv) Mixed

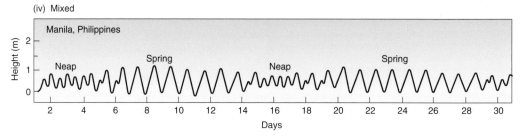

Figure 7.14 Coastlines throughout the world experience a variety of tidal signatures. (a) The major types include: 1. Diurnal tides, one tide daily; 2. Semidiurnal tides, two tides daily; 3. Semidiurnal tides with strong inequality, 4. Mixed tides, combination of diurnal and semi-diurnal tides. (b) Geographic distribution of tidal types. (*Source:* R. A. Davis, 1977).

bulges (the tidal wave) have unequal magnitudes, producing different tidal ranges and high and low tides reaching different elevations. This condition of unequal tides is called a semidiurnal inequality.

Mixed Tides

This type of tide occurs extensively throughout the world. As the name implies, mixed

tides have elements of both diurnal and semidiurnal tides. Its signature varies during a lunar cycle from a dominant semidiurnal tide with a small inequality to one that exhibits a very pronounced inequality. At some sites, including San Francisco, California, Seattle, Washington, and Port Adelaide, Australia during part of the lunar month one of the two daily tides manifests itself as a very

(b)

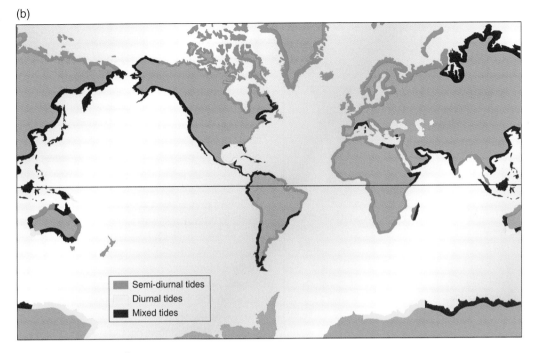

Figure 7.14 (Continued)

small vertical excursion measuring no more than 0.1–0.3 m. These tides have a distinct diurnal signature. Along other coasts, such as Los Angeles, Honolulu, and Manila in the Philippines, the second daily tide essentially disappears and the tide becomes totally diurnal.

Thus, the complexities that produce and modify the Earth's tides are revealed by the variability in their tidal signature throughout the world and even temporally as viewed during a lunar cycle.

7.8 Tides in Shallow Water

Continental Shelf Effects

In the middle of the ocean the tidal wave travels with a speed of 700 km h⁻¹. In these regions the tidal range is only about 0.5 m. The tidal wave that reaches the coast travels from the deep open ocean across the continental margin to the shallow inner continental shelf. Similar to wind-generated waves, shoaling of the tidal wave along this pathway causes it to slow down. The tidal wave that traverses the entire continental margin is reduced in speed to about 10–20 km h⁻¹. Like wind waves, the tidal wave also steepens, which is reflected in an increase in tidal range. For example, the tidal wave in the north Atlantic is estimated to be 0.8 m in height (tidal range) at the edge of the continental shelf. The wave steepens as it propagates through the Gulf of Maine, producing a tidal range of 2.7 m along the coast of Maine. On a worldwide basis using an average shelf width of 75 km, it is estimated that the tidal wave will increase in height from 0.5 m in the deep ocean to about 2.4 m along the coast after it traverses the continental shelf. Variations from this value are due to differences in shelf width and slope, and variability in the configuration of the coast. It is of interest to note that along the east coast of the United States, the continental shelf is relatively wide off Georgia, where tidal ranges reach 2.6 m, whereas north and south of this region the shelf narrows and tidal ranges correspondingly reduce to 1.1 m at Cape Hatteras and less than a meter along central Florida (Figure 7.15).

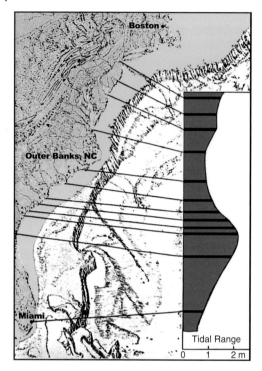

Figure 7.15 Tidal range along the East Coast of United States is controlled in part by the width of the continental shelf. As the tidal wave travels from the deep ocean across the shallow continental shelf, its speed slows and its crest steepens. The amount of wave steepening, which increases the tidal range, is proportional to the length and gradient of the shelf. Thus, the wide, gentle shelf off the Georgia coast produces a tidal range greater than 2 m whereas the narrow, steep shelves bordering south Florida and Cape Hatteras in North Carolina generate tidal ranges less than 1 m. (*Source:* From Nummedal et al. 1977).

Coriolis Effect

Just as Coriolis produces a counterclockwise rotation of the tidal wave in large amphidromic systems in the northern hemisphere it also influences the propagation of the tidal wave into gulfs and seas from the open ocean. This phenomenon is demonstrated well in the North Sea, where the Coriolis effect dramatically modifies tidal ranges. The North Sea is a shallow (<200 m), rectangular-shaped basin approximately 850 km long from the Shetland Islands southeastward to the German Friesian Islands and 600 km wide from Great Britain eastward to Jutland, Denmark. Tides in the North Sea are forced

by the North Atlantic amphidromic system (Figure 7.12). The tidal wave approaches through the northern open boundary of the North Sea and propagates southward. Ultimately, the wave partially reflects off the southern margin of the basin and interacts with next incoming wave. The resulting oscillations combined with the Coriolis effect produce three amphidromic systems, two of which are displaced toward the coasts of Norway and Denmark. As illustrated by the co-range lines in Figure 7.16, tidal ranges are much higher along the east coasts of England and Scotland (~4 m) as compared to the Norwegian and Danish coasts (<1 m). This disparity is caused by the tidal wave being deflected to the right as it moves into the North Sea. Water is piled up on the western side of the basin and diminishes the tidal ranges along the eastern side.

A similar situation occurs in the English Channel. Here the tidal wave approaches from the southwest and propagates eastward through the channel. Coriolis deflects water away from the English side of the channel and toward the coast of France. Tidal ranges along the coasts of Brittany and Normandy are greater than 4–5 m, whereas along the southwest English coast they are less than 3 m.

Funnel-Shaped Embayments

The configuration of the coast can also have a pronounced influence on tidal ranges. Rotary tidal waves do not exist in funnel-shaped embayments due to their narrowness. Instead, the tidal wave propagates into and out of funnel-shaped bays. The wave is constricted by the seabed, which shallows in a landward direction, and by the ever-narrowing confines of the embayment. Although frictional elements serve to decrease the energy of the propagating wave, the overall steepening of the tidal wave causes an amplification of the tidal range (Figure 7.17). For example, the 2.0-m tidal range at the entrance to the Saint Lawrence estuary increases to over 5 m during spring tidal conditions at Quebec City some 600 km upstream. Funnel-shaped embayments are found all over the world including the Bay

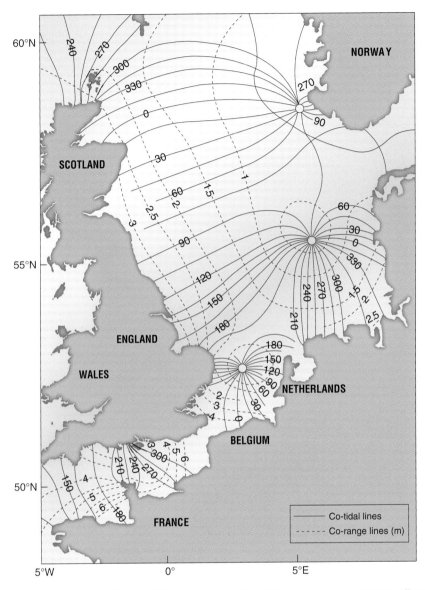

Figure 7.16 The distribution of tidal ranges in the North Sea illustrates how Coriolis affects tidal wave propagation into a shallow sea. Tides in the North Sea are forced by the North Atlantic amphidromic system. As the tidal wave travels southward into the shallow basin, Coriolis deflects the tidal wave toward the coasts of Great Britain and Scotland, creating large tidal ranges. Conversely, the tidal wave is deflected away from Norway and Denmark producing small tidal ranges. (*Source:* from Huntley 1980).

of Fundy in Canada (see Box 7.1), the Gironde and Seine Estuaries in France, the Wash and Severn Estuary in England, Cambridge Gulf in Australia, Cook Inlet and Bristol Bay in Alaska, the Gulf of Cambay in India, the head of the Gulf of California, and the Rio de Plata in South America.

Tidal Bores

In some estuaries large tidal ranges lead to the formation of tidal bores (Figure 7.18). A tidal bore is a steep-crested wave or breaking wave that moves upstream with the rising tide. It is a product of a large funnel-shaped estuary having a tidal range exceeding 5 m

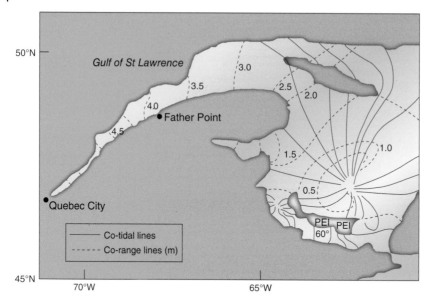

Figure 7.17 The Gulf of St. Lawrence narrows from a width of 150 km at the entrance to the St. Lawrence River to less than 15 km wide just downstream of Quebec City. Gradual constriction of the tidal wave in this funnel-shaped embayment increases spring tidal ranges from 1.0 m at the entrance to over 5.0 m at Grosse Île near Quebec City.

Box 7.1 Bay of Fundy: The Largest Tides in the World

The Bay of Fundy in eastern Canada is perhaps the most famous funnel-shaped embayment in the world. It has record tides, equal to the height of a five-story building. The bay connects to the northern end of the Gulf of Maine and separates the provinces of New Brunswick and Nova Scotia (Box Figure 7.1.1). Formation of the bay is related to the opening of the Atlantic Ocean, which occurred about 180 million years ago when North America began separating from Europe and northern Africa. The bay is part of a rift valley that developed within a broad sandy arid plain. Remnants of basaltic eruptions associated with early rifting can still be found at several locations along the bay's margin. During repeated episodes of Pleistocene glaciation, ice sheets scoured and deepened the basin and then deposited a thick carpet of glacial sediment. Following deglaciation, isostatic rebound (see Chapter 17, Glaciated Coasts) caused the sea to retreat from the Bay of Fundy exposing the basin to riverine reworking. Approximately 6000 years ago, rising eustatic sea level inundated the bay and allowed waves to erode the soft sedimentary rocks that surround most of the Bay of Fundy shoreline. Sandstone cliffs, intertidal marine platforms, and vast sand shoals are a product of this erosion.

The Bay of Fundy is 260 km long and 50 km wide at its opening gradually narrowing into two separate bays; Chignecto Bay to the northeast and the Minas Basin to the east. It is deepest at its mouth and progressively shoals toward its eastern end having an average depth of about 32 m along its length. The immense tidal range in the Minas Basin (Box Figure 7.1.2) leads to more than 14 km^3 of water flushing the bay twice daily (actually every 12 hours and 25 minutes). Tidal currents at the entrance to the Minas Basin exceed 4 m s^{-1} (~9 mph). These strong currents mold the sandy bottom into giant sand ripples called sandwaves that have heights of more than 10 m. Flow in the 5-km-wide tidal channel leading into the Minas Basin is equal to the combined discharge of all the streams and rivers on Earth. The weight of the huge volume of water entering the Minas Basin actually causes a slight tilting of western Nova Scotia.

Although the funnel shape of this coast contributes to very large tides, it is **tidal resonance**

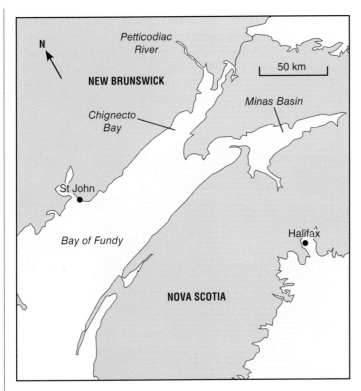

Box Figure 7.1.1 Location map of Bay of Fundy.

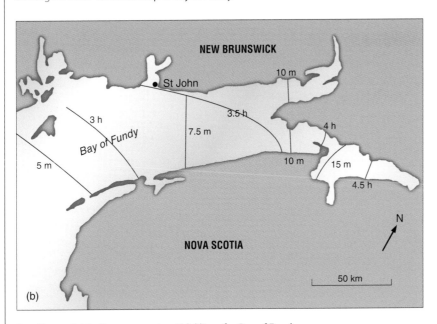

Box Figure 7.1.2 Co-range and co-tidal lines for Bay of Fundy.

that generates the world's largest tidal ranges. The length and depth of the Bay of Fundy promote the development of a standing wave (similar to the sloshing back and forth of water in a bathtub or a coffee cup) that is constructively perturbed by the tide-generating forces in the Atlantic Ocean. Resonance occurs in elongated embayments when the advancing tidal wave reflects off the head of the bay back toward the bay's entrance. A standing wave is produced when the geometry of the bay is of the correct dimensions such that the reflected wave arrives at the bay entrance at the same time as the next incoming tidal wave. Each incoming wave amplifies the standing wave until the energy that is added balances the energy lost due to friction. Tidal ranges at the mouth to the Bay of Fundy are a modest 3 m but resonance and funneling effects gradually increase the range to an amazing 16 m near Wolfville at the eastern end of the Minas Basin (Box Figure 7.1.3).

Box Figure 7.1.3 Picture illustrating large tidal ranges showing high and low tide. (*Source:* Courtesy of Vik Pahwa).

(a)

Tidal crest

River flow

Incoming tide ➡

(b)

(c)

Figure 7.18 Tidal bores occur in funnel-shaped estuaries having tidal ranges greater than 5 m. (a) They are produced when shoaling and constriction of the landward moving tidal wave oversteepens and may begin to break. (b) View of tidal bore in the Salmon River in the Minas Basin in the Bay of Fundy, Nova Scotia. (c) Close-up view of tidal bore approximately 30 cm in height.

and a channel that progressively shallows upstream. The height of most bores is less than 0.4 m; however there are some spectacular bores that adventurers surf on as the wave advances upriver. A tidal bore is formed when the propagating tidal wave oversteepens and breaks due to a constriction of the channel and retarding effects of the river's discharge. Tidal bores are best developed during spring tide conditions when tidal ranges are near maximum. Bores are found in the Severn and Trent in England, the Seine in France, the Truro and Petitcodiac Rivers that discharge into the Bay of Fundy, the Ganges in Bangladesh and several rivers along the coast of China. Some of the largest tidal bores in the world occur in the Qiantang River in northern China and in the Pororoca River, a branch of the Amazon. Their heights have been reported to approach 5 m and travel at speeds close to 20 km h^{-1}. Fisherman and shippers will often travel upstream by riding the ensuing strong currents that follow the passage of a tidal bore.

Tidal Currents

Tidal currents are most readily observed in coastal regions where the tidal wave becomes constricted. As the tidal wave approaches the entrance to harbors, tidal inlets, and rocky straits, the tide rises at a faster rate in the ocean than it does inside the harbor or bay. This produces a slope of the water surface and, just like a river system, the water flows downhill, producing a tidal current (Figure 7.19). The water moving through a tidal inlet and flooding a bay is called a flood-tidal current. The water emptying out of a bay and moving seaward is referred to as an ebb-tidal current. In a slight over-simplification, when the tidal waters in the ocean and bay are at the same elevation, there is slack water at the tidal inlet. This condition usually occurs at high tide and low tide. Likewise, the strongest current velocities are produced when the water surface through the inlet achieves the steepest slope, which commonly is near mid-tide but may also occur closer to high or low tide. During spring tide conditions when the maximum volume of water is exchanged between the ocean and bay, tidal currents can reach velocities of 3 m s^{-1}.

Along non-sandy shorelines, tidal currents can achieve strong current velocities, particularly in regions with large tidal ranges, large bay areas, and narrow constrictions. One such location is along the Norwegian coast north of Bodø where Vestfjord connects to the Norwegian Sea. The fierce tidal currents that flow through the straits reach speeds greater than 4.0 m s^{-1}. This creates strong whirlpools that make travel through the strait

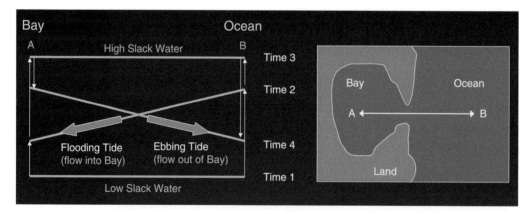

Figure 7.19 Tidal currents occur at the entrance to bays, harbors, and tidal inlets and are due to a constriction of the tidal wave. As the ocean tide rises and falls, this condition creates a water surface slope between the ocean and bay. In simplest terms tidal currents are a manifestation of water flowing downhill under the influence of gravity.

extremely dangerous during peak current flow. The Norwegians call these whirlpools the Maelstrom and fisherman time their passage to avoid these perilous eddies.

The strong tidal currents that are generated in funnel-shaped estuaries and elsewhere along the world's coastlines can be harnessed to provide a source of energy. For example, in the Gulf of Saint-Malo along the Brittany coast of France the tidal range in the Rance Estuary can exceed 12 m. This exceptionally large tidal fluctuation produces very strong tidal currents. A 750 m-wide barricade has been constructed across the river to house 24 hydroelectric power generators. The reversing tidal currents of the Rance have been providing electricity since 1966.

7.9   Summary

The Moon and Sun's force of attraction exerted on Earth's hydrosphere causes ocean tides. The Moon's tide-generating force is about twice that of the Sun because it is much closer to Earth. The Moon and Earth revolve around a common center of mass inside Earth, which produces a centrifugal force that balances the forces of attraction. In the equilibrium tide model, two tidal bulges are developed because masses on Earth's surface are acted on unequally by gravitational and centrifugal forces. One bulge faces the Moon and the other is directed away from the Moon. The tidal period is 12 hours and 25 minutes rather than 12 hours (half of Earth's rotation) because it takes Earth an additional 50 minutes each day to catch up with the Moon in its orbit. As the Moon revolves around its common center of mass with Earth its orbit makes excursions north and south of the equator. When the Moon is over the tropics, Earth passes through unequal successive tidal bulges, producing different elevations in successive high and low tides and unequal tidal ranges. This tidal condition is called a semidiurnal inequality.

The effects of the Sun can enhance or retard the Moon's tide-generating force. When the Moon, Earth, and Sun are aligned (position called syzygy) the Sun's effects are additive and we experience spring tides and large tidal ranges. When the Moon, Earth, and Sun are at right angles (quadratic position), the Sun's effects diminish the Moon's tide-generating forces and we have neap tides and relatively small tidal ranges. Mean tides and average tidal ranges occur between syzygy and quadratic positions. Due to the elliptical orbits of the Moon and Earth, the height and range of the tides increases when the Moon is proximate to Earth (perigean tides) and Earth is close to the Sun (perihelion tides).

The continents and island archipelagos partition Earth's hydrosphere into several interconnected large and small ocean basins. Based on their dimensions, Coriolis and tide-generating forces cause the tidal wave to rotate around one or more amphidromic points within these basins, counter-clockwise in the Northern Hemisphere and clockwise in the Southern Hemisphere. Tidal range increases with distance from the amphidromic point, but ranges in the open ocean are generally quite low (<0.6 m). When the tidal wave propagates across the continental margin the wave slows down and the crest steepens, resulting in an increase in the tidal range (1.0–2.0 m at the coast). The tidal signature along the coast reflects the geometry of the basin and shoaling behavior of the tidal wave. Most open-ocean coasts experience semidiurnal tides (two tides daily) or mixed tides, which is a tidal signature that exhibits periods of semidiurnal tides and distinctly diurnal (one tide daily) tides during other periods of the lunar month. Diurnal tides are most common in restricted basins where the tidal wave resonates with a period close to 24 hours and 50 minutes.

The tidal wave can undergo dramatic distortions as it moves into restricted ocean basins and through straits due to Coriolis effects and shoaling effects. Deformation of the tidal wave can result in dramatic differences in tidal range over distances less than 50 km. This is particularly apparent in funnel-shaped embayments where steepening of the advancing tidal wave can increase the tidal range by 2–4 m at the head of the bay. Even greater tidal ranges can result if a standing wave is produced in the bay that is constructively interfered by the incoming tidal wave. This phenomenon is best developed in the Bay of Fundy where tidal ranges (up to 16 m in the Minas Basin) are the largest in the world. In some estuaries with very large tidal ranges, the advancing tidal wave steepens forming a steep-crested wave or breaking wave called a tidal bore that moves upstream with the rising tide. Tidal currents are produced in coastal settings when the tidal wave becomes constricted such as at the entrance to a bay or tidal inlet.

References

Huntley, D.A. (1980). Tides on the north-west European continental shelf. In: *The North-West European Shelf Seas: The Sea Bed and the Sea in Motion. II Physical and Chemical Oceanography and Physical Resources* (ed. F.T. Banner, M.B. Collins and K.S. Massie). Amsterdam: Elsevier.

Nummedal, D., Oertel, G., Hubbard, D.K., and Hine, A., 1977. Tidal inlet variability – Cape Hatteras to Cape Canaveral, *Proc. of Coastal Sediments '77.* ASCE, Charleston, SC, pp. 543–562.

NOAA, n.d. Tides & Currents; Chapter 3, Detailed Explanation of the Differential Tide Producing Forces. https://tidesandcurrents. noaa.gov/restles3.html

Suggested Reading

Defant, A. (1958). *Ebb and Flow: The Tides of Earth, Air, and Water.* Ann Arbor, MI: University of Michigan Press.

Fischer, A. (1989). The model makers. *Oceanus* 32: 16–21.

Greenberg, D.A. (1987). Modeling tidal power. *Scientific American* 247: 128–131.

Lynch, D.K. (1982). Tidal bores. *Scientific American* 247: 146–157.

Open University (1989). *Waves, Tides, and Shallow-Water Processes.* Oxford, UK: Pergamon Press.

Redfield, A.C. (1980). *Introduction to Tides.* Woods Hole, MA: Marine Science International.

Sobey, J.C. (1982). What is sea level? *Sea Frontiers* 28: 136–142.

von Arx, W.S. (1962). *An Introduction to Physical Ocenaography.* Reading, MA: Addison-Wesley.

8

River Deltas

The Source of Most of our Coastal Sediments

The rock cycle shows us that most of the sediment that is eroded from land is carried by streams and rivers to a water body that receives their discharge, both the water and the sediment that it carries. This is usually one of the oceans or a *mediterranean* associated with an ocean, such as the Gulf of Mexico or the Mediterranean Sea. In some situations, large lakes may be the final destination of streams and rivers. The distance over which this sediment is carried by the river may be only a few tens of kilometers such as on much of the west coast of the United States, or it might extend thousands of kilometers such as in the Amazon and Mississippi river systems. These major differences in river length are related to the concept of global tectonics. The drainage system may be developed on a leading edge of a plate such as the west coast of both North and South America, or on a trailing edge such as the coastal plains of the United States and the stable crustal shield of Brazil.

Once the sediment is discharged at the mouth of the river along the coast, it might be carried out into deep water if it is fine-grained and suspended, as is the case with silt and clay, or it might be sand that is transported along the coast to be included in various coastal features such as nearshore bars, beaches or dunes, or it might come to rest at or near the mouth of the river in the form of a large and complex sediment accumulation called a river or fluvial *delta* (Figure 8.1). In this chapter we will discuss these river deltas, how they develop, their characteristics, and what causes them to change. Deltas are one of the coastal environments that is greatly influenced by human activity, and this will also be considered in this discussion.

Although we do not know how he recognized its shape, Herodotus is commonly given credit for coining the term delta in the fifth century BC in his description of the famous Nile Delta on the Mediterranean coast of Egypt. Given the date of his account it is a real puzzle as to how he determined that the sediment accumulation at the mouth of the Nile was in the shape of the Greek capital letter delta. River deltas have historically been the site of human settlement because of their proximity to the sea and their abundant food supply in the form of waterfowl, finfish and shellfish. In the past century ancient portions of deltas have become a major source of petroleum. The latter is a major economic aspect of many deltas and has led to extensive research on a wide range of environmental topics in addition to their geology.

Virtually all our scientific knowledge of deltas as a geologic sedimentary environment has been acquired during the past century. G.K. Gilbert (1890), a famous geologist with the U. S. Geological Survey, investigated much of northern Utah including what is known as ancient Lake Bonneville, a large lake of the Pleistocene Epoch from which the Great Salt Lake was evolved. He recognized thick deltaic accumulations from rivers that emptied into it.

Beaches and Coasts, Second Edition. Richard A. Davis, Jr. and Duncan M. FitzGerald.
© 2020 John Wiley & Sons Ltd. Published 2020 by John Wiley & Sons Ltd.

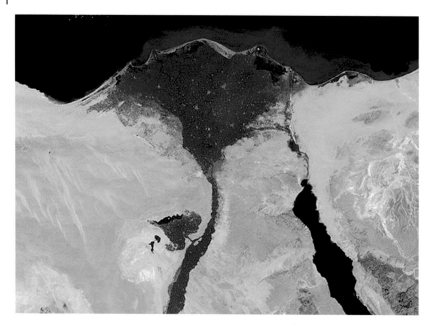

Figure 8.1 Satellite photo of the Nile Delta showing its generally triangular shape. (*Source:* Jeff Schmaltz, NASA Visible Earth).

Modern river deltas did not attract the attention of geologists until after the beginning of the twentieth century. Joseph Barrel of Yale University is generally given credit for writing the first research paper on the Mississippi Delta in 1914. Little was done on the research of river deltas until the extensive work of H.N. Fisk on the Mississippi Delta that began in the 1940s. His research, combined with the interest in deltas generated by the production of oil and gas, initiated an explosion of activity beginning in the 1950s. This began on the Mississippi and then expanded to the Niger Delta in Africa and the Orinoco Delta in Venezuela as well as others. All are important oil-producing deltas. Because of the extensive research and interest in the Mississippi Delta, it became the primary model for interpretations of deltas throughout the world. In fact, however, Delta is an extreme case, essentially one of a kind, and is a poor example with which to compare other river deltas. This is largely because the Mississippi is an extreme river-dominated delta.

8.1   How Deltas Develop

The presence of a river delta along any coast is an indication that the river is providing more sediment than can be removed and redistributed by coastal processes. This accumulation of sediments in riverine deltas may be quite temporary or it may be permanent. Some small deltas may be seasonal, appearing only in the spring when water and sediment discharge is at its highest. As the year proceeds, coastal processes remove that sediment and the delta is gone by the next spring when it is re-formed. There is considerable interaction of the riverine processes of sedimentation with open coastal marine processes; especially waves, longshore currents and tidal currents. The interaction of these processes, along with the sediment load of the river and the physical setting at and near the river mouth, determine the presence and the nature of the delta. The most important requirement for the formation of a delta is the discharge of sufficient sediment to produce a net accumulation above that amount removed and redistributed by waves and currents. The amount required is quite different from one coastal location to another. River mouths where the wave climate is characterized by large waves and/or where strong tidal currents persist, require considerably more sediment to produce a delta than those locations where waves and tidal flux are small. An equally important factor is the geologic and bathymetric setting on the continental margin adjacent to the coast. The sediments discharged at the coast must have a place to accumulate and form a delta.

Plate tectonic history, regional geologic setting, and sea level change are quite important in the development of large river deltas, as shown by their global distribution. Trailing-edge or passive margins foster the development of deltas but leading edges or active margins are difficult places for deltas to form. Extensive drainage basins typically form in areas where there is little relief with no mountain ranges or other high-relief landforms blocking the path of rivers to the coast. Two of the best examples are the Mississippi River drainage system in the United States and the Amazon River in South America. The Mississippi system drains most of the country

between the Rocky Mountains and the Appalachian Mountains. It includes the Missouri and Ohio River systems as well as other rivers that empty directly into the Mississippi. Most of the terrain in this system is the stable mid-continent area known geologically as a craton and the coastal plain that begins near St. Louis. The Amazon River system drains most of the northern part of South America from the continental divide in the Andes as the western boundary of the system where many small tributaries flow toward the east. Both of these river systems drain huge, relatively stable continental regions and deliver their sediment load onto a stable, trailing-edge continental margin with a broad, gently sloping shelf; ideal geologic settings for the development of river deltas (Figure 8.2).

Marginal sea coasts are also good places for deltas to form. The marginal sea that receives the greatest volume of river-borne sediment is the South China Sea. Here the Yangtze and Wang Ho rivers of China form large muddy deltas. In this geologic setting the combination of huge volumes of sediment delivered to a fetch-limited basin provides good conditions for delta development. Other examples of marginal sea coasts where deltas have developed

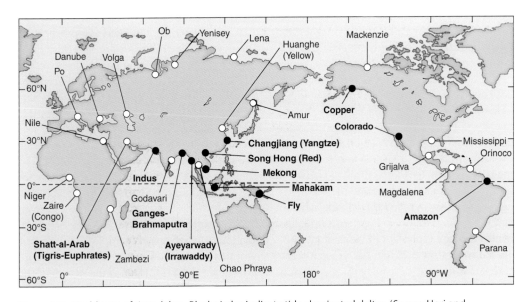

Figure 8.2 World map of river deltas. Black circles indicate tide-dominated deltas. (*Source:* Hori and Saito (2007)).

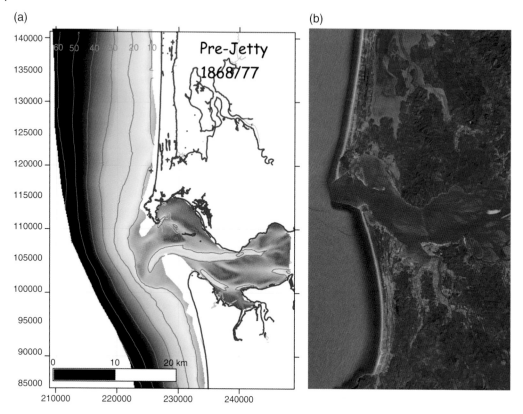

(a)

Pre-Jetty 1868/77

(b)

Figure 8.3 Image of the mouth of the Columbia River (a) before structures were added in the late nineteenth century and (b) currently with structures that were added in the 1950s. (*Sources:* (a) Kaminsky et al. (2010), (b) Image © 2018 Google Earth).

include the north coast of Alaska (McKenzie), and the north coast of the Mediterranean Sea (Rhone and Ebro), although these rivers and their sediment load pale in comparison with those of China.

On the other hand, leading-edge coastal settings do not permit the development of even modest-sized river deltas, e.g. the Columbia River mouth (Figure 8.3). The first reason for this is the absence of large drainage systems in this type of geologic setting. These leading edge, active margins tend to be next to high relief, mountainous areas with drainage divides that are typically only tens of kilometers from the coast. While the gradients are steep and therefore there is significant erosion through down-cutting of the flowing water, there is typically not much soil development

that would produce sediment for transport by the rivers. A second major problem with delta development on leading-edge coasts is the absence of a proper site for sediment accumulation. Typically the continental margin is narrow and steep; in some cases, it has multiple faults that create small basins. Further, this steep and narrow margin allows large oceanic waves to move very close to the coast without significant loss of energy because they do not feel bottom until almost at the shoreline. As a result, sediment can readily be removed from the mouth of a river thus inhibiting delta formation. A pretty good example of a river that empties into a leading-edge margin where no significant delta is formed is the mouth of the Columbia River on the west coast of the United States (Figure 8.3). In summary, with rare exceptions, large deltas can only develop on trailing-edge coasts because they provide abundant sediment, proper site for accumulation, and appropriate physical conditions for their maintenance. Global distribution of the major deltas shows this relationship with plate tectonics quite well (see Figure 8.2).

There are some exceptions to this generalization about delta formation and leading-edge coasts. On the west coast of North America we have two pretty good examples; the deltas of the Fraser River near Vancouver, Canada, and the delta of the Copper River on the south coast of Alaska. The Fraser River has its tributaries near the continental divide in the Canadian Rockies in the province of Alberta. It flows for a few hundred kilometers and empties into somewhat protected waters on the coast of British Columbia, near Vancouver. Tides here are in the macrotidal range but the combination of sediment discharge and protection from the high wave energy of the Pacific Ocean has allowed a modest size delta to develop in spite of the geologic setting.

The Copper River Delta (Figure 8.4) is located between Anchorage and Juneau on the south coast of Alaska. Special circumstances have permitted the development of a river delta along this coast even though it is one of very high wave energy and very severe winter storms. Most of the water and sediment-discharge from the Copper River is derived from melting glaciers. This condition has produced a huge sediment discharge that has permitted the development of a significant tide-dominated delta. In addition to the unusual presence of a river delta on this coast, there are also several barrier islands. Both of these features are typical of trailing edge tectonic settings but the huge volume of sediment has compensated for the high-energy, tide-dominated conditions and other leading-edge characteristics.

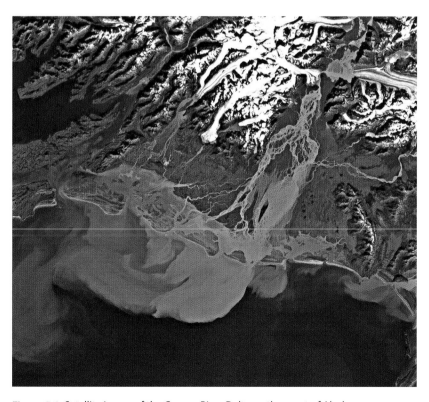

Figure 8.4 Satellite image of the Copper River Delta on the coast of Alaska.

Table 8.1 Some large modern deltas.

River	Land Mass	Receiving Basin	Size (km^2)	Annual Sediment Discharge (tons × 10^6)
Amazon	S. America	Atlantic	467,000	1200
Chao Phraya	Asia	Gulf of Siam	25,000	5
Danube	Europe	Black Sea	2700	67
Ebro	Europe	Mediterranean	600	?
Ganges–Brahmaputra	Asia	Bay of Bengal	106,000	1670
Huang	Asia	Yellow Sea	36,000	1080
Irrawaddy	Asia	Bay of Bengal	21,000	285
Mahakam	Borneo	Makassar Strait	5000	8
Mekong	Asia	S. China Sea	94,000	160
Mississippi	N. America	Gulf of Mexico	29,000	210 (469)
Niger	Africa	Gulf of Guinea	19,000	40
Nile	Africa	Mediterranean	12,500	0 (54)
Orinoco	S. America.	Atlantic	21,000	210
Po	Europe	Adriatic Sea	13,400	61
Rio Grande	N. America	Gulf of Mexico	8000	17
Sao Francis.	S. America	Atlantic	700	?
Senegal	Africa	Atlantic	4300	?
Yangtze	Asia	E. China Sea	66,700	478

8.2 Deltas and Sea Level

Sea level is an important factor in the development and maintenance of river deltas. Deltas that we see around the world today are geologically quite young. Virtually all are postglacial in age. Although these deltaic systems range from a few thousand to about ten thousand years in age, the currently active delta lobe is typically at the young end of this spectrum. Deltas cannot exist without sediment supply from rivers. The sediment-discharge of the river is partly a consequence of sea level position and the rate of change of sea level.

At the time of widespread glaciers during the Pleistocene Epoch, sea level was much lower than it is at the present time. The large rivers flowed across what is now the continental shelf and discharged their sediment load near the edge of the continental shelf. The consequence of this was widespread and large density currents of suspended sediment called turbidity currents along with other sediment gravity processes that transported most of the sediment-discharge of the rivers directly to the continental rise where it accumulated in thick wedge-shaped deposits.

Under these sea-level conditions deltas were not being formed and those deltas that existed from previous sea level highstands were being bypassed as rivers flowed across the continental shelf. The great ice sheets began to melt about 18,000 years ago causing a rapid rise of sea level across what is now the continental shelf. The river mouth was essentially retreating across the shelf so rapidly that there was not enough time for deltas to accumulate. As the rate of sea-level rise slowed about 6000–7000 years ago the shoreline migration slowed, permitting deltas to actively accumulate large quantities of sediment as it was no longer being vigorously dispersed by waves or tidal currents.

This is not to imply that all deltas are geologically very young, some have existed for millions of years. They have not, however, been continuously receiving sediment from their associated rivers because they were

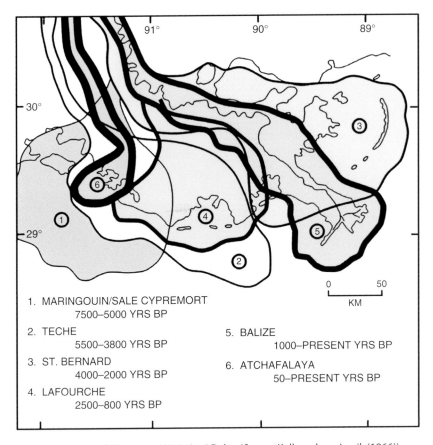

1. MARINGOUIN/SALE CYPREMORT
   7500–5000 YRS BP

2. TECHE
   5500–3800 YRS BP

3. ST. BERNARD
   4000–2000 YRS BP

4. LAFOURCHE
   2500–800 YRS BP

5. BALIZE
   1000–PRESENT YRS BP

6. ATCHAFALAYA
   50–PRESENT YRS BP

Figure 8.5 Lobes of the young Mississippi Delta. (*Source:* Kolb and van Lopik (1966)).

abandoned by the shoreline as it moved in association with sea-level change. The Mississippi Delta and the Niger Delta in Africa are good examples among many old deltas. Both deltas have been reactivated and are currently active. They are underlain by deposits that are at least ten million years old.

The young portion of the Mississippi Delta is 6–7000 years old, coincident with the slowing of sea level rise. This part of the Mississippi Delta consists of numerous recognizable lobes. Each of these lobes represents sediment accumulation at the mouth of a different geographic location of the river. These different lobes are abandoned when the location of river-mouth deposition shifts due to channel switching, *avulsion,* or other natural causes. Although many lobes of sediment accumulation have been recognized, they can be combined into only a few (Figure 8.5) based upon radiometric dating and location. The present lobe of the

Mississippi Delta began to form only about 600 years ago, not much before Columbus' first voyage to the New World. Most of the active portion has developed since the settlement of New Orleans by Europeans. The rate of sediment accumulation at the mouth of the Mississippi has been so great that nearly one-half of the State of Louisiana has been formed by the river since sea level rise slowed about 6000 years ago.

We can also see older deltaic deposits at the mouth of the Niger River (Figure 8.6). Like the Mississippi Delta, exploration for petroleum on the Niger has provided a wealth of information on the age and development of the delta. From these data it is possible to recognize sediment strata at least Miocene in age, which is up to 15 million years ago. Differences of climate in western Africa over this extent of time have provided great quantities of sediment as the result of more humid conditions and associated rainfall.

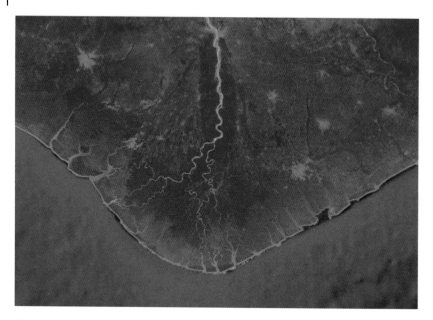

Figure 8.6 Satellite image of the Niger Delta on the coast of Africa. (*Source:* Image © 2018 Google Earth).

8.3   Delta Environments

Deltas are a transitional coastal environment located between terrestrial and marine conditions. Distinct landward or seaward boundaries do not exist on deltas; they grade continuously in both directions. This gradual transition is primarily due to the change from fresh water to seawater, and to the differences in sediment accumulation from the river through the open marine environment. The discharge of the river is carried from the main channel through a series of smaller channels that split off from the river channel into multiple *distributaries* that actually distribute the discharge of the river, both water and sediment, across the delta and into the marine basin. The result is a condition of overall progradation of sediment accumulation into the basin of deposition that could be a lake, estuary, or other standing body of water as well as the ocean itself.

As a consequence of this type of setting, the delta includes subaerial, intertidal, and subaqueous sedimentary environments as well as freshwater, brackish, and marine conditions. For purposes of discussing sedimentary environments and their processes, we can best subdivide the delta into three major parts, each of which has its own specific environments. From landward to seaward these are the *delta plain*, the *delta front* and the *prodelta*. This discussion will emphasize the first two; the prodelta is strictly subtidal and extends into the fairly deep water of the outer continental shelf.

The delta plain is primarily influenced by the river and its processes with tides and waves playing a minor role overall. There is, however, an increase in the influence of marine processes toward the seaward portion of the delta plain. The delta front is dominated by marine processes and tends to be subtidal with a small intertidal portion in some deltas.

8.4   Delta Plain

We can think of the delta plain (Figure 8.7) as the coastal extension of a river system. The delta plain is dominated by channels and their deposits, and the associated overbank environments that receive sediment during flooding. This scenario is parallel to that of a

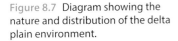

Figure 8.7 Diagram showing the nature and distribution of the delta plain environment.

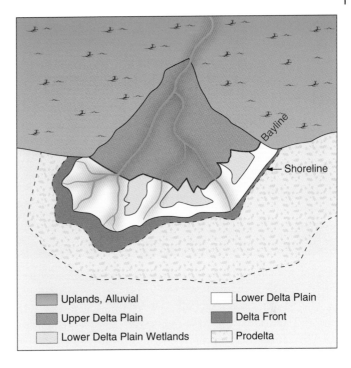

typical meandering river system. In fact, all of the specific elements of a meandering river complex are typically present on many deltas. There are some deltas, however, that have only a portion of this spectrum of environments.

The distributary channels on a delta plain contain point bars formed as the channel migrates. In doing so, they form broad meander loops that may be cut off, leading to formation of *oxbow lakes*. As the channels migrate across the delta plain they produce scars of their former location that leave subtle but recognizable geomorphic and vegetation patterns, another parallel with the fluvial system. Adjacent to the channels are three major types of overbank or flooding deposits; natural levees, crevasse splays, and floodplains, in order away from the channel.

Natural levees
(Figure 8.8) are produced during flooding when the river overtops its banks and immediately deposits much of its sediment load. The confinement of the channel coupled with a high discharge volume and rate, causes

the river to carry considerable sediment. Sudden loss of this confining characteristic as overflow occurs results in a sudden loss of speed and carrying capacity, causing much sediment to be deposited at the edge of the bank as natural levees. Breaches in these levees produce sediment accumulation in the form of *crevasse splay* deposits (Figure 8.9). These are fan-shaped deposits that can cover up to many square kilometers with a sediment thickness that is typically less than that of the adjacent natural levee. These splays may be reactivated multiple times during successive flooding conditions and thereby they can grow significantly in elevation and extent. This condition takes place each time a channel floods and the levees build vertically. Although the natural levees may be only a meter or so high, they are important features of the distributary channel system.

The most widespread but the thinnest of the overbank accumulations are the *floodplain* sediments. Even after losing sediment to natural levees and splay deposits, there is substantial fine sediment in suspension during flooding conditions. The spreading

Figure 8.8 Natural levees produced during flooding from rapidly deposited sediment load.

of the floodwaters beyond the channel causes important loss of water speed and thereby capacity, resulting in the deposition of fine and extensive floodplain deposits. Commonly, such floodplain sediments are draped over vegetation or other materials that occupy this environment. We have all seen many examples in the media of mud deposited by flooding, covering cars, carpets, and furniture in houses. The floodplain in the delta plain may take on a variety of characteristics. These include subtidal environments such as interdistributary bays, intertidal marshes, swamps and tidal flats or, in the most landward areas, even subaerial environments of various types.

The upward and lateral growth of the delta plain portion of the delta is dependent upon flooding periods for sediment distribution to the overbank environments. The typical situation is that the channel and its associated levee extend seaward at the outer limit of the delta. The levees may even be subaqueous at the most distal end of the channel. The initial subaerial portion of this distributary channel is the natural levee, followed by small splay deposits. Continued flooding will enlarge the splays until at least a portion is subaerial. Continued accumulation of these splays along with the slower but more extensive floodplain deposits will eventually lead to the interdistributary area

being filled, producing a continuous delta plain system.

8.4.1 Delta Front

The seaward edge of the delta plain merges with the generally continuous subtidal portion of the delta called the *delta front*. It is this part of the delta that is most affected by marine processes, especially the waves. Sediment empties out of the mouth of the distributary channels as both suspended load and as bed load. The finer suspended sediments tend to be carried away from the mouth of the channel by currents, whereas much of the coarser bed load tends to accumulate near the channel mouth. The vast majority of the coarse sediment is sand that comprises the delta front system.

The nature of the sand accumulations in the delta front depends upon the volume of sand transported to the distributary mouth and the relative roles of the interacting river currents with the waves and tidal currents. A common sand body is the distributary mouth bar (Figure 8.10) that accumulates just seaward of the channel mouth and typically causes the channel to bifurcate. The isolated distributary bar with little or no sand on either side is not generally common because of the influence of waves that spread the sand along the delta front on most deltas. As the

(a)

(b)

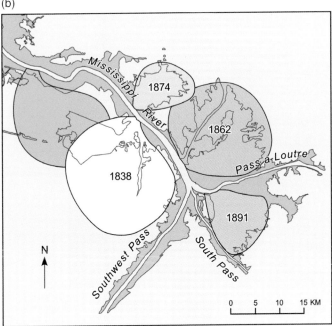

Figure 8.9   (a) Oblique aerial photo of a flooded delta plain showing numerous natural levees. (b) Map of crevasse splay deposits and dates of deposition. (*Source:* After Coleman and Gagliano (1964). Reproduced with permission of GCAGS).

waves approach the shallow part of the delta they refract and generate longshore currents, in the same fashion as they would along a beach. These currents carry the sand away from the mouth of the channel and distribute it along the outer delta plain, forming a nearly continuous delta front system. The degree to which this takes place is dependent upon the wave climate and the abundance of sand-sized sediment.

(a)

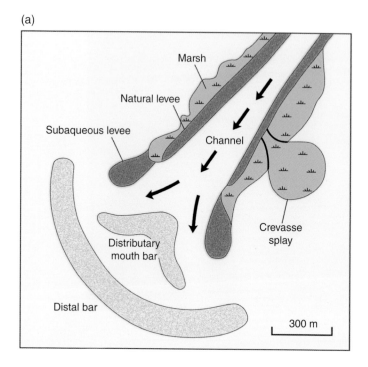

(b)

Figure 8.10 (a) Diagram of individual elements of the outer distributary showing a mouth bar and (b) an aerial oblique photo of a distributary mouth bar that is dominated by sand. (*Source:* After Coleman and Gagliano (1964). Reproduced with permission of GCAGS).

There is a wide range in the nature of the delta front sand bodies that comprise the outer part of the upper delta. In some deltas where there are several distinct distributaries, as on the Mississippi Delta, the delta front tends to be rather subtle, with distinct sand bars near the channels. By contrast, on some river deltas there is considerable redistribution of the sand from the channel mouths across the outer delta plain margin. In these situations there may be beaches and dunes on this part of the delta due to an abundance of sand and the appropriate wave climate to redistribute it. The Sao Francisco River in southern Brazil is a good example of this type (Figure 8.11).

Figure 8.11 Image of the Sao Francisco Delta on the coast of Brazil where the outer portion of the delta is dominated by sand. (*Source:* Courtesy of Landsat).

8.5 Delta Processes

The interaction of riverine processes with the wave- and tide-generated marine processes is quite complicated and results in a wide variety of deltaic forms and features. River-generated processes include both confined flow in open channels and unconfined flow during flood conditions. Wave-generated processes include the waves themselves in a variety of scales along with the currents developed by wave refraction. Intense storms also produce very high-energy wave conditions that impact deltas. Tides produce important currents that not only distribute sediment but also influence the discharge from the distributary channels of the delta.

8.6 River Processes

The fundamental role of the river in the delta system is that of providing the sediment. In doing so the river is at the mercy of climatic conditions and seasonal changes in discharge. The variables that influence the nature, amount and rate of sediment delivery include the geology, geomorphology and climate of the drainage basin. More recently, humans have played an important role in many river systems and have caused many problems on the delta; human influences include agriculture, navigational structures and dams (see Section 8.9).

The combination of rock type and climate are major controls on the sediment provided to the river. Rainfall and its distribution over time are the fundamental factors of river discharge and therefore on sediment provided to the delta. Prolonged periods of drought place serious constraints on delta formation or maintenance. The typical climatic influence is in the seasonal distribution of temperature and precipitation. There are at least two important aspects to these cycles, including the annual distribution of rainfall. In areas where *monsoon* conditions exist during the summer, such as Southeast Asia, there is tremendous discharge and typically devastating flooding during this two- to three-month period. The flooding in Bangladesh of the Ganges–Brahmaputra Delta area is probably the most consistently dramatic case of flooding in the world. Most rivers, however, experience flooding during the rainy season and in some places this is

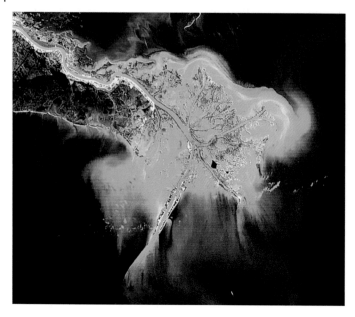

Figure 8.12 Infra-red satellite image of a huge amount of sediment being discharged from the present active lobe of the Mississippi Delta. (*Source:* Courtesy of Landsat).

especially problematic because the spring, wet season coincides with the spring melting of snow. The Mississippi River is such an example (Figure 8.12). Most of the midwestern part of the United States has high rainfall in the spring during the same time that the snow in the Rocky Mountains and northern latitudes of the basin is melting. Flooding can be very severe for the people living along the river, especially in agricultural regions. Similar phenomena may take place in the delta area. On the other hand, these floods are highly beneficial to the delta in that these conditions provide the highest rate of sediment delivery to the delta, a crucial source of mineral nutrients. Flooding also washes out soluble salts which have built up in delta sediments.

The annual distribution of sediment to a river delta varies greatly at each river because of the dependency on climatic conditions. Desert rivers tend to have little discharge of water and therefore transport little sediment. However, when there is rainfall, it is typically heavy and intense, a condition that delivers considerable sediment to the river; it

is essentially a flood condition. Rivers like the Ganges–Brahmaputra that experience monsoon conditions discharge many times the normal rate during this season. Even the Amazon and Mississippi show a marked difference in discharge during the wet season compared to the rest of the year.

8.7 Delta Classification

Both wave and tidal processes on the delta are essentially competing with the riverine processes to leave their imprint on the delta morphology. These competing processes and the resulting configuration of the delta provide a framework for classifying deltas. The three major processes that influence deltas provide convenient end-members for a comprehensive organization of deltas by shape. This classification was first presented in 1975 in published form by William Galloway of the University of Texas–Austin, and has become a standard.

The classification consists of a triangle-shaped diagram with riverine processes,

Figure 8.13 Diagram showing the classification of river deltas as originally proposed in 1975 by William Galloway. (*Source:* Galloway (1975)).

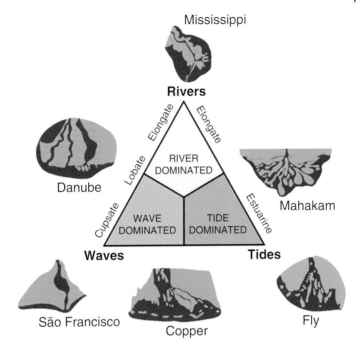

Figure 8.14 Satellite image of the tide-dominated Fly River Delta on the coast of Papua New Guinea. (*Source:* Image © 2018 Google Earth).

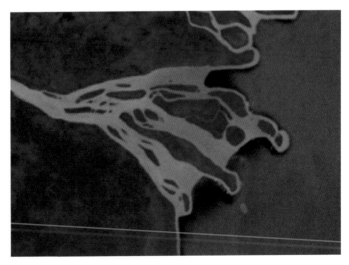

waves, and tides at the three apices (Figure 8.13). A delta that is clearly dominated by any one of the three processes is placed at the appropriate apex. The Mississippi Delta is quite distinctly dominated by river processes in the form of sediment input due to the large volume of sediment-discharge and little marine reworking. This gives it a so-called "bird's foot" configuration. By contrast, the Fly Delta in Papua New Guinea (Figure 8.14) and the Ord Delta in the Cambridge Gulf of northwestern Australia are fine examples of domination by tidal flux. At each of the sites sediment bodies of the delta are oriented essentially perpendicular to the trend of the coast. Lastly, the Sao Francisco Delta in Brazil (see Figure 8.11) and the Senegal Delta in Africa (Figure 8.15) show distinct domination by wave processes. Both contain smooth

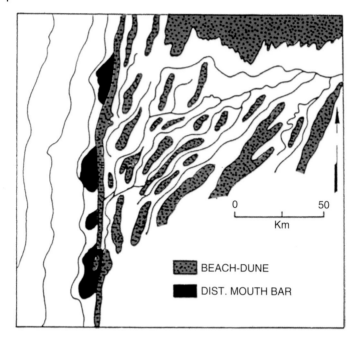

Figure 8.15 Diagram of the asymmetrical wave-dominated Senegal River Delta on the west coast of Africa. The dominant sediment transport direction is to the south. (*Source:* Wright (1985)).

outer margins caused by distribution of sediment along the coast as the result of wave-generated longshore currents.

Most deltas fall somewhere nearer the middle of the classification than these examples because all deltas experience some influence from all three types of processes. The morphology of each tends to reflect these influences. In general, river influence produces a finger-like morphology with a well-developed delta plain having several distributaries. *Tide-dominated deltas* display a strong shore-perpendicular trend and have extensive tidal flats with little mud. *Wave-dominated deltas* typically have well-developed beach and dune systems at their outer limits with few distributaries. Neither the absolute values of the processes nor the size of the delta are important in determining the position of a given delta in the overall classification scheme. It is the relative influence of the interactive processes that gives the delta its character.

8.7.1 River-Dominated Deltas

Conditions that foster *river-dominated deltas* include a high water and sediment-discharge with small waves and low tidal ranges in the receiving basin. A broad, gently sloping continental shelf provides the typical resting place for this large volume of sediments. These conditions are best fulfilled by trailing-edge and marginal sea, tectonically stable coasts that are sheltered from large waves and have small tidal ranges. The Gulf of Mexico is a perfect setting, and it hosts the Mississippi Delta. Other similar settings and their example deltas are the Black Sea with the Danube, the Adriatic Sea with the Po and the Yellow Sea with the Huang Ho. All have good sediment supplies and low tidal ranges, and all are sheltered from large waves. The Mississippi Delta does experience exceptional wave energy when hurricanes pass through the area. Some of the most devastating of these storms have eroded large areas of the delta.

Figure 8.16 Satellite image of the Ganges–Brahmaputra River Delta on the coast of the Bay of Bengal. (*Source:* Courtesy of Google Earth).

8.7.2 Tide-Dominated Deltas

Strong tidal currents in the absence of a substantial wave climate and strong river influence will produce a delta that is tide-dominated. Large tidal channels with intervening tidal sediment bodies dominate the delta and are generally more numerous than the river distributaries. Intertidal environments are widespread and are commonly partly covered with vegetation such as salt marsh or mangroves.

The Ganges–Brahmaputra (Figure 8.16) is the largest of the tide-dominated deltas and is an area of great interest because of the common and devastating floods it experiences. This is a large river system that supplies huge quantities of sediment to a coast where the tidal range exceeds three meters and waves are modest. One of the major factors in the development of this delta is the annual variation in discharge. During the monsoon season the amount of sediment delivered to the delta is orders of magnitude greater than during the rest of the year. Strong tidal currents redistribute the sediment in elongate bands that are separated by numerous large tidal channels.

Tidal currents usually have at least a shore-perpendicular component and move in and out of the delta complex as flooding and ebbing occurs. The stronger the current the more sediment is redistributed in this shore-perpendicular fashion. Notable deltas that have developed on coasts with high tidal ranges, include the Colorado (Figure 8.17) in the United States. (3.5 m), the Ganges–Brahmaputra in Pakistan (4.0 m) and the Ord River in Australia (7.0 m). These and similar deltas experience sediment being carried inland and deposited by flooding currents as well as large volumes of sediment being carried offshore even beyond the delta, such as on the Amazon Delta.

8.7.3 Wave-Dominated Deltas

Some deltas do not really look like deltas because of the strong influence of waves. They may mimic the appearance of barrier

(a)

(b)

Figure 8.17 (a) Oblique aerial photo of the Colorado River Delta that has no vegetation and is essentially without discharge at the present time. (*Source:* R. A. Davis). (b) Main channel of delta for Colorado River. (*Source:* R. A. Davis).

island systems with beaches, dunes and wetlands landward of them. These wave-dominated coasts are deltaic in nature because the sediment is supplied directly by the river and then reworked by the waves and wave-generated currents. The distributary channels do not protrude into the basin, thus providing a smooth outer shoreline. Typically, wave-dominated deltas are small. In fact, they grade into conditions of no delta when wave processes are strong enough to carry away all of the sediment supplied by the river.

There are different styles taken by wave-dominated deltas depending upon the nature of the longshore current patterns. The Sao Francisco Delta in Brazil and the Brazos River delta on the coast of Texas are fairly symmetrical about a single large distributary with a smooth overall cuspate shape. This is due to the absence of a strong littoral drift in either direction caused by the wind patterns and related direction of wave approach. By contrast, the Senegal River in western Africa (see Figure 8.15) displays a very strong change in direction of its course due to longshore currents and resulting littoral drift. The river course is shifted over 50 km by the longshore currents. The mouth of the river is marked by a distinct spit that mimics a coastal barrier, and extensive wetlands cover the delta plain.

8.8   Intermediate Deltas

Intermediate types of deltas display features of both river influence and marine processes. The Mahakam Delta on the coast of Borneo is small but has a shape that shows important influence of both river and tidal currents. It has numerous distributaries and a well-developed delta plain with distinct lobes that protrude into the receiving basin. Spring tides range up to 3 m and have currents of $1\,\text{m s}^{-1}$ that form distinct tidal channels between the distributary mouths.

The Nile Delta (see Figure 8.1) is a good example of a delta intermediate between river- and wave-dominated form. Tides in the Mediterranean Sea are nominal and waves are modest because of the fetch-limited basin. River input has historically been fairly high until the construction of the Aswan Dam in the 1960s that captured much of the river's sediment load. The delta plain is traversed by a modest number of well-defined distributaries each protruding into the sea. Between the distributary mouths the delta displays a relatively smooth outline with beaches and other wave-dominated features.

Probably the best example of an intermediate delta is the Niger Delta on the coast of Nigeria (see Figure 8.6). It falls in the middle of the classification showing equal influence of the river, waves and tides. It has a well-developed delta plain with a complex network of distributaries and a spring tidal range of up to 2.8 m, and it is exposed to the waves of the southern Atlantic Ocean. The result is a delta that incorporates some features of each of the major processes that influence deltaic coasts.

8.9   Human Influence

As people began to populate drainage basins of major rivers and the banks along the courses of these rivers, they profoundly influenced the delta in several ways. Most, but not all, of these influences have had detrimental effects. In most territories, the earliest important human activity was agriculture and forestry. Both tended to benefit the growth of the delta although they produced some important negative effects in the drainage basin. Cultivation and deforestation increase erosion of the soil and provide the river, and therefore the delta, with a high rate of sediment-discharge. This has resulted in the accelerated growth of many deltas, with the prime examples being the Mississippi Delta in the nineteenth century and, at the present time, the Amazon Delta. As the rapid diminution of the rainforest in Brazil takes place, vast quantities of sediment are provided to the delta. In the case of the Mississippi Delta the increase in sediment was the consequence of the expansion of agriculture throughout its drainage basin.

Box 8.1 Sea Level Rise in the Mississippi Delta Area of Louisiana

The combination of decreased discharge of sediment from the Mississippi River, eustatic sea level rise, compaction of sediment on the delta, and withdrawal of fluids associated with the petroleum industry has caused sea level to rise at catastrophic rates on the delta; about four times the global average. This condition is resulting in the loss of tens of square kilometers of land each year; Louisiana is literally drowning in the Gulf of Mexico.

Most river deltas are the sites of very rapid rates of sediment accumulation. This is typically fine sediment, much of it clay-sized. These clay particles have shapes like small sheets of paper or small sticks; about 2–4 µm in maximum dimension. When these particles come to rest they trap a tremendous amount of water in the sediment; it is very soft mud! As hundreds, or in some cases thousands, of meters of this sediment pile up along the continental margin there is significant compaction because of the mass of the sediment involved. This compaction causes water to be driven from the sediment and the particles to be better organized or layered. The result is that there is a significant decrease in the volume of the sediment which causes the surface of the delta to sink. When the surface sinks that means that the water is getting deeper; relative sea level is rising.

This sinking did not occur in the early days of human habitation of the Mississippi Delta area, in New Orleans and other nearby communities, because the river was continually providing new sediment to fill the areas where compaction was taking place. This became even more true during the mid-nineteenth century when agriculture first became extensive throughout the midwest and Great Plains, the primary drainage area of the Mississippi. Removal of grasses and trees for agriculture caused increased rates of erosion and sent increased volumes of sediment down the river.

As commerce developed in this area, boat traffic on the river increased greatly. Flood control and navigation gave rise to numerous dams along the entire Mississippi system, especially during the first half of the twentieth century. These dams are helpful for shipping but they stop sediment from moving down the river. Hence, the amount of sediment being delivered to the mouth of the Mississippi has been cut by about half over the past century. The result has been the "starvation" of the delta.

In the middle of the twentieth century the petroleum industry discovered the great oil and gas resources of the Mississippi Delta and exploration and exploitation took off. Now there are thousands of wells on and adjacent to the delta from coastal lands out to many hundreds of meters of water. Billions of barrels of oil and trillions of cubic feet of gas have been removed from under the delta surface. Removal of such volumes caused tremendous compaction of the delta sediments which has added to the rate of relative sea-level rise.

When combined, the compaction of sediments (about $4\,mm\,year^{-1}$), the withdrawal of oil and gas (about $4\,mm\,year^{-1}$), and the eustatic sea level rise (about $3\,mm\,year^{-1}$) causes a centimeter of sea level rise each year. This occurs on a coast that is dominated by wetlands that are intertidal to less than a meter above mean sea level.

Small and low barrier islands along this coast have been destroyed or greatly reduced during historical times. Predictions are that some will be completely gone by the end of the twenty-first century. Marshes, which are among the most productive environments, are being drowned because of lack of sediment to nourish them. In order to sustain the marsh environment, it is necessary for the rate of sediment being delivered to the marsh to be the same as the rate of sea level rise. That is not even close to happening on the Mississippi Delta at the present time.

One positive note that gives some hope is that the Corps of Engineers is planning to return significant portions of the river to its natural state, especially in the delta area. This will permit floods to take place and thereby provide the wetlands with the sediment needed to maintain them during this time of rapid sea level rise.

The more widespread human influence is the reverse situation, i.e. a reduction in the sediment supply thereby causing the delta to shrink in size. There are three important ways by which this occurs: diverting water from the river; navigation controls on the river; and damming the river. All reduce the discharge of the river and the latter two physically trap sediment and keep it from moving down the river.

There are several major cities that take large percentages of the discharge of rivers to use in the municipal water supply. The southern California area is dependent on water from various rivers in the southwest for its water, both for irrigation and for domestic use, notably the Colorado River that flows through the Grand Canyon. Both activities greatly decrease water and thereby diminish the sediment provided to

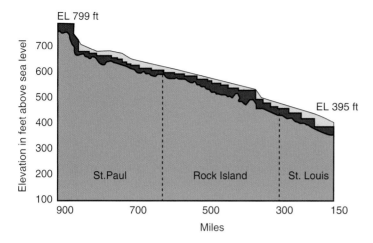

Figure 8.18 Diagram showing all of the dams on the Mississippi River between Minneapolis and St. Louis. (*Source:* Courtesy of U.S. Army, Corps of Engineers).

maintain the river delta. They also decrease the frequency of flooding across the delta, causing delta sediments to become increasingly salty, which may affect the type of vegetation that grows there. The present Colorado River Delta (Figure 8.17) is not experiencing significant discharge of either water or sediment.

Locks for navigation on major rivers invariably have dams associated with them. On the navigable portion of the Mississippi River that begins near Minneapolis, Minnesota there are many such structures (Figure 8.18). The small amount of water impounded is typically not a big problem but the sediment that is trapped behind the dam is literally stolen from the system and eventually, from the Mississippi Delta. More important, related impoundments are the huge dams built for reservoirs and/or hydroelectric power. They are extremely good sediment traps and some also serve as sources for water diversion. The bottom line is that the amount of water and sediment that the river has available and can transport is not being delivered to the delta.

Two good examples of this problem are the Colorado River that empties into the Gulf of California (see Figure 8.17) and the Nile River in Egypt. The headwaters of the Colorado are in the Rocky Mountains in the state for which it is named. Along the course of over 1000 km there are numerous dams and reservoirs as well as places of diversion. The result is that virtually no water nor sediment is being provided to the Colorado River Delta and it is rapidly being eroded by strong tidal currents.

The case of the Nile River has become similar since the Aswan Dam has been in place, constructed to make the desert fertile. The dam has been quite successful in trapping virtually all of the sediment being carried by the Nile that was destined for the delta. As a consequence, the outer margin of the delta is being eroded rapidly by waves in the eastern Mediterranean.

We cannot continue to rob our rivers of their water and sediment load without experiencing the consequences for the deltas that they feed (Figure 8.19). At present the most viable alternative appears to be stopping development of any kind on deltas. Waves, tides and their resulting currents interact with the riverine processes to prevent, mold or destroy the deltas depending upon the specific local circumstances. At those river mouths where waves and tides carry all the sediment away there is no delta. At many places the delta is allowed to accumulate and prograde but at some, such as the previously mentioned examples, the processes are now resulting in overall erosion. The relative role

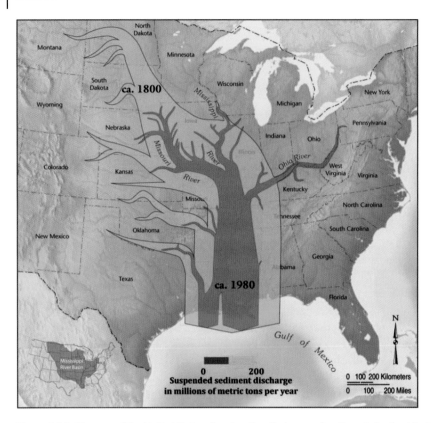

Figure 8.19 Diagram of the United States showing the discharge of sediments by the Mississippi River decreasing considerably from 1800 to 1980. (*Source:* Courtesy of USGS).

of the waves and the tides is also an important factor in the formation, maintenance, and the overall morphology of deltas.

The primary direct marine processes are the waves, wave-generated currents and tidal currents; the rise and fall of the tide has little direct effect on redistributing deltaic sediment. Waves impart energy along the delta and cause sediment to go into temporary suspension, whereas longshore and/or tidal currents transport it. During storms this distribution or removal of sediment reaches its maximum. The overall influence of the waves ranges from mostly longshore transport along the delta front providing sediment to various parts of the delta, to actual offshore or alongshore removal of sediment from the delta proper. Waves and wave-generated processes work toward a smoothing of the outer delta shape.

Another very important anthropogenic influence on deltas is the petroleum industry. This is true for many deltas, including the Niger and Orinoco, but the Mississippi is tremendously affected. There are literally hundreds of wells that have been drilled on it, and they have had two major impacts. Firstly the delta plain is covered with channels that have been dredged to enable access the location where drilling takes place. These channels destroy wetlands and the dredge spoil is dumped along the channel, producing a levee that prevents flooding and sediment delivery to the wetlands (Figure 8.20). Construction of infrastructure to support the petroleum industry also inflicts a negative impact on the delta plain (Figure 8.21).

The other aspect of this industry is the removal of huge volumes of fluids from

Figure 8.20 Aerial photo of numerous canals with dredge spoil levees produced for the petroleum industry.

Figure 8.21 Photo of large infrastructure complex built on the delta plain of the Mississippi.

beneath the delta surface. The removal of all of this fluid causes the surface to subside and, along with eustatic sea-level rise, contribute to the high rate of relative sea-level rise. The combination of the subsidence produced by the production of oil and gas coupled with the numerous natural and human-constructed levees is a major problem for the environment. The levees prevent sediment from reaching the wetlands and subsidence causes the wetlands on the delta plain to drown (Figure 8.22).

Figure 8.22 Photo of a portion of the Mississippi Delta plain showing dredge canals and drowning wetlands due to subsidence produced by fluid withdrawal.

Most deltas are in great jeopardy due to a combination of sea-level rise and human impact.

8.10 Summary

In some ways river deltas may be considered as the most important of all coastal environments because they are the site of sediment introduction for most of the other parts of the coast. On the other hand, people rarely spend any time visiting a delta on vacation or while going to and from places of work or play. Deltas tend to be remote, without traffic arteries, and are generally inhospitable due to the plethora of insects they support. Because of their critical role in the overall scheme of the coastal zone it is important that we have an understanding of river deltas and their characteristics.

Deltas are among the most productive and valuable environments in the world. They contain fruitful ecological niches where a wide variety of both plants and animals thrive. Their marshes are among the most extensive and productive environments anywhere. They are very important nursery grounds for juvenile fish and marine invertebrates. Their marshes also are important filters that trap contaminants and pollutants during flooding of distributaries.

The size and shape of the delta is a consequence of the interplay between the river and the sediment it provides with the wave and tidal processes of the marine coast. Deltas have developed quite rapidly in the context of geologic time and they can be destroyed just as rapidly. As we influence our environment more and more, we need to do a better job of considering the long-term consequences of our actions. The role of human intervention is critical to the maintenance of this coastal environment, as evidenced by what has happened to the deltas of the Nile and Colorado as well as others.

References

Coleman, J.M. and Gagliano, S.M. (1964). Cyclic sedimentation in the Mississippi deltaic plain. *Trans. Gulf Coast Assoc. Geol. Soc.* 14: 67–80.

Galloway, W.E. (1975). Process framework for describing the morphological and stratigraphic evolution of deltaic depositional systems. In: *Deltas*, 2e (ed. M.L. Broussard), 87–98. Houston, TX,: Houston Geological Society.

Gilbert, G.K. (1890). *Lake Bonneville: U.S. Geological Survey Monograph I*, 438. Washington, D.C.

Hori, K. and Saito, Y. (2007). Classification, architecture and evolution of large-river deltas. In: *Large Rivers: Geomorphology and Management* (ed. A. Gupta), 75–96. New York: John Wiley and Sons.

Kaminsky, G.M., Ruggiero, P., Buijsman, M.C. et al. (2010). Historical evolution of the Columbia River littoral cell. *Mar. Geol.* 273: 96–126.

Kolb, C.R. and van Lopik, J.R. (1966). Depositional environments of the Mississippi Delta plain, southeastern Louisiana. In: *Deltas in Their Geologic Framework* (ed. M.L. Shirley), 17–61. Houston, TX: Houston Geological Society.

Wright, L.D. (1985). River Deltas. In: *Coastal Sedimentary Environments* (ed. R.A. Davis), 1–76. New York: Springer-Verlag.

Suggested Reading

Broussard, M.L. (ed.) (1975). *Deltas: Models for Exploration*. Houston, TX: Houston Geological Society.

Giosan, L. and Bhattacharya, J.P. (eds.) (2005). *River Deltas – Concepts, Models and Examples*, SEPM Special Publication 83. Tulsa, OK: SEPM.

Oti, M.N. and Postma, G. (eds.) (1995). *Geology of Deltas*. Rotterdam: A.A. Balkema.

Schmidt, P.E. (ed.) (2011). *River Deltas: Types, Structures and Ecology,*. Hauppauge, NY: Nova Science Publishers.

Wright, L.D. (1985). River Deltas. In: *Coastal Sedimentary Environments* (ed. R.A. Davis), 1–76. New York: Springer-Verlag.

9

Estuaries

Most coasts have embayments of various sizes, shapes, and origins. The differences in the morphologies of these embayments is the result of a wide range of origins. The different types of embayment depend primarily on the nature of their interactions with various sources of water and their circulation. The two primary types are estuaries and lagoons. An estuary experiences freshwater inflow from the mainland and tidal influx from the marine environment; a lagoon has no significant freshwater inflow and no tidal circulation (lagoons will be discussed in the next chapter). There are some bays that have tidal influx but no freshwater. They are called tidal estuaries. All of these bays come in a wide range of sizes, shapes and origins. There are two basic origins for coastal bays; rising sea level and tectonic activity. Bays of tectonic origin tend to be only in crustal plate collision zones, whereas those of a sea-level change origin are associated with climate change. Some may have a combination of both, as along the Alaskan coast.

The bays that have been formed as a direct consequence of tectonic activity are generally located on leading-edge margins like the west coast of both North and South America. Here, faults and movement along these faults may produce bays that are typically long and narrow, such as Tomales Bay in California (Figure 9.1), where the San Andreas Fault system provides the geologic setting for a coastal bay. The most common type of coastal bay is typical of trailing-edge and mediterranean coasts with broad coastal plains and well-developed river systems. These are generally drowned fluvial systems (Figure 9.2). Other varieties include: fjords, which are elongate embayments excavated by glaciers (Figure 9.3); bays formed by barriers such as coral reefs and barrier islands (Figure 9.4); and embayments constructed by human activity, more commonly called harbors (Figure 9.5). This brief list provides some idea about the origins of coastal bays and the shapes that are related to those origins.

The definition for estuaries given above makes one wonder about the difference between these and river deltas, which also have important freshwater input and which experience tidal influence. The simple difference is that an estuary is a coastal embayment but a delta protrudes into the ocean or other adjacent water body. There are several factors common to these two coastal environments: both typically have tidal flats and wetlands, e.g. marshes or mangrove swamps; both are influenced by rivers, waves and tides; both are important sites of sediment accumulation; and both are geologically young features.

Estuaries are sediment sinks, that is, places where sediment tends to accumulate and stay for long periods of time. It is this characteristic that limits the geologic lifetime of an estuary. The runoff from rivers as well as tidal flux transports sediment from both landward and seaward directions into an estuary. The embayment provides a local basin for sediments to come to rest. Estuaries tend, therefore, to be filled in from the margins toward the middle in a manner that is commonly referred to as

Beaches and Coasts, Second Edition. Richard A. Davis, Jr. and Duncan M. FitzGerald.
© 2020 John Wiley & Sons Ltd. Published 2020 by John Wiley & Sons Ltd.

Figure 9.1 Tomales Bay, California, an example of a fault-generated estuary in the San Andreas complex north of the San Francisco area. (*Source:* Courtesy of NASA).

Figure 9.2 Baffin Bay on the south Texas coast is a bay that still reflects its origin from a fluvial drainage system. (*Source:* Image © 2018 Google Earth).

Figure 9.3 A fjord on the Alaskan coast. (*Source:* Courtesy of G. Ashley).

(a)

(b)

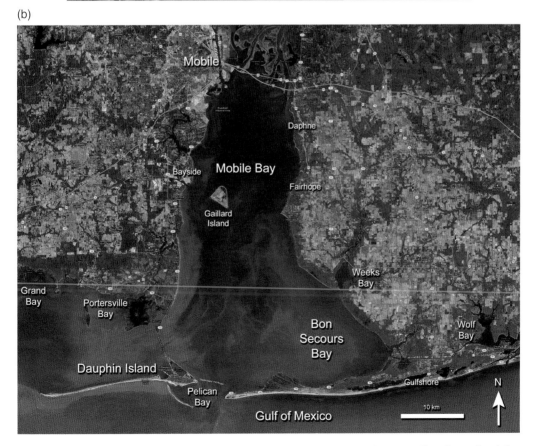

Figure 9.4 (a) A bay on the Texas coast behind the city of South Padre Island. (b) is a satellite photo of Mobile Bay, Alabama showing a bay formed by a barrier island.

Figure 9.5 Photograph of Aberdeen Harbour, Hong Kong. (*Source:* Tokyo Metro at Chinese Wikipedia, https://commons.wikimedia.org/wiki/File:Aberdeen_Harbour_view.jpg. Licensed under CC BY SA 3.0).

progradation. If sea level rises during this infilling process then the space available for sediment continues to increase. On the other hand, if sea level falls, the estuary is drained and the river(s) flow across its prior location leaving the "basin" essentially "high and dry."

9.1 Estuarine Hydrology

In addition to the size, shape and origin of these bays, their character also includes their hydrology. This is comprised of the characteristics of the water coming from both the land runoff and the marine environment, coupled with its circulation within the estuary. The hydrology controls the water chemistry, the biota and the sediment that forms the substrate. It is the hydrologic characteristics that provide the best criteria for classifying coastal bays into broad categories.

The fact that both rivers and tides flow into the estuary means that fresh water and seawater are being mixed. This interaction of different water types gives the estuary one of its most important characteristics along the coast; brackish salinity. Runoff from a river is generally continuous but the discharge may vary greatly depending upon the season, the overall climate, and other factors, just like the situation for deltas. In contrast, the tidal influence from the open marine environment to the estuary is typically regular and predictable. The range in tidal fluctuation changes with spring and neap conditions but the periodicity remains fairly constant.

The salinity of seawater is about 35 parts per thousand, or 3.5 %, whereas fresh water is essentially zero. This contrast in salinity produces a significant difference in the density of the two water types. Remember how much easier it is to float in salt water as compared to a freshwater pool? In absolute numbers the densities are 1.000 g cc^{-1} for fresh water and 1.026 g cc^{-1} for normal marine salt water; a relatively small difference but a very important one. In the absence of waves or strong currents, these different water types are layered, with the lighter fresh water "floating" on the heavier salt water. This phenomenon demonstrates the potential complications that the merging of these water types present to the estuary in terms of circulation into and out of the embayment.

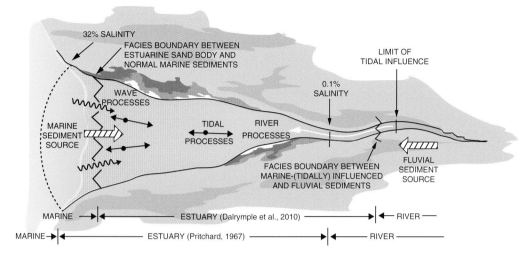

Figure 9.6 Diagram showing the boundaries of an estuary according to Pritchard (1967) and Dalrymple (2010).

The boundaries of estuaries depend on salinity and sediment facies. There have been two different definitions proposed and utilized by different estuarine scientists. Pritchard defined the boundaries of an estuary based on salinity, ranging from 32‰ on the seaward limit to 0.1‰ on the landward side (Figure 9.6). By contrast Dalrymple used sediment facies as the limits of an estuary. The seaward limit was the presence of marine sediments and the landward limit is the presence of fluvial facies. The Pritchard definition is the easier of the two to resolve simply by taking salinity measurements but the Dalrymple definition is geologically significant and can be resolved in the ancient stratigraphic record.

9.1.1 Classification of Estuaries

A common classification of estuaries is based on the way that fresh water and salt water interact. In the 1950s, Donald Pritchard, a scientist at the Chesapeake Bay Institute, recognized three types of circulation conditions in estuaries: stratified; partially mixed; and mixed (Figure 9.7). In a stratified estuary there is essentially complete separation between the freshwater and saltwater masses due to lack of mixing caused by waves or strong currents. Estuaries dominated by

rivers may display stratified water masses, for instance the Hudson River estuary in New York, where the saltwater wedge extends tens of kilometers up the river. In some estuaries some of the salt water mixes with the fresh water to produce a transition zone of intermediate salinity between the fresh water and salt water. Places where tidal currents influence part, but not all, of the estuary, such as Chesapeake Bay, display this characteristic. Totally mixed estuaries produce a vertically homogenized water column with a gradient of increasing salinity toward the ocean. This could be the result of waves in a shallow estuary such as Pamlico Sound, North Carolina or Mobile Bay, Alabama, or it could be due to strong tidal currents such as in the Bay of Fundy, Canada or Delaware Bay on the Atlantic coast of the United States. Large and complicated estuaries such as Chesapeake Bay or San Francisco Bay can experience different conditions in different locations and at different times in the lunar tidal cycle.

Many estuaries move from one hydrologic type to another depending upon seasonal variations in runoff, changes in wave climate, topographic variations of the estuary floor, or other phenomena that lead to variations in the amount of mixing. For example, a large but shallow estuary such as Mobile Bay or

TYPES OF ESTUARIES

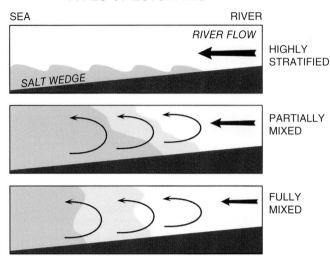

Figure 9.7 Diagram showing the three basic types of estuarine circulation between seawater and fresh water.

Pamlico Sound is susceptible to waves mixing the water column thereby destroying any layering of water mass types. Waves tend to be absent or small during the summer, conditions that foster stratification. Near the other end of the spectrum, tidal currents in the Bay of Fundy are always strong enough to mix the water in this estuary completely.

9.1.2 Estuarine Processes

Estuaries tend to be influenced primarily by river or tidal processes, with wave influence being dependent upon the size and depth of the estuarine basin. Fresh water and sediment are provided to the estuary by river discharge. The amount of both of these and the rate at which they are delivered are important to the character and longevity of the estuary. Some estuaries are supplied by a single river and therefore the sediment supply is essentially at one point. This situation tends to form a bayhead delta, where much of the sediment delivered by the river accumulates (Figure 9.8). Some of the large estuaries on the Texas coast have this characteristic; San Antonio Bay is fed by the Guadalupe River, Corpus Christi Bay by the Nueces River, and Galveston Bay by the Trinity River. Another good example is Mobile Bay in Alabama, where the Tensaw

River forms such a delta (Figure 9.9). These estuaries tend to be river-dominated because of the strong influence of the stream processes and the absence of strong tidal currents and/or big waves. The tidal influence in these Gulf Coast estuaries is diminished by the presence of the barrier islands across the mouth of the bays. It is also limited by the small tidal range around the Gulf; less than 1 m spring tide for all of the barrier islands.

Some estuaries have multiple rivers emptying into them with little or no development of a bayhead delta. Probably the best example is Chesapeake Bay (Figure 9.10) which receives input from numerous large rivers but which has no significant bay-head deltas. Here the digitate nature of the many river valleys leading to the estuary traps most of the relatively coarse sediment before it reaches the open portion of the estuary. Many of the small west-coast estuaries have similar conditions, though they generally have only a single stream feeding them. Although this type of estuary can develop any of the three hydrologic styles mentioned above, nearly all are stratified or partially mixed. In addition to the presence of the bayhead deltas, terrigenous sediment accumulation in these river-dominated estuaries tends to be dominated by mud. (See Box 9.1.)

Figure 9.8 Aerial photograph of the Guadalupe River Delta as it is positioned at the head of San Antonio Bay on the central Texas coast. (*Source:* Courtesy of Google Earth).

At the other end of the spectrum are tide-dominated estuaries. This type is typically funnel-shaped and has no barrier or other constriction at its mouth (Figure 9.11). Such a configuration not only eliminates the dampening effect that barriers have on tidal flux but commonly amplifies the progressing tidal wave during flooding, producing high tidal ranges. Both conditions result in maximizing the influence of tidal flux, and they create fully mixed hydrologic conditions in the estuary. The combination of high tidal range with strong tidal currents generally results in a sand-dominated estuary floor because the mud tends to be carried out to sea in suspension or is trapped at the low-energy, landward limits of the estuary. Good examples of tide-dominated estuaries are the Bay of Fundy and also the Bay of St.-Malo on the north coast of France (Figure 9.12).

The strong tidal currents in these tide-dominated estuaries move much sediment into the estuary, and they also move sediment that is already in the estuary back and forth during each flood and ebb of the tide.

Sand can be moved by currents of as low as 20–30 cm s^{-1} depending on the size of the sand particles. These conditions are typically achieved or exceeded for several hours during each flood and ebb tidal cycle. As a result, the tidal currents produce numerous bedforms on the floor of the estuary; basically the same ones that we see exposed on tidal flats (Figure 9.13). These bedforms are developed because of the shear between the bed (sediment) and the water column, thereby causing turbulence and producing these regular irregularities on the substrate. These bedforms range in size from ripples to sand waves. The size and shape of these features is controlled by grain size and by the speed of the tidal currents.

9.1.3 Time–Velocity Relationships

The graphic record of the rise and fall of the tides shows the change in water level over each tidal cycle. This curve is essentially symmetrical. If we plot the velocity of tidal currents that are produced by this rise and

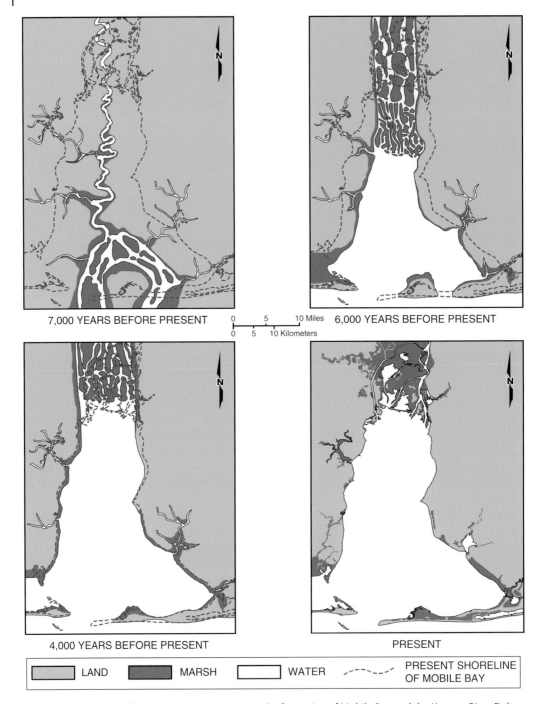

7,000 YEARS BEFORE PRESENT

0   5   10 Miles
0   5   10 Kilometers

6,000 YEARS BEFORE PRESENT

4,000 YEARS BEFORE PRESENT

PRESENT

LAND   MARSH   WATER   PRESENT SHORELINE OF MOBILE BAY

Figure 9.9 Sequence of diagrams showing stages in the formation of Mobile Bay and the Kensaw River Delta as sea level rose during the Holocene period. (*Source:* Courtesy of Geological Survey of Alabama).

Figure 9.10 Map of Chesapeake Bay showing the numerous rivers that empty into this estuary. (*Source:* Landsat/NASA, https://commons.wikimedia.org/wiki/File:Chesapeakelandsat.jpeg).

Box 9.1 Chesapeake Bay

Chesapeake Bay is one of the largest estuaries in the world. It is supplied by a fluvial system that covers part of six states in the U.S. and includes more than 100 streams, 11 of which are major rivers (Box Figure 9.1.1). This bay is 320 km long and 48 km wide at its widest, with 18,800 km of shoreline, and has an area of 11,600 km^2. The average depth is 6 m and the average discharge is 2200 m^3 s^{-1}.

This estuary is surrounded by a population of several million. As a result, the anthropogenic impact has been substantial, especially in terms of water quality. The basin in which the bay resides is the result of a bolide impact about 35.5 million years before present. As sea level rose in the post-glacial sea-level rise it flooded the Susquehanna River valley to form the present fluvial system.

Like most estuaries this one has been extremely productive, although it has had some periods of low levels. It is best known for oyster production (Box Figure 9.1.2) although

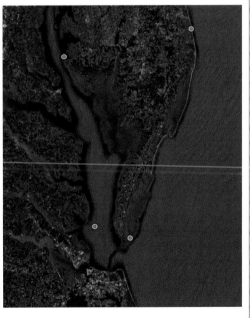

Box Figure 9.1.1 Map of the drainage basin that feeds into Chesapeake Bay. (*Source:* Pennfuture).

Box Figure 9.1.2 Close-up photo of a large oyster reef at low tide. (*Source:* Nature Conservancy).

Box Figure 9.1.3 Aerial photo of a large concentration of suspended sediment. (*Source:* Jane Thomas, courtesy of the Integration and Application Network, University of Maryland Center for Environmental Science.)

this has also seen a range of productivity. Blue crabs and clams have also been significant. Sport and commercial fishing are widespread on Chesapeake Bay. Sediments in the bay are a combination of sand and mud (Box Figure 9.1.3). Like most estuaries this bay is surrounded by a combination of wetlands (Box Figure 9.1.4) and fine beaches (Box Figure 9.1.5).

Box Figure 9.1.4 Aerial photo of marsh with numerous tidal channels. (*Source:* EPA. https://www.epa.gov/sites/production/files/styles/large/public/2015-03/what_is_the_bay_tmdl.jpg).

Box Figure 9.1.5 Photo of a beach with several groins at low tide. (*Source:* Chesapeake Bay Program, https://www.chesapeakebay.net/news/blog/eight_reasons_the_chesapeake_bay_is_an_exceptional_estuary).

fall of the tides we find that the curve has a very different shape. There is considerable asymmetry to the velocity data and the duration of flood and ebb may be different. This graph is called a time–velocity curve (Figure 9.14) because it is a record of the velocity of the tidal current over time. The lack of symmetry is called time–velocity asymmetry.

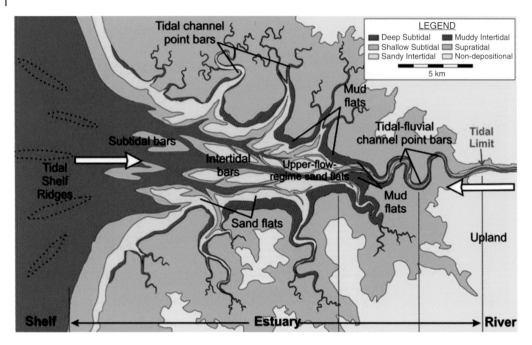

Figure 9.11 Diagrammatic map of a tide-dominated estuary showing the various sedimentary environments. (*Source:* Dalrymple et al. (2012)).

Each location in an estuary displays its own characteristic time–velocity curve showing its own asymmetry. Changes in the asymmetry will occur within the lunar cycle from neap to spring conditions. If the estuary floor or tidal channels are changed, all of which strongly influence the flow of tidal current, then time–velocity curves may show steep or more gradual slopes or they may show distinct differences in the duration of the flood and ebb portion of the tidal cycle. This time difference exceeds an hour in many cases. These conditions may produce either flood-dominated locations or ebb-dominated locations and these different locations may be adjacent to one another. It is common for example, for a channel to be ebb-dominated but the adjacent tidal flats to be flood-dominated.

We can show how sediment is transported by looking at the time–velocity relationships. There is a threshold velocity above which sediment is transported. By looking at the length of time that the velocity is above that value it is possible to show the amount of sediment that is moved and to determine if

the location of this time–velocity plot is flood- or ebb-dominated (Figure 9.15).

9.1.4 Model Estuary

A good way to conceptualize an estuary is through the use of a simple model. The estuary can be subdivided into three main parts: the landward area of river influence; the middle, truly estuarine area; and the seaward area of marine influence (Figure 9.16). The relative proportions of each vary with individual estuaries and with the influences of the major processes. Tides commonly diminish in their influence landward, although there are exceptions where the shape of the estuary enhances tidal range, the Bay of Fundy being an example. Wave influence tends to be directly proportional to the size of the estuary or to the fetch of a particular portion of it. Riverine influence is likewise proportional to the amount and rate of river input relative to tidal flux.

There is nearly always significant overlap in the sediment supply from the river and

(a)

(b)

Figure 9.12 Examples of major tide-dominated estuaries include (a) the Bay of Fundy in Nova Scotia, Canada, and (b) the Bay of St.-Malo on the north coast of France. (*Source:* Tessier (2012)).

from marine sources, and the nature of the contribution is commonly different between these sources. River sediments are generally sand and mud, whereas marine sediments tend to be dominated by sand with some shell gravel; mud is rare (see Figure 9.16). The estuarine transport of sediments includes both bed load and suspended load. The latter is particularly important in low-energy estuaries where it forms the bulk of the sediment that accumulates; less is carried out by ebbing tides into the open marine environment.

(a)

(b)

Figure 9.13 Bedforms are widespread on the surface of tide-dominated estuaries such as (a) an aerial photo of multiple sets at Cobequid Bay in the Bay of Fundy, and (b) on a sandy tidal flat on the coast of Queensland, Australia. (*Source:* R. A. Davis).

The zone of fresh water and salt water mixing has a strong influence on suspended sediments because it is a place where water density changes significantly. Flocculation of fine clay mineral particles (<2 μm in diameter) takes place here, and floc size may reach up to 30 μm. This is also the zone of the turbidity maximum in both partially and fully mixed estuaries. Here, suspended sediment concentrations are highest. This phenomenon is controlled by the mixing of fresh water with the leading edge of the salt water. Some

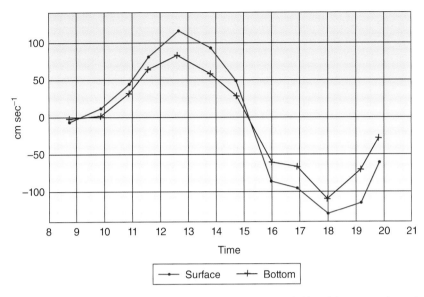

Figure 9.14 Time–velocity curve showing how flood-tidal and ebb-tidal current velocity is distributed through a tidal cycle.

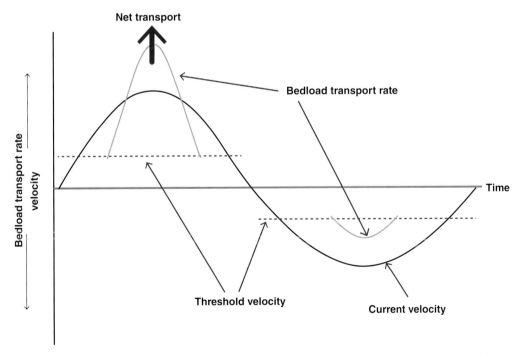

Figure 9.15 Model showing sediment transport over a tidal cycle as based on the time-velocity curve for currents. (*Source:* Wang (2012)).

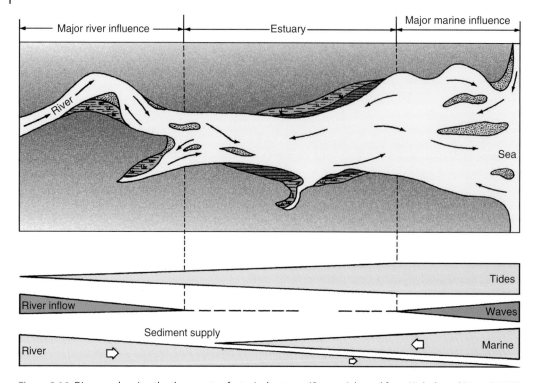

Figure 9.16 Diagram showing the three parts of a typical estuary. (*Source:* Adapted from Nichols and Biggs (1985)).

of the particles suspended in the overlying fresh water mass settle as currents diminish and are then entrained by the lower, denser salt water and carried landward to the turbidity maximum. This process produces the high sediment accumulation rate associated with the turbidity maximum.

There is a third source of sediment in estuaries; the biogenic material that is produced or modified in the estuary itself. This sediment tends to be most abundant and accumulates most rapidly in the middle zone (see Figure 9.16). There are numerous organisms that thrive on the brackish salinities that characterize the central portion of most estuaries. The typical salinity range here is from about 5–20 ‰. Ostracods, foraminifera, various mollusks, and worms are the most common animals. In addition various types of algae are present and some subtidal grasses on the fringe where water clarity is sufficient. Poor water clarity strongly inhibits photosynthesis in most estuaries, a consequence of the abundance of fine sediment, and the waves

and currents that can cause it to become suspended. Skeletal carbonate material from many of these animals makes an important contribution to the sediment of the estuary. Oyster reefs are particularly abundant and widespread in the middle section of many of the low-energy, muddy estuaries (Figure 9.17). In some areas mussels are also common.

The other aspect of the contribution made by bottom-dwelling organisms is by their pelletization of suspended sediments by filter feeders and to a lesser extent by grazers. The biggest contributors to pelleted muds are oysters, mussels and worms, all of which filter their nourishment from fine suspended particles provided by currents. The organisms pass these particles through their digestive tracts and excrete pelleted mud in large quantities. Much of the accumulated estuarine sediment is actually in the form of these pellets – sand-sized, cohesive aggregates of mud-sized particles. Because they are soft they appear to be simply an accumulation of fine mud. Filter feeders greatly increase the

Figure 9.17 Oyster reefs on the Gulf Coast of Florida.

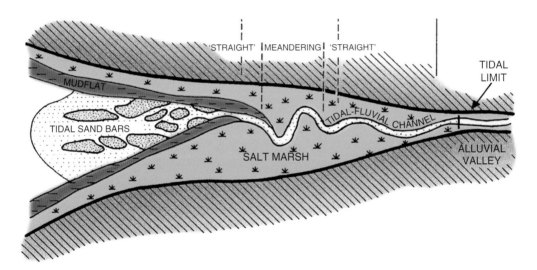

Figure 9.18 General model of a tide-dominated estuary. (*Source:* Courtesy of DalrymplChapte 2012).

rate of benthic sediment accumulation by taking suspended sediment that might otherwise be carried to sea out of the water column and converting it to larger particles that settle to the floor of the estuary.

9.1.5 Estuary Types

We can place most estuaries into one of two general morphodynamic types; wave-dominated and tide-dominated. If we consider the

general estuarine model shown in Figure 9.16 as a general morphologic model, it is possible to place it into either of these two types. The following discussion will consider the differences and similarities along with various examples of each type.

9.1.5.1 Tide-Dominated Estuaries

The general configuration of the tide-dominated estuary is a funnel shape (Figure 9.18) with tidal processes having a strong influence over most of the estuary. This type of estuary develops along coasts that have some combination of high tidal range and large tidal prism (the volume of water in an estuary between mean high tide and mean low tide) with an absence of an energetic wave climate. As a consequence, we have an estuary that tends to be fully mixed throughout due to the influence of tidal currents. In addition, the sediments that accumulate in a tide-dominated estuary tend to be sand; most of the mud being swept away by the strong currents. This sandy estuary floor is characterized by linear sediment bodies aligned along the estuary by the flow of tidal currents (Figure 9.19). The mobility of the substrate caused by these currents tends to inhibit colonization by benthic organisms.

The Gironde estuary along the west coast of northern France is a classic example of a funnel-shaped tide-dominated estuary. Other examples include the Minas and Chignecto basins of the Bay of Fundy, the Wash on the east coast of England, and to a lesser extent, Delaware Bay on the east coast of the United States (Figure 9.20).

9.1.5.2 Wave-Dominated Estuaries

If we take the same funnel-shaped estuary as discussed above and greatly reduce the tidal flux while at the same time increasing the incident wave energy along the coast, the result is some type of barrier across the mouth of the estuary (Figure 9.21). This is the general morphology of the wave-dominated estuary. The three parts can be seen as the marine portion that includes the flood tidal delta, the central basin that is the true estuary and the fluvial-dominated portion that is the bayhead delta. In some of these estuaries the barrier is detached from the mainland, as along the Texas coast (Figure 9.22a), and in others, it is attached to the mainland, generally some type of headland (Figure 9.22b). This is typical of the estuaries along the west coast of the United States, especially in the states of Oregon and Washington. Compare these photographs with a tidal estuary type of coastal bay (Figure 9.23).

Such estuaries develop along coasts where waves and wave-generated currents dominate over tidal processes. The barriers between the estuary and the open marine environment inhibit tidal flux but still permit enough marine influx to produce brackish conditions when combined with fresh water runoff via rivers. Sediments in wave-dominated estuaries tend to be dominated by mud or muddy sand. The absence of strong currents permits the fine sediment to settle to the bottom and remain there. In addition, filter feeders are common on a relatively stable substrate

Figure 9.19 Linear bedforms on the coast of Maine, USA.

Figure 9.20 Map of Delaware Bay a tide-dominated estuary showing sand bodies in its bathymetry on the Atlantic coast of the United States. (*Source:* Courtesy of NOAA).

further contributing to the muddy sediments through the production of pellets. This environment tends to be a very productive oyster habitat.

These estuaries can have any of the three main hydrologic characteristics discussed above (see Figure 9.7), but are likely to be in the layered or partially mixed types more that in the fully mixed category. They will, however, become fully mixed in the event that waves are generated over extensive shallow estuarine water bodies. Such an example

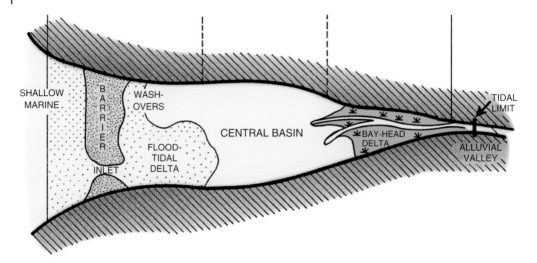

Figure 9.21 Simplified model of a wave-dominated estuary with the vertical lines dividing the three main parts. (*Source:* Courtesy of Dalrymple 2012).

of this condition would be Pamlico Sound behind the Outer Banks of North Carolina and most of the estuaries of the Gulf Coast.

9.2 Human Impact on Estuaries

The growth of population on the coast has had considerable effect on estuaries. Most of the impact is negative and almost all is related to various types of construction. Among the most significant of these impacts are those associated with commercial harbors (Figure 9.24). Here dredging of channels is common, large industrial docks have been constructed as have seawalls. In addition to the construction work, populated regions contribute considerable quantities of pollutants to bays. As a consequence of this, water quality can become quite poor limiting the ability of many species to live or reproduce.

Two other types of construction are prevalent in many of the bays associated with coastal plains. First, the dredging of shallow water and wetlands to produce upland environments for residential construction (Figure 9.25). This approach to the modification of bays generates multiple problems:

it produces canals with little to no circulation, leading to poor water quality and oxygen deficiency; it destroys important wetland environments; finally, it reduces the area of the bay, which reduces the tidal prism and causes circulation problems (Figure 9.26). The other type of construction that causes problems is fill-type causeways connecting the mainland to barrier islands (Figure 9.27). These structures are essentially a dam across the bay; no circulation can pass through them except at an intracoastal waterway channel.

Pollution can be a major factor in an estuary. One of the most serious events situation took place on the southern coast of Spain, associated with the Rio Tinto mining district in the upland area of the province of Andalusia (Figure 9.28). This was the site of some of the original precious-metal mines that were used for coinage by the Roman Empire. The tailings from the mine were included in the runoff into the Rio Tinto which polluted the estuary to the point that all life in it was destroyed. Indeed, its pH is still only 3; very acidic. The mining has ceased but the estuary is now subjected to pollution from a phosphate beneficiation plant (Figure 9.29).

(a)

(b)

Figure 9.22 Aerial photo of a wave dominated estuary on the (a) Texas coast and (b) the Oregon coast in the United States. (*Source:* Images ©2018 Google Earth).

Figure 9.23 Example of a tidal estuary type of coastal bay on the coast of England. (*Source:* Courtesy of Southwest Coastal Group).

(a)

(b)

Figure 9.24 Aerial photos showing (a) development of a harbor and (b) dredge spoil islands in the estuarine environment. (*Source:* Images © 2018 Google Earth).

Figure 9.25 Aerial photo of finger canals and upland development that originated on coastal wetlands in the Gulf Coast of Florida.

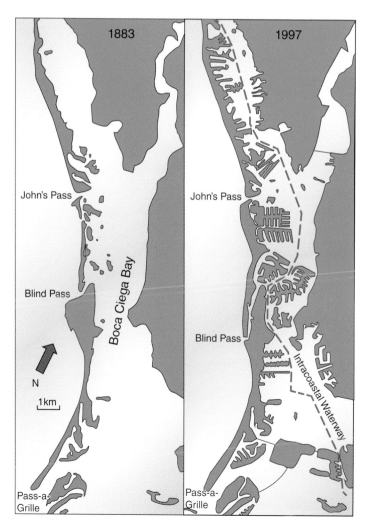

Figure 9.26 Map of Boca Ciega Bay on the Florida Gulf coast showing the changes in the area of the estuary that resulted from dredge and fill construction. (*Source:* Davis and Barnard (2003)).

Figure 9.27 Aerial photo of a fill causeway on the Florida Gulf coast. Such a dam-like structure causes significant changes in the tidal circulation of an estuary.

(a)

(b)

Figure 9.28 Photos of the (a) mining area where serious pollution is generated and (b) stream carrying pollution in the Rio Tinto in the Andalusia area of southern Spain. (*Source:* R. A. Davis).

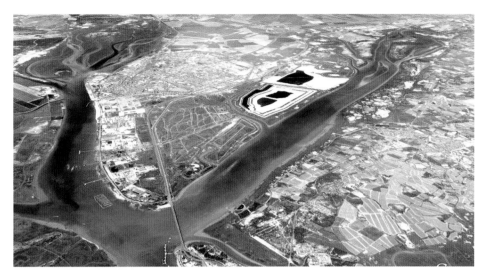

Figure 9.29 Aerial photo of the Rio Tinto (right) estuary near its mouth. The white area is a phosphate beneficiation area, another source of pollution for this estuary. The estuary on the left is the Odiel. (*Source:* Courtesy of J.A. Morales, Director, The Research Group).

9.3 Summary

Estuaries are among the most complex of all coastal environments because of their wide range of hydrologic, biotic, and sedimentologic conditions. They can be formed by a variety of different factors, although most are the result of sea-level rise over the past few thousand years. The rivers emptying into estuaries contribute both fresh water and sediments. The tidal flux from the marine environment contributes sediments and salt water. The estuary is the site of the mixing of these elements into a transition between marine and freshwater conditions. Organisms tend to be controlled by the nature of the water and by the substrate conditions.

References

Dalrymple, R.W., Mackay, D.A., Ichaso, A.A., and Choi, K.S. (2012). Processes, morphodynamics and facies of tide-dominated estuaries. In: *Principles of Tidal Sedimentology* (ed. R.A. Davis and R.W. Dalrymple), 79–108. New York: Springer.

Davis, R.A. and Barnard, P. (2003). Morphodynamics of the barrier-inlet system, west-Central Florida. *Mar. Geol.* 200: 77–101.

Nichols, R. and Biggs, R.L. (1985). Estuaries. In: *Coastal Sedimentary Environments*, 2e (ed. R.A. Davis), 77–186. New York: Springer.

Pritchard, D.W. (1967). What is an estuary? A physical viewpoint. *Science* 83: 3–5.

Tessier, B. (2012). Stratigraphy of tide-dominated estuaries. In: *Principles of Tidal Sedimentology* (ed. R.A. Davis and R.W. Dalrymple), 109–128. New York: Springer.

Wang, P. (2012). Principles of sediment transport applicable in tidal environments. In: *Principles of Tidal Sedimentology* (ed. R.A. Davis and R.W. Dalrymple), 19–34. New York: Springer.

Suggested Reading

Dyer, K.R. (1979). *Estuaine Hydrography and Sedimentation*. Cambridge: Cambridge University Press.

Dyer, K.R. (1998). *Estuaries: A Physical Introduction*, 2e. New York: John Wiley and Sons.

Eisma, D. (1998). *Intertidal Deposits; River-Mouths, Tidal Flats and Coastal Lagoons*. Boca Raton, FL: CRC Press.

Isla, F.I. (1995). Coastal Lagoons. In: *Geomorphology and Sedimentology of Estuaries*, Developments in Sedimentology No. 53 (ed. G.M.E. Perillo). Amsterdam: Elsevier.

Hardisty, J. (2008). *Estuaries: Monitoring and Modeling the Physical System*. New York: John Wiley and Sons.

Kennish, M.J. (ed.) (2016). *Encyclopedia of Estuaries*. New York: Springer.

Nelson, B.W. (ed.) (1972). *Environmental Framework of Coastal Plain Estuaries*, Geological Society of America, Memoir No. 133. Boulder, CO.

Perillo, G.M.E. (ed.) (1995). *Geomorphology and Sedimentology of Estuaries*, Developments in Sedimentology No. 53. Amsterdam: Elsevier.

10

Coastal Lagoons

Lagoons are very restricted coastal bays. They occur as the result of specific climatic, geologic and hydrographic situations. They have specific hydrologic characteristics, they host their own special fauna and flora and they have their own sediment signature. Lagoons are not particularly common globally but they are important because they provide a special environment with unique characteristics. In general coastal lagoons represent what is considered to be a stressed environment because of the extreme conditions that prevail.

This chapter will consider the various characteristics of coastal lagoons and what distinguishes them from other coastal water bodies. Some important examples will be discussed and compared.

10.1 Definition

Most definitions of the term lagoon are non-specific. Many authors apply the name lagoon to any water body that is landward of a barrier. Typically, lagoons are indicated as being parallel to the coast and separated from the open marine water by a natural barrier. The major problem with this loose use of the term lagoon is that there is no restriction or limitation on the nature of the circulation and water characteristics within the water body.

Some coastal bays are simply open embayments of the sea (Figure 10.1a) with similar tides, salinities and other characteristics to

the open marine environment, but with lower wave energy. The other end of the coastal bay spectrum is the lagoon, where there is an absence of significant freshwater influx and where there is no significant tidal flux because of the presence of an efficient barrier blocking interaction between the bay and the open marine environment (Figure 10.1b). Lagoons, therefore, will be expected to have an elevated salinity due to a general excess of evaporation over precipitation which is the primary mode of introducing water to the lagoon, although storms may wash over low barriers and introduce water to the lagoon. Such coastal water bodies with high salinities have very different characteristics than open bays and estuaries. For these reasons it is important to maintain strict definitions for the various types of coastal bay.

10.2 Morphology and Setting

Lagoons display various shapes, but most are elongate parallel to the coastline and virtually all are separated from the open marine environment by a barrier island, though in some places by a reef, such as in the Persian Gulf. Because the wave-dominated nature of the coast produces long barrier islands and coral reefs, many lagoons extend for tens of kilometers along the coast. Atoll reefs also have a lagoon in their center. Lagoons may form in a variety of geologic settings as long as there is some embayment from the open marine

Beaches and Coasts, Second Edition. Richard A. Davis, Jr. and Duncan M. FitzGerald.
© 2020 John Wiley & Sons Ltd. Published 2020 by John Wiley & Sons Ltd.

(a)

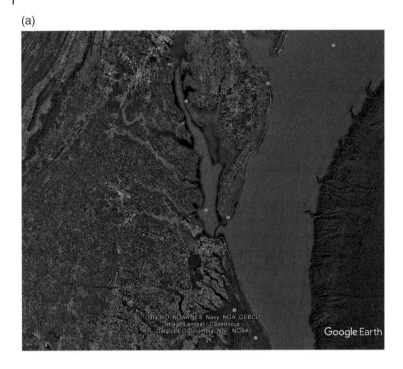

(b)

Figure 10.1 Various settings of different types of coastal bays including (a) an open embayment of the sea and (b) a lagoon which lacks streams and tidal flux. (*Source:* Courtesy of (a) South West Coastal Group and (b) Wikipedia).

environment and a mechanism for isolating it, such as development of a barrier or spit. Lagoons may develop along high-relief areas such as the Pacific coast of Mexico where high wave energy and pronounced longshore drift develop barrier spits that are wave-dominated and become closed to significant tidal flux, but most lagoons are developed along coastal plains where barriers have effectively separated a coastal bay from the open ocean.

Climate is another factor that plays a role in most coastal plain or prograding strandplain lagoons. In order to eliminate freshwater runoff into the coastal bay it is necessary for rainfall to be significantly limited. Typically this occurs along coasts where near-desert conditions prevail. In fact, many of the world's lagoons have developed along such coasts: southern Australia, northern Africa, the Persian Gulf, south Texas, much of Mexico, southern Brazil and southeast Africa.

The nature of coastal lagoons makes them typically quite shallow; a meter or less is common. Some are actually ephemeral, having dry beds during the dry season and water standing in the lagoonal basin only during the wet season.

10.3   General Characteristics

Although there are significant differences among various coastal lagoons, there are many similarities. We can characterize them by their salinity, organisms, processes and sediments.

Salinity

Lagoons are commonly schizohaline or hypersaline, at least in the broad senses of these terms. Schizohaline water bodies are those which display great change in salinity from brackish to hypersaline, generally in response to seasonal rainfall or some cyclic phenomena. Some shallow coastal lagoons fall into this category. An extreme example is Lake Reeve in Victoria, Australia, which is a long, shallow lagoon (Figure 10.2). During the wet season it is almost fresh, whereas in the dry season it dries up locally and becomes quite saline in those areas where water remains.

This type of salinity pattern would characterize lagoons of the mid-latitudes with strong seasonality to precipitation, and with high evaporation:precipitation ratios. Hypersaline lagoons are those in which salinities are

Figure 10.2 Vertical aerial photo of Lake Reeve, Victoria, Australia showing some dry portions. (*Source:* Image © 2018 Google Earth).

continually above normal marine concentrations. They are characteristic of semiarid and arid coastal areas where little or no freshwater influx occurs. Salinity commonly increases away from the connections with the open sea, if any are present. Laguna Madre in Texas and Shark Bay in Western Australia are examples, and will be discussed in detail later in the chapter.

Organisms

Hypersaline or schizohaline conditions cause serious problems for organisms. Typical marine or estuarine species cannot tolerate either of these salinity situations so we find special communities present in lagoons. It is usual for such extreme environments is to have very few species because of the special adaptations that are required for such severe salinity conditions. Generally, however, the numbers of individuals within these specially adapted species is very high.

A good example of an organism that can tolerate major fluctuations in salinity is the killifish, *Funulus*, which is common in parts of Laguna Madre, Texas. This small fish inhabits many of the isolated or nearly isolated ponds and embayments of this lagoon where salinity may change from nearly fresh water (<5 parts per thousand (5 ppt, or 5 ‰)) to as much as 200 ppt. This is almost six times higher than normal marine concentrations which are around 35 ppt (=35 ‰). Evaporation will raise salinity to near the high end of this range during summer and one fall of rain can lower it to brackish levels of 10–15 ppt. Killifish can tolerate such extreme and rapid changes in salinity through its ability to rapidly osmoregulate its body fluids to match those in its aquatic environment.

Another excellent example of an organism that can live under severe salinity conditions can be found in both of the southeastern Australian lagoons; the Coorong and Lake Reeve. Here a single species of cerithid gastropod, *Rosiella*, appears in huge numbers along the shallow and exposed wet margins of the lagoons. These detritus feeders graze while moving slowly over the surface. They also contribute large numbers of small pellets to lagoonal sediment. Their numbers reach such high concentrations that is common for the small beaches along the lagoons shores to be made exclusively of the shells from this single species (Figure 10.3).

Figure 10.3 Cerithid snails which abound in many schizohaline lagoons.

Chemical Precipitates

High concentrations of dissolved elements and ions in lagoons may result in actual precipitation of minerals. There is a generally predictable order of mineral precipitation depending on concentration. The typical salinity of open lagoons ranges from about 40 through 70 ppt most of the time. These levels of salinity do not result in formation of any of the typical evaporite minerals but carbonate minerals may precipitate. These carbonate minerals (aragonite and calcite) are different in that they do precipitate from solution in elevated salinities but they are also influenced by photosynthesis, pH and other factors.

The true evaporite minerals that may appear in lagoons do not precipitate until a salinity of about 200 ppt is reached and gypsum (calcium sulfate) forms. The next common mineral to precipitate is halite, common salt, which appears at levels of 300 ppt. In very extreme situations, other evaporite minerals may form but they are typically restricted to saline lakes and intermittent lakes in inland environments. Even gypsum and halite are uncommon in coastal lagoons, and are generally restricted to local ponds and embayments of the major lagoonal water body where salinities can reach extreme levels.

10.4 Lagoonal Processes

Physical Processes

Typical coastal processes, such as tides and waves, are not prevalent in lagoons. The definition of the term and the nature of lagoons limits any tidal influence. Tidal flux is absent or very local adjacent to typically small inlets. Wind tides do occur but without any regularity. Thus, there is no tidal mechanism for transporting sediments into or out of coastal lagoons. Waves are limited in their influence by the short fetch that is typical of the normally elongate lagoons. The width of these coastal bays is generally only a kilometer or so, which produces short and steep waves if strong winds are present. These waves do not have much influence on the lagoon floor but they do develop narrow beaches and small beach ridges (Figure 10.4). These choppy waves can also cause coastal erosion along lagoonal shorelines.

Figure 10.4 Small beach and related beach ridges in Lake Reeve, Australia; both wave-formed features of this lagoon.

There are other physical processes that have a major influence on coastal lagoons. All are derived from some influence of wind. This can be the result of storm activity or from ambient onshore winds that prevail along most coasts. Wind can produce circulation in lagoons that homogenizes any stratification of salinity (density) that might develop from very low energy conditions.

Storms, especially hurricanes, produce large waves and elevated water level or storm surge. It is common for this combination to breach dunes of barrier islands and transport both water and sediment into the lagoon. Storm surges of over 3 m are fairly common in severe hurricanes and when combined with large waves they can develop large washover fans that extend well into the lagoon (Figure 10.5). This phenomenon transports large volumes of sediment into the lagoon in a very short time. The same is true for water. Such high-intensity hurricanes are most common in fairly low latitudes. None the less, many coastal lagoons do not receive this input of either water or sediment. This is especially true if the barriers contain dunes of several meters elevation.

The other storm-related process that affects lagoons is the wind tide that results from setup in the water body itself. When wind blows over these shallow water bodies, water is pushed toward the downwind side of the lagoon. Because lagoonal shorelines generally have a gentle gradient, there may be significant flooding by this elevated, wind-blown water; causing a wind tide. When this happens on Laguna Madre, it is common for sediment to be transported onto the wind tidal flats (Figure 10.6) on the landward side of Padre Island, Texas which borders the lagoon, and along the mainland shoreline. It is common for fine-grained sediment, suspended by the turbulence of the storm, to accumulate in a thin veneer on these normally sandy wind tidal flats.

Another important wind-related phenomenon that influences lagoonal environments is the prevailing wind transporting dry and unstabilized sediment into the lagoon. This dry sediment might be from the backbeach environment but is most typically from dunes. Its transport might also be associated with the daily sea breeze cycle. As a result, large amounts of sediment are transported into lagoons. This transport can take place in

Figure 10.5 Washover fans on Padre Island extending into Laguna Madre as the result of a severe storm which overtopped this barrier island.

Figure 10.6 Extensive wind-tidal flats along the barrier island side of Laguna Madre, Texas.

two ways: by individual grains being picked up and transported in suspension by the wind; and by large numbers of grains being carried along the sediment surface as the dunes migrate into the lagoon.

Most people who have been to the beach have experienced the situation where wind blows sand in your eyes, onto your blanket and into your sandwich. This is the same phenomenon that transports sand grains from the front part of the barrier island to the backbarrier and into the lagoon. This condition results in a persistent but slow rate of sediment transport to the lagoon.

Unvegetated dunes are vulnerable to wind transport because of the absence of any stabilizing plants. The absence of vegetation may be the result of arid conditions, loss due to overgrazing of the barrier islands, deforestation by human activity, or the catastrophic loss of vegetation due to storms or disease. As a wind with an onshore component blows, there is essentially a mass transport of sediment across the dune surface and down the leading edge or slipface of the dune. This is how dunes migrate and without any stabilization there can be an almost catastrophic rate of sediment transport. In many locations there are very high and steep slipfaces on dunes that are migrating landward

(Figure 10.7) including some that are actually migrating over forests. In several situations this migration of dunes has moved huge amounts of sediment into coastal lagoons. The best example is probably that of Laguna Madre in Texas where wind-blown sand has filled the entire width of the lagoon and dredging is necessary to maintain a navigable channel for the Intracoastal Waterway (Figure 10.8).

In summary, although there is no significant sediment introduction to the lagoon from either river discharge or tidal flux, a significant amount of sediment is delivered. The rates of sediment introduction are low in many situations and the methods of delivery are dominantly the result of wind-generated processes. The result is that most lagoon sediment is derived from the seaward direction and is carried to the lagoon throughout its length; there are no detectable point sources of sediment.

10.5 Lagoonal Sediments

The nature of lagoonal sediments is quite diverse because there are diverse sources and mechanisms for their presence. There are three types of sediment that accumulate in

Figure 10.7 Active dunes on Padre Island due in part to cattle grazing on the vegetation. (Reprinted by permission of the Bureau of Economic Geology, The University of Texas at Austin, 2018).

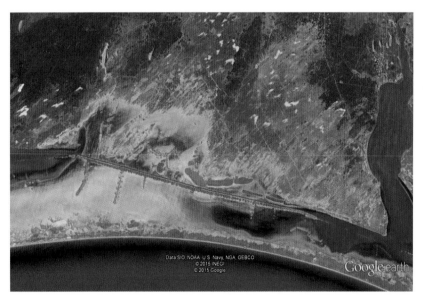

Figure 10.8 Area called the "Landcut" in Laguna Madre, Texas where windblown sand closes the entire lagoon, and a channel must be dredged to keep the Intracoastal Waterway open. (*Source:* Image © 2018 Google Earth).

lagoons: chemical precipitates; sediment particles carried in by various wind-related processes; and skeletal material from organisms that live in the lagoon.

Chemical precipitation of evaporite minerals is typically limited to local sites where high evaporation rates are present and where salinities are extreme. The most common situation for this type of sediment accumulation is in small ponds and lakes that have been separated from the main lagoon. This might occur during the dry season or in

Figure 10.9 Thick lenses of calcium carbonate mud (micrite) in Lake Reeve, Australia.

situations where a longer isolation of the pond or lake takes place.

The more common chemical precipitate is various species of calcium carbonate. This usually occurs as clay-size particles of aragonite or calcite; both types of $CaCO_3$. The precipitation of these minerals is commonly the result of photosynthesis which in turn alters the pH of the shallow aquatic environment of the lagoon. The carbonate mud, commonly called micrite, generally occurs in thin and discontinuous layers in the open lagoons (Figure 10.9), but also may be present as extensive layers (Figure 10.10). In many situations the carbonate mud accumulates in association with filamentous blue-green algae also called cyanobacteria. These microorganisms typically develop so-called algal mats (Figure 10.11) along the periphery of coastal lagoons. It is the photosynthesis of these cyanobacteria that assists in the precipitation of the carbonate mud. Algal mat

layers below the present surface represent earlier wind-tidal flat surfaces.

The sediment that is introduced by wind activity is typically well-sorted, fine to medium sand. Although it is generally quartz in composition, virtually any composition is possible depending on the nature of the overall barrier sediment. This sediment appears in a variety of forms. It may be isolated sand grains in a muddy lagoonal sediment. Extreme storms that generate washover processes can transport thin layers of sand into the lagoon with a thickness up to tens of centimeters. In the case of migrating sand dunes, the entire dune may become part of the lagoon (Figure 10.12).

Fine sediment that settles out of suspension from the water column during and just after storms, is most recognizable along the margins of the lagoon where it accumulates in thin layers only a few millimeters thick. After high water conditions cease, the thin

Figure 10.10 Very extensive carbonate mud covering stromatolites accumulation in a lagoon of the Coorong, South Australia.

Figure 10.11 Algal mats (dark layers and surface) formed by large populations of cyanobacteria (blue-green algae). Layers below the present surface represent earlier surfaces.

mud layer is exposed and dries quickly. Shrinkage takes place causing the layer to become a series of curled up flakes (Figure 10.13) that can easily be removed by wind or the next period of high water.

Although the variety of mollusks, ostracods and other skeleton-bearing organisms is not great, each species present is usually very abundant, thereby producing a significant amount of gravel and sand-sized sediment in the lagoon. This skeletal sediment component may be scattered throughout the lagoonal sediment or it may be concentrated along the shoreline by wave action.

Figure 10.12 Dune migrating landward, encroaching on the lagoon at the Coorong, Australia.

Figure 10.13 Flakes of a thin mud that has been desiccated due to its position on a wind-tidal flat; Baffin Bay, Texas. (*Source:* R. A. Davis).

10.6 Example Lagoons

There are many good examples of lagoons throughout the world. Some are large and others are limited in their extent. Because of the nature of the lagoon and its high evaporative environment, the examples included here will represent a spectrum of increasing salinity and therefore, increasing amount of chemical precipitate. The three lagoons considered, in order of increasing salinity, are the Laguna Madre/Baffin Bay complex in south Texas, Lake Reeve in Victoria, Australia and the Coorong in South Australia. All are more than 100 km long and have only one inlet at one end of the lagoon.

Laguna Madre

This lagoon extends from near Corpus Christi along the entire south Texas coast to near Brownsville at the Texas–Mexico border (Figure 10.14). It has some circulation at its north end and is connected to the open Gulf via an artificial inlet called Mansfield Pass. There is no significant perennial stream that enters Laguna Madre or adjacent Baffin Bay. Baffin Bay is a drowned fluvial system that carries very little fresh water at the present time. The lagoon is pristine, with significant development only at each end. Most of the adjacent barrier is a National Seashore and the adjacent mainland is an area of cattle

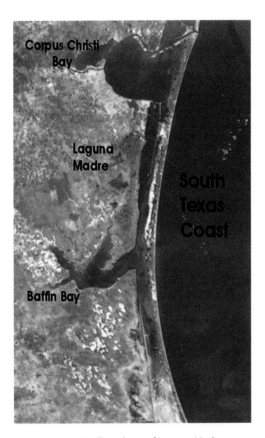

Figure 10.14 Satellite photo of Laguna Madre, Texas, the longest coastal lagoon in North America.

grazing and cotton fields. Oil exploration has taken place in various locations throughout the lagoon.

The salinity ranges from about the low 40s to near 90 ppt from north to south. There are seasonal variations and local places of higher concentrations. Evaporite precipitation is not present in surface waters of the open lagoon but does occur in the pore waters of the sediment along the coast and probably beneath it as well. Carbonate mud occurs only in small patches in the southern part of the lagoon and ooids also occur.

Laguna Madre has extensive wind-tidal flats (see Figure 10.6), most of which are covered by algal mats. In some places it is possible to see multiple layers of these mats interspersed between sandy layers produced by washover events (see Figure 10.11). Back island dunes are widespread and are mobile due to the absence of vegetation as a consequence of overgrazing (see Figure 10.7).

Landward of Laguna Madre and connected with it, is a separate lagoonal basin, Baffin Bay. This irregularly shaped lagoon is a relict estuary that has lost its freshwater influx due to a change in climate. No significant streams are present in adjacent south Texas. This shallow, hypersaline water body has an additional element of interest –small but numerous serpulid worm reefs. These reefs rise nearly a meter above the surrounding waters (Figure 10.15) and contribute considerable skeletal material to the sediment of the lagoon floor. The reefs do display some asymmetry that indicates the direction of high energy, in this case waves.

Lake Reeve, Australia

The southeastern coast of Victoria, Australia is characterized by a continuous barrier island known as 90-Mile Beach. This barrier has one small opening on the northeast end at Lakes Entrance. The backbarrier aquatic environment is Lake Reeve (Figure 10.16), a shallow, coast-parallel lagoon that is schizohaline in its character. This lagoon displays a wide range of salinities from nearly fresh at its northeast end to near 100 ppt in various of

(a)

(b)

Figure 10.15 (a) Map showing serpulids in Baffin Bay as shown in red and (b) close-up of serpulid worms that comprise the reefs. (*Source:* (a) and (b) Courtesy of Alex Simms).

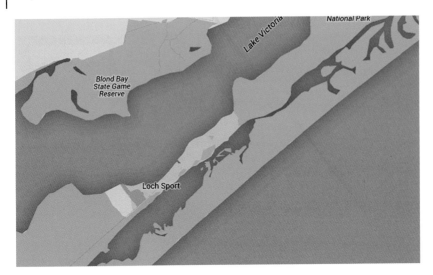

Figure 10.16 Map of Lake Reeve, Australia located landward of the 90-mile Beach Barrier showing the only opening to the open ocean at the northeast end, Lakes Entrance. (*Source:* © State of Victoria. Licensed under CC BY 4.0).

Figure 10.17 Aerial photo of a portion of the 90-mile barrier showing relic tidal inlets that are now covered with carbonate mud. (*Source:* Image © 2018 Google Earth).

the small and sometimes isolated basins within the lagoon. Salinity also varies greatly with the season; the winter is wet and salinity is low; the summer is dry and salinity high.

It is apparent from the geomorphology that the barrier here was formed by significant northeast-to-southwest sediment transport.

This can be seen by the shape of the closed inlets (Figure 10.17). These relict inlet surfaces are covered by carbonate mud.

Because the dunes on the 90-Mile barrier rise 10 m or more above sea level, there is no washover of the barrier and there has not been for much of its history. Additionally, the

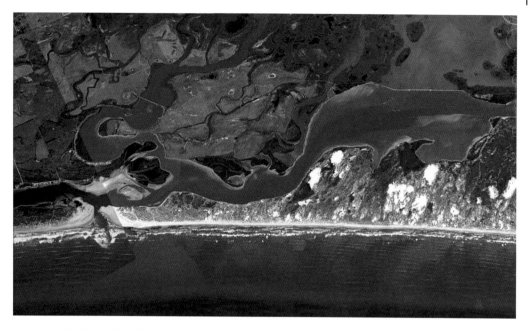

Figure 10.18 Photo of the Coorong in South Australia with the opening at the north end where the Murray River discharges. (*Source:* Image © 2018 Google Earth).

dunes are well stabilized by vegetation. This means that there has been virtually no introduction of sediment from the seaward direction. The floor of most of the lagoon is comprised of shelly sand with very little mud. Radiocarbon dating of the shells shows that they are more than 4000 years old, testimony to the absence of sediment introduction.

Lake Reeve is accumulating two primary types of sediment at the present time. One is small pellets of mud produced by small cerithid snails that inhabit the lagoon in huge numbers. These snails graze over the typically algal-mat-covered margin or dried portion of the lagoon where they feed on the cyanobacteria and produce large numbers of these pellets. The other type of sediment is calcite which precipitates directly from the lagoonal water as the result of high salinity and photosynthesis. The mud may be in thin lenses or it may extend over large areas of the lagoon floor.

The Coorong

Another very long coastal lagoon in the southern part of Australia is the Coorong, south of Adelaide. It has an inlet at its southern end where the Murray River discharges into the Southern Ocean (Figure 10.18). This lagoon and its associated barrier represents the latest of several similar coastal systems that developed as this part of South Australia formed during the Pleistocene Epoch. The barrier is dominated by very large, mobile dunes. The combination of the climate and the lack of freshwater and marine influx except at the northernmost end produces a high-salinity lagoon with increased concentrations to the north. There are also several small lakes that are relict coastal lagoons. Unlike Lake Reeve, the Coorong is receiving significant sediment from the barrier and it is precipitating a variety of minerals throughout much of the lagoon and the related small lakes.

Onshore wind from the Southern Ocean can be quite strong and causes individual grains to be removed from the unvegetated dunes and carried into the lagoon. There is also extensive landward migration of the dunes into the Coorong.

Chemical precipitates include thick carbonate mud (Figure 10.19), and both gypsum

Figure 10.19 Extensive and thick carbonate mud that has precipitated in one of the many small isolated ponds in the Coorong complex.

Figure 10.20 Halite deposits formed as the result of evaporite precipitation.

and halite (Figure 10.20). The halite or common salt was mined in the nineteenth century by Chinese immigrants who accidently discovered these deposits on their way to the gold mines further inland in Australia.

10.7 Summary

Lagoons are quite special coastal bays. Their typical hypersaline conditions result from lack of significant freshwater influx and lack of tidal flux. Because climate tends to be a factor in lagoonal development, most are associated with arid coastal conditions. These characteristics result in unusual biota with few species but abundant numbers. Limited methods of sediment introduction result in slow rates of influx except in the case of major storms. Additional accumulation of material in lagoons is produced by chemical precipitation of carbonate and evaporite minerals.

Suggested Reading

Eisma, D. (1998). *Intertidal Deposits; River-Mouths, Tidal Flats and Coastal Lagoons.* Boca Raton: Florida, CRC Press.

Isla, F.I. (1995). Coastal Lagoons. In: *Geomorphology and Sedimentology of Estuaries*, Developments in Sedimentology No. 53 (ed. G.M.E. Perillo), 241–272. Amsterdam: Elsevier.

Kjerfve, B.J. (ed.) (1994). *Coastal Lagoon Processes*. Amsterdam: Elseriver.

Warren, J.K. (2006). *Evaporites: Sediments, Resources, and Hydrocarbons*. New York: Springer.

11

Tidal Flats

Unvegetated intertidal environments that accumulate sediment and are occupied by specially adapted organisms are located around the margins of most coastal embayments and some open coasts. These environments are called intertidal flats or simply tidal flats. The width and extent of tidal flats are directly related to tidal range and to the morphology of the bay or other environment in which they are located. This term is typically reserved for those intertidal environments that are not exposed to significant wave energy. For example, many beaches have extensive intertidal components that are not vegetated but they are not considered to be tidal flats.

In some embayments, especially those that experience macrotidal conditions and are tide-dominated, much of the bay may be intertidal except for tidal channels that dissect the flats. The Wadden Sea along the Dutch and German coast of the North Sea is such a place as are the Bay of Fundy in Canada and the Bay of St.-Malo on the north coast of France, the two places with the highest tidal ranges in the world. Each of these is discussed in a later section of this chapter.

Tidal flat surfaces and their associated tidal channels may be composed of mud, sand or more typically, a combination of both. Some channels may have high concentrations of shell debris on the floor. Most tidal flats and tidal channels have various types of bedforms, regular undulations, on their sediment surface. The nature and rigor of the tidal currents and waves tend to be the controlling factors in both the rate and nature of sediment accumulation, and the types of bedforms that develop on the sediment surface.

In this chapter the morphology, sediments, and processes that characterize the tidal flat complex are discussed. This environment is by far most common along mesotidal and macrotidal coasts where low gradients characterize the shore zone.

11.1 Morphology of Tidal Flats

The standard appearance of a tidal flat is a gently sloping and fairly broad surface of unconsolidated sediments that is alternately inundated and exposed as the tide floods and ebbs. Typically, the width is directly related to the tidal range but the underlying geology and regional geomorphology can cause variations. Along broad coastal plains or other flat-lying areas the tidal flats tend to be very gently sloping (Figure 11.1), but along some coasts the tidal flat slope tends to be relatively steep (Figure 11.2).

Another factor in the size of tidal flats is the extent to which the estuary has been filled with sediment. Those that have not experienced significant infilling of sediment tend to have narrower tidal flats than those that have a great deal of sediment.

The most pronounced interruption of the nearly flat and featureless intertidal flat is the presence of tidal channels (Figure 11.3)

Beaches and Coasts, Second Edition. Richard A. Davis, Jr. and Duncan M. FitzGerald.
© 2020 John Wiley & Sons Ltd. Published 2020 by John Wiley & Sons Ltd.

Figure 11.1 Photograph of a wide, gently sloping tidal flat in Westernport Bay, Victoria, Australia.

Figure 11.2 Photograph of a narrow and steep tidal flat in the Bay of Fundy, Nova Scotia, Canada.

Figure 11.3 Photograph of a typical tidal channel that cuts into the tidal flat.

that dissect most tidal flats. These channels range from small and ephemeral ones that may be closely spaced to those that are large and deep. The latter commonly have water in them throughout the tidal cycle, even during spring tide conditions. These channels serve as major conduits for sediment during both flood- and ebb-tidal cycles. The pattern of channel development on many tidal flat systems is quite similar to a typical river system, with small tributary channels merging to serve a single major channel.

11.2 Sediments

Sand and mud with some scattered shells form the typical sediment package on tidal flats. There is generally a regular and predictable pattern of sediment distribution related to both physical energy and position within the intertidal flat. Highest energy occurs at the base of the intertidal zone, the lowest part of the tidal flat covered by water for the longest period. Here, typically, is a concentration of sand with a specific grain size dependent on the spectrum of sizes available in the specific estuarine system. Grain size decreases landward and upward across the intertidal flat (Figure 11.4) with mud at the top or landward side. This is a contrast to most shoreline environments, where sediment grain size decreases from land to sea. Grain size decreases from the source; because tidal flats receive their sediment from the ocean side, grain size decreases toward land.

Generally tidal flat sediments are quite well-sorted at any specific location because they are subjected to similar conditions on a regular basis. The only common exceptions to this generalization is the presence of shells that may be scattered over various grain sizes because they are indigenous to the tidal flat environment, and mud that can settle from suspension during still water conditions. Some tidal flat environments may be adjacent to bedrock exposures that, when eroded, will provide large rock fragments to adjacent tidal flats.

11.3 Organisms

Although tidal flats are a rather harsh environment because of the regular and continual exposure and inundation, they do support a community of abundant organisms. This discussion will be restricted to the benthic portion of the community because that is the only portion that is truly restricted to the tidal flats. There are two different living habits here: vagrant benthos that move about, and sessile benthos that are fixed in their position. The mobility of sediment dictates that infaunal organisms are the most abundant.

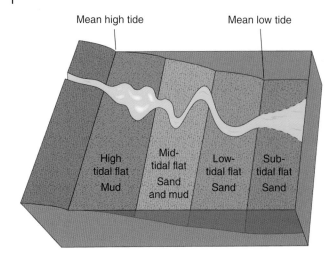

Figure 11.4 Diagram showing the trend in grain size over a tidal flat. (*Source:* Adapted from Klein (1972)).

11.3.1 Vagrant Organisms

There are various animals that move over the tidal flat surface, such as snails, worms and amphipods. The small snails feed on detritus that accumulates on the sediment surface, and are important fecal pellet producers. They are especially abundant on the upper part of the intertidal zone. Pellets can be an important constituent of sediments in intertidal and shallow marine environments.

Other vagrant types of snails are carnivorous and live on burrowing bivalves and oysters. These creatures have the ability to bore holes in the shells and then ingest the soft parts of the organism for food. So-called oyster drills may wipe out an entire oyster population.

Some burrowers are also quite mobile. The best example is the fiddler crab (Figure 11.5) that scatters over tidal flats quickly and feeds on detritus. It gets its name from the single large claw that resembles a violin or fiddle. This appears only on males; females have two like-sized claws.

11.3.2 Sessile Organisms

The most important and most abundant creatures on the tidal flat do not move about. They include both epifaunal organisms that live on the sediment surface, such as oysters or cyanobacteria, and burrowers, such as some worms and bivalves. Oysters (Figure 11.6) and mussels are both intertidal and subtidal. They tend to occur in clusters of many individuals. As sessile filter feeders, they produce many pellets which become part of the tidal flat sediment accumulation.

Cyanobacteria or blue-green algae exist as microscopic filamentous organisms that produce a mat-like coverage of the upper parts of intertidal flats in lower latitudes. These mats (Figure 11.7) are important sediment stabilizers in that they protect otherwise non-cohesive sediments from deflation by wind or from wave erosion under non-storm conditions.

Filter-feeding infaunal bivalves and worms (Figure 11.8) produce abundant fecal pellets on tidal flats. Worms are typically much more abundant than bivalves and they cause the destruction of most laminations in the sediment (Figure 11.9) due to their burrowing activities. They may have tubes (Figure 11.10) or they may not.

11.3.3 Limiting Factors

Exposure to the atmosphere and subsequent desiccation is a problem, especially for soft-bodied organisms like worms. Most can only tolerate exposure for a short period of time. The higher the position in the tidal flats, the fewer sessile organisms will be present

(a)

(b)

Figure 11.5 Tidal flat with (a) numerous fiddler crabs, and (b) a close-up photograph of a fiddler individual. (*Source:* NOAA, https://commons.wikimedia.org/wiki/File:Fiddler_crab.jpg).

because this is where exposure can last for at least several hours during each tidal cycle. Most of the shelled invertebrates such as oysters, clams and snails are able to seal their soft parts from the atmosphere and can withstand fairly long periods of exposure during each tidal cycle.

Another limiting factor to benthic organisms is a large concentration of suspended sediment particles in the water column. Most of the infaunal and some epifaunal organisms that live on the tidal flat obtain their nourishment through filtering organic debris from the water column. These filter

Figure 11.6 Intertidal oyster reefs near Cedar Key, Florida.

Figure 11.7 Surface and subsurface mats of cyanobacteria on wide tidal flats of Padre Island, Texas.

feeding organisms do not have the ability to select specific suspended particles for ingestion into their filtering system. As a consequence, when large concentrations of suspended sediment are present the organisms will ingest too much indigestible material, their siphons will become clogged, and they will die.

The other important limitation is a mobile sediment bed caused by waves and/or strong tidal currents. Many burrowing organisms,

especially sedentary ones, need a reasonably stable sediment base in which to burrow and maintain existence. It is obvious, therefore, that numerous problems confront benthic organisms in estuaries. Waves may cause significant sediment mobility in the large estuaries, and tidal currents can mobilize the bottom sediment in many locations. The floor of tidal channels is probably the most hostile environment for benthic organisms because sediment is moving almost

Figure 11.8 (a) Burrowed surface, and (b) subsurface burrow of *Arenicola*, a worm that prefers sandy substrates.

 (a)

 (b)

Figure 11.9 Shallow trench in tidal flat sand showing a total absence of stratification due to burrowing organisms.

Figure 11.10 Abundant worm tubes protruding above the tidal flat surface on Martens Plate, East Frisian Wadden Sea, Germany.

throughout the tidal cycle. In many areas, the lowest portion of the intertidal flats can experience vigorous substrate mobility during most of the tidal cycle.

11.3.4 Bioturbation

Many of the numerous benthic organisms that live in estuaries are infaunal; they burrow into the sediment both for protection and for feeding. It is these same animals that take in suspended particles and produce most of the pellets that accumulate in estuaries, but the activity of interest here is the actual burrowing process. As a bivalve or worm burrows into and through sediment it destroys layering by essentially homogenizing the sediment (Figure 11.9). This churning of sediment by burrowers, which may number hundreds of individuals per square meter, destroys the characteristic structure of sediments deposited in a tidal environment. Extensive tidal flats of the Georgia and South Carolina coasts fall into this category. In fact, in the German Wadden Sea the effects of waves and bioturbation combine to destroy the laminations in tidal sediments.

It is only in places where there are few burrowing organisms that stratification is typically preserved. This lack of benthic infaunal organisms can result from a variety of conditions such as exposure, too much suspended sediment, and substrate mobility.

11.4 Sedimentary Structures

As might be expected, this distinctive environment also contains some special types of sedimentary structures, and also some that are not so special. That is, there are some sedimentary structures that are present, or even common, on tidal flats which are not unique to that environment, for instance ripples, megaripples and sand waves (Figure 11.11).

The features that are indicative of the tidal-flat environment are of more importance because they become key factors for geologists in their identification and interpretation of sedimentary depositional environments from the ancient stratigraphic record. Included among these features are both physical structures and biogenic structures (see Box 11.1). The many burrowing organisms leave characteristic markings on and in tidal flat sediments, especially such organisms as the lug worm, *Arenicola*, which is

Figure 11.11 Aerial photograph of tidal flats in the Bay of Fundy, Canada showing various scales of bedforms at low tide.

Figure 11.12 Mud surface showing desiccation cracks resulting from extended exposure on a tidal flat in the Wadden Sea.

widespread on sandy tidal flats (see Figure 11.8). Mudcracks or desiccation features are important physical structures that occur on the upper, muddy portions of the intertidal zone. These develop as the result of significant exposure in high places on tidal flats such as between neap and spring high tide (Figure 11.12).

The type of bedding on tidal flats is varied and can be summarized in a diagram proposed by H. E. Reineck from his work on the German coast. This classification is based on the ratio of sand to mud. There are only three categories: flaser bedding, wavy bedding and lenticular bedding (Figure 11.13). Flaser bedding is the most common; in it sand

Box 11.1 Preservation Potential of Tide-Dominated Sediments

One of the main reasons that geologists study modern coastal environments is so that when they examine the stratigraphy of ancient sedimentary strata they can interpret the depositional environments of those strata. This is an illustration of an important axiom of geology, the Law of Uniformitarianism: the present is the key to the past. More specifically, studying and understanding what is happening in modern environments prepares you to interpret ancient sediment layers.

Tidal-dominated environments are among the depositional environments that have a fairly good potential for preservation in the stratigraphic record. Most coastal environments are not in this category. If we consider barrier islands as an example, only the washover deposits in this system are reasonably likely to be preserved. If we consider tidal environments we know from examining the stratigraphic record that they are preserved throughout most of geologic time. Let's take a look at both modern and ancient tidal flat deposits to demonstrate this situation.

There are several features of intertidal environments that permit their recognition in the stratigraphic record. We are here considering both tidal flats and tidal channels. The most obvious feature to recognize is sediment accumulation that shows bi-directional bedforms or a portion thereof, thus reflecting the flood and ebb cycles of the tidal cycle. This feature tends not to be common in either modern or ancient strata because near-symmetrical currents are required and we know that such conditions are not common. Both situations do exist however (Box Figures 11.1.1 and 11.1.2).

Other recognizable tidal signatures include reactivation surfaces and tidal bedding. Reactivation surfaces show the bidirectional nature of tidal currents but lack bidirectional cross-strata. These surfaces reflect the asymmetry of tidal currents (Box Figures 11.1.3 and 11.1.4).

Box Figure 11.1.1 Photograph of a box core from a tidal channel in Martens Plate on the Wadden Sea of Germany, showing bidirectional cross-stratification.

Box Figure 11.1.2 Photograph of bidirectional cross-stratification from 1.7 billion-year-old Baraboo Quartzite in Wisconsin.

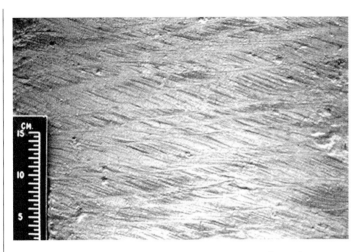

Box Figure 11.1.3 Example of modern reactivations surfaces that separate bedform cross-strata.

Box Figure 11.1.4 Photograph of Precambrian reactivation surfaces from the Baraboo Quartzite of Wisconsin.

They are at acute angles with the cross-strata and dipping in the same direction. Tidal bedding is a very important element of tide-dominated environments (Box Figures 11.1.5–11.1.8). It can commonly be seen in couplets of relatively fine and coarse sediment grains but also occurs in what appears to be uniform grain size.

Other tidal signatures are found in tidal bundles where lunar tidal cycles are recorded. Individual cycles range from sand beds that are relatively thick to thin mud deposits that represent suspension deposits that have come to rest (Box Figures 11.1.9 and 11.1.10).

Box Figure 11.1.5 Tidal bedding from a macrotidal coast in China.

Box Figure 11.1.6 Tidal bedding from Pleistocene strata on the coast of Washington state.

Box Figure 11.1.7 Tidal bedding from the Miocene in Florida.

Box Figure 11.1.8 Tidal bedding from the Precambrian Baraboo Quartzite.

Box Figure 11.1.9 Tidal bundles from a tidal channel in Martens Plate on the German Wadden Sea.

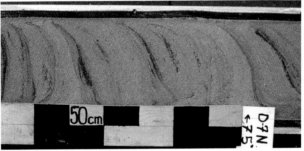

Box Figure 11.1.10 Exposure of Cretaceous strata in the San Juan Basin of New Mexico that contain tidal bundles.

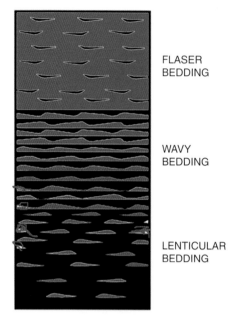

FLASER BEDDING

WAVY BEDDING

LENTICULAR BEDDING

Figure 11.13 Diagram proposed by Reineck showing the three primary bedding types, flaser, wavy and lenticular, that characterize tidal flat sediments. (*Source:* Courtesy of H.E. Reineck).

certain times and places. The rise and fall of the tides as the tidal wave is forced into estuaries and then back creates significant tidal currents, particularly in the tidal channels. Currents may range from only a few centimeters per second near slack tides on the flats to about a meter per second in the channels. These currents distribute sediment throughout the tidal flats. For purposes of explanation, the tidal flat is best viewed as a smooth and gently sloping surface. Certain key horizons may be noted on this surface based on the position of the water level at given tidal stages. Spring high tide is the highest position of regular and predictable inundation of the sediment surface by water and spring low tide is the lowest. Wind tides may cause water to be pushed up to the supratidal environment or down below low tide depending upon the direction and strength of the wind. Neap tidal range may be only about half of the spring range thereby causing considerable variation in the intertidal zone depending upon the lunar condition.

dominates over mud, which only occurs is small lenses (Figure 11.14). Flaser bedding is also easily preserved in the stratigraphic record because it commonly is buried rapidly (Figure 11.15).

Probably the most characteristic feature of tidal flats is a sedimentary structure called tidal bedding (Figure 11.16). This is a special type of sediment accumulation that is a consequence of of tidal cycles. The alternating energy levels produced by the flooding and ebbing of the tide lay down alternating thin layers of sand and mud (Figure 11.17). These rhythmites as they are called, record the rise and fall of the tide along with slack tide conditions (Figure 11.18). They also preserve neap and spring cycles with neap conditions producing thinner layers than spring.

11.5 Tidal Flat Processes

Tide-generated processes tend to dominate most sediments that accumulate on tidal flats, although waves can be important at

11.5.1 Tides

The scheme of sediment transport on tidal flats has been best described from studies of the tidal flats on the Wadden Sea. The model produced shows a combination of the settling-lag effect and the scour-lag effect on the paths of sediment particles as they are transported up on to the tidal flat surface (Figure 11.19) and how, over a long time, sediment builds up and out into the estuary. The distance–velocity curves are asymmetrical and show where a given particle is entrained, transported and deposited.

During flooding tides, the sediment particle at location 1 on curve $A–A'$ is picked up at a tidal current velocity shown at point 2 and carried landward until that velocity is again reached at point 3 on the curve. The particle then begins to fall and reaches the tidal flat surface at point 5 when the current velocity is at location 4. The difference in the velocity of entrainment, which is greater than at settling, is quite important. This provides the settling lag effect.

(a)

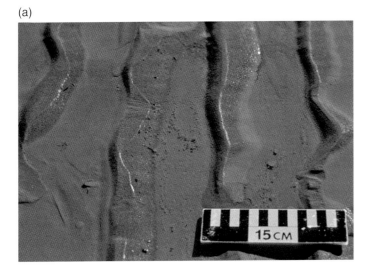

(b)

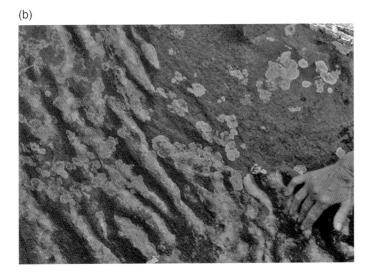

Figure 11.14 Photographs of (a) a modern tidal flat surface with mud in ripple troughs, and (b) an ancient example 1.7 billion years old in the Precambrian Baraboo Formation from Wisconsin.

Ebbing tidal currents follow curve $B–B'$, which represents a more landward water mass and achieves a lower maximum velocity at this position on the tidal flat. The same particle is picked up at location 5 when the velocity at point 6 on the curve is reached. It is carried until point 7 when it begins to fall and eventually settles to the bottom at point 9. This diagram shows that the net result is movement of a sediment particle from location 1 to location 9 during a single tidal cycle. This is obviously an oversimplified model because there are many perturbations on the tidal flat that interfere with the processes. It does, however, serve to illustrate the basic mechanism by which sediment particles are transported up on to the tidal flats.

Size of the sediment particles is an important variable in the above scheme. Bigger particles will be transported a shorter distance over a given tidal cycle than smaller particles. The more times an area is covered and the deeper the water, the more tidal energy is expended on a given location on the

Figure 11.15 Photograph of flaser bedding from the sediments in the Bay of St. Malo, France.

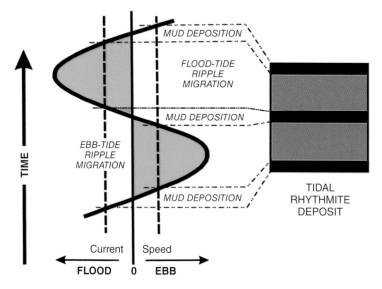

Figure 11.16 Diagram showing how tidal bedding is formed by alternating sand and mud during a tidal cycle as originally described by G.D. Klein (1977). (*Source:* Adapted from Dalrymple et al. (1992)).

tidal flat. As a consequence, there is a regular decrease in sediment particle size up the tidal flat toward the spring high tide level.

Conditions in some areas produce discontinuous mud or sand layers due to some combination of sediment availability and tidal current strength. These can be formed through tidal processes and represent what is essentially discontinuous tidal bedding, or they may represent alternations in current energies that are not produced by flooding

and ebbing tides. These are the conditions that produce flaser bedding (see Figure 10.15) which commonly form in the troughs of bedforms. Discontinuous sandy lenses within a mud sequence are referred to as lenticular bedding, a feature generally associated with limited sand availability but that may also reflect variations in tidal current velocities.

Another type of tidally produced stratification is tidal bundles, a type of stratification typically associated with tidal channels or

Figure 11.17 Photograph showing example of tidal bedding from the coast of China.

relatively strong tidal currents and large bed-forms (Figure 11.20). These types of stratification are generally found in the tidal channels where sand dominates and bedforms are at least bigger than ripples. The alternation of flooding and ebbing tide is generally accompanied by significant differences in current velocity. This commonly produces pulses in the migration of large bedforms that are characterized by medium- to large-scale cross-stratification. The dominant current moves the bedform and the recessive current commonly deposits a mud drape over the bedform, producing a muddy seam between each sand cross-stratum. In many tide-dominated areas the sequence contains readily distinguishable sets of cross-strata that change in thickness and sand : mud ratio in packages of 14 bundles. These tidal bundle packages (Figure 11.20) represent a spring and neap tidal cycle, and when preserved, they are found on the floors and margins of tidal channels (Figure 11.21).

11.5.2 Waves

We associate tidal processes with tidal flats, but there are some locations where waves play an important role in the dynamics of tidal flats, sufficient to destroy all of the signatures of tidal processes. In order for this to happen the energy imparted by waves onto the tidal flat must exceed that of tides. The most common conditions under which this can occur are in places where extensive shallow water covers the tidal flat for long portions of each tidal cycle. Waves formed in this shallow environment will move large quantities of sediment through the back and forth motion they produce, thereby destroying any tidal signature or preventing it from happening in the first place.

The Wadden Sea area on the German coast of the North Sea is a good example of such wave-influenced tidal flats. Here broad, sandy intertidal flats, several kilometers wide, cover most of the area between the barrier islands and the mainland. The muddy tidal flats are mostly near the mainland. The sandy flats are flooded for four to six hours of each tidal cycle, and the fetch of several kilometers permits the commonly strong wind to generate modest-sized waves. This combination of strong wave action coupled with relatively weak tidal currents prevents tidal bedding from forming over much of the tidal flat environment. It also concentrates sand on the intertidal surface. In contrast, the tidal

(a)

(b)

Figure 11.18 Examples of ancient tidal bedding from (a) the Miocene of Florida, and (b) the Precambrian in Wisconsin.

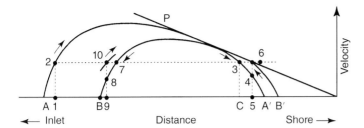

Figure 11.19 Diagram of settling lag and scour lag showing how sediment grains are transported up an intertidal flat, as proposed by H. Postma (1961).

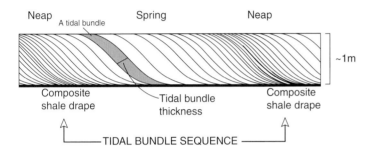

Figure 11.20 Schematic diagram showing the nature of tidal bundles as proposed by Visser (1980). (*Source:* Adapted from Visser 1980).

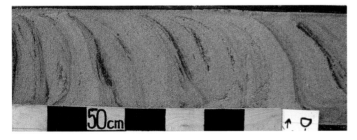

Figure 11.21 Vibracore of tidal bundle sequences taken from a channel on Martens Plate, East Frisian Wadden Sea, Germany.

channels that dissect the tidal flats display tidal bundling, showing that tidal currents are dominant in this environment.

11.6 Tidal Channels

Much like the flat upland plain environments of the U.S. Midwest or the high plains, tidal flats develop a drainage network that contains a range of small to relatively large channels. Unlike the upland environments, however,

these tidal channels carry water in two directions; the flood and ebb as the tide rises and falls. These channels are narrow and shallow in the upper reaches of the tidal flats where muddy sediments are more common, and extend to the lower elevations of the intertidal zone where they might be large and have floors in the subtidal zone.

Slight and subtle undulations in the tidal flat surface cause water to be concentrated in the low areas after emergence of the tidal flat surface begins during the ebb phase of the tidal cycle. These small channels increase in size as each tidal cycle passes until an equilibrium condition is established between the tidal flat/channels surface and the tidal flow. Where these channels are cutting through muddy and cohesive sediments, they tend to have relatively steep channel walls (Figure 11.22) but where they cut through sand sediments the channels are broad with more gently sloping walls.

The floors of these tidal channels are the sites of important sediment transport. This sub-environment is characterized by sand-sized sediment that lacks cohesion and is, therefore, susceptible to movement by both flood and ebb currents during each tidal cycle. As the tide floods over a tidal flat complex, the forced wave of the rising tide produces currents that may be strong enough to move the sand on the channel floor. Once the water level has risen above the level of the channel margin then conditions are essentially like those of a flooding river. Water and suspended sediment spill over the channel walls onto the tidal flats forming natural levees and causing tidal currents to slow.

During ebbing conditions of the tide there is a slow current as the water flows over the tidal flats in response to gravity. As soon as the tidal level is so low that part of the tidal flat surface becomes exposed, then there is some channeling of the rest of the ebbing waters. These waters are fed into the tidal channels in large volumes, causing the ebbing currents to be rapid. It is typical that flow in tidal channels, like most main channels in tidal inlets, is ebb-dominated.

The tidal currents that persist in these channels during the early parts of the flood cycle and the later parts of the ebb cycle, transport considerable sediment, and in doing so, they develop a spectrum of bedforms along the channel floor. The size of these bedforms is partly related to the grain size of the sediments but is mostly due to the flow strength of the currents. Most channels with sand floors display what are called megaripples or small subaqueous dunes. These are asymmetrical bedforms that have a wave length of about 1–5 m with a wave height of 20–50 cm (Figure 11.23). The asymmetry of these bedforms is the result of the direction of current flow such that we can tell the direction of the current that formed a particular group of bedforms by looking at them. Surveys of tidal channels show the nature of

Figure 11.22 Channel on tidal flat in The Wash on the North Sea side of England.

Figure 11.23 Bedforms about a meter or so in wave length from the Bay of Fundy, Nova Scotia, Canada.

these bedforms; their wave length and wave height as well as their orientation.

Most channels display bedforms that show that they were formed by ebbing tidal currents, further indication of the ebb-domination of the channels. In some instances these bedforms are modified by incoming flood currents but retain their ebb orientation. However, near-equal flood and ebb tidal currents lead to the bedforms reversing their orientation during each ebb and flood of the tides.

The migration of bedforms produces another feature that is a signature of tidal influence: the reactivation surface (Figure 11.24), a result of time–velocity asymmetry during the change of tides in a channel or on a tidal flat with medium to large bedforms. The dominant current moves the bedform while the recessive current only removes some of the sediment. The current then reverses and the bedform is again moved; it is reactivated. This set of conditions produces a contact that dips in the direction of the bedform movement but at a lower angle. Reactivation surfaces are among the best indicators of tidal environments in the stratigraphic record (Figure 11.25).

Another important but uncommon tidal signature is the presence of bidirectional cross-strata. This feature forms only when the flood and ebb portions of the tidal cycle are almost equal in tidal flux and energy. Such conditions are present in some parts of the tidal channels on the German coast (Figure 11.26). Because of their rapid burial and their occurrence in tidal channels, bidirectional cross-strata are commonly preserved in the stratigraphic record (Figure 11.27).

11.7 Some Examples

There are several places in the world where entire estuaries or large portions of estuaries are intertidal. As might be expected, these tend to be macrotidal or at least high in the mesotidal range. In this section we will take a look at some well-known examples in order to demonstrate the profound influence tides have on estuaries and on tidal flats.

11.7.1 German Wadden Sea and Jade Bay

The north coast of Germany includes some of the most studied tidal flat complexes in the world. This coast is a continuation of the north coast of the Netherlands and comprises

(a)

Water surface

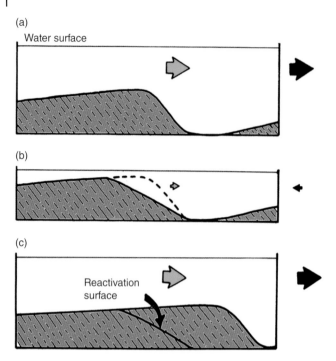

Figure 11.24 Diagram showing how reactivation surfaces form with dominant tidal flux followed by recessive tidal flux and then a return to the dominant one as tides change.

(b)

(c)

Reactivation surface

Figure 11.25 Photograph of ancient reactivation surfaces from the ancient stratigraphic record.

15 CM

short barrier islands with very large tidal inlets. The barrier system is separated from the mainland by the Wadden Sea at the southern end of the North Sea (Figure 11.28) near the apex of the German Bight where the coast ranges from mixed-energy to tide-dominated.

The Wadden Sea is essentially all intertidal flat except for subtidal channels that feed the tidal inlets (Figure 11.29). This extensive intertidal complex is subjected to spring tidal ranges of about 3 m. Drainage divides are located behind the middle of each of the barrier islands and represent locations of muddy sediments as compared to the intermediate areas where sand dominates. There is also a fining of grain size toward the mainland, where a fringe of marsh borders the extensive

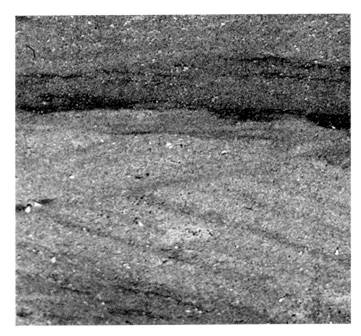

Figure 11.26 Photograph of a vibracore from a tidal channel in the German Wadden Sea showing the presence of bidirectional cross-strata.

Figure 11.27 Photograph of bidirectional cross-strata from the Baraboo Quartzite in the Precambrian of Wisconsin.

tidal flats. Wadden Sea tidal flats actually tend to be dominated by wave action instead of tidal currents as might be expected. This is because the barrier islands are several kilometers from the mainland meaning there is a fairly long fetch during high tide. Jade Bay (Figure 11.30a) is just around the corner from the Wadden Sea and is an enclosed tidal estuary that is largely intertidal. Here mud-dominated tidal flats (Figure 11.30b) extend for several kilometers due to the combination of the low relief and the spring tidal range of

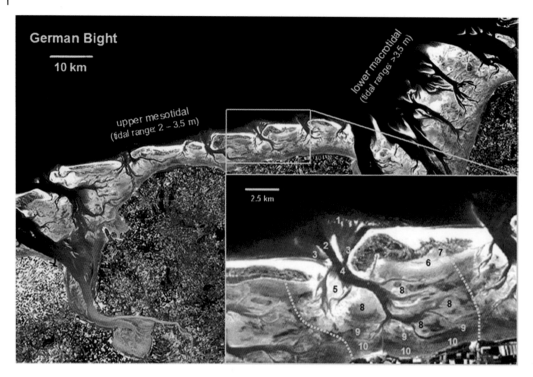

Figure 11.28 Satellite image of the German Wadden Sea barrier islands. (*Source:* B.W. Flemming).

Figure 11.29 Example of extensive tidal flats and intervening tidal channels on Martens Plate in the German Wadden Sea.

nearly 4 m. The widespread tidal rhythmites are soft, thinly laminated accumulations of fine sediments. This tidal estuary was the location of some of the first detailed investigations of tidal flat sedimentology, particularly the interactions of organisms with sediments. The many burrowing species provided a wide range of these relationships. This work took place at the Senckenberg Institute at Wilhelmshaven, Germany.

(a)

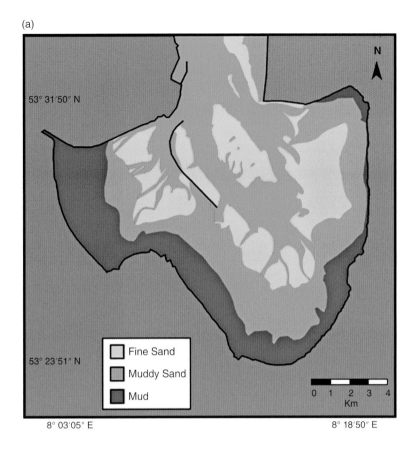

53° 31′50″ N

N

53° 23′51″ N

Fine Sand
Muddy Sand
Mud

0 1 2 3 4
Km

8° 03′05″ E

8° 18′50″ E

(b)

Figure 11.30 (a) General map of sediment distribution on Jade Bay (*Source:* Courtesy of H.E. Reineck), and (b) extensive muddy intertidal flats on Jade Bay in the German Wadden Sea.

11.7.2 The Wash

Although much of the coast of the British Isles has fairly high tidal range, the embayment known as The Wash on the northern Norfolk and southern Lincolnshire coast (Figure 11.31) has spring ranges of about 7 m. This primarily tidal estuary displays the typical transition from sand in the outer parts to mud in the landward fringes. Marshes dominated by *Salicornia* extend across much of the upper muddy levels; these are dissected by tidal channels (see Figure 11.22) that carry mud clasts eroded from the adjacent marsh areas.

As one proceeds in a seaward direction, there is an increase in the size of these tidal channels as they cut through the cohesive mud, and the bottom sediments become sandy. Eventually the entire intertidal system becomes sand-dominated and channels lose

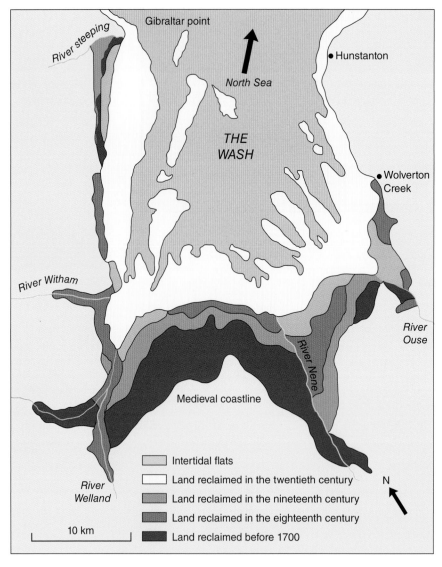

Figure 11.31 Map of The Wash, a large tide-dominated embayment on the Norfolk and Lincolnshire coast of England.

Figure 11.32 Aerial photograph of the abbey at Mont St.-Michel on the Bay of St.-Malo on the coast of France.

their definition. Here the sand is reworked and mobilized by the strong tidal currents associated with the macrotidal conditions of this tidal estuary.

11.7.3 Bay of St.-Malo

The Bay of St.-Malo is a large embayment along the northern Brittany coast of France. This tidal estuary is one of the two areas in the world where spring tidal range is near 15 m. The inner portion of this huge estuary, called the Bay of Mont St.-Michel after the well-known monastery and abbey that carries that name (Figure 11.32), is a major tourist attraction in northern France. Here, just down the coast from the area of the Normandy invasion in World War II, the tides are so large that detailed study of the lower part of the tidal flat system has not yet been completed. Only the upper, muddy areas are accessible to investigation because the intertidal flats extend for tens of kilometers (Figure 11.33).

Like most tidal estuaries there is a marsh fringe. Here it is dominated by succulent plants of the *Salicornia* type. The sediments under these plants and on the unvegetated tidal flats are typically muddy and show cyclic patterns of accumulation (Figure 11.34). These rhythmites display neap and spring cycles depending on the combination of the thickness of each lamination and the grain size of the sediments.

11.7.4 Bay of Fundy

The Bay of Fundy is an elongate coastal bay whose presence is the result of faulting in the lithosphere producing a structural basin called a graben. This bay is split into two tide-dominated, macrotidal basins at its landward end (Figure 11.35), the Minas Basin and Chignecto Bay. Here, huge tidal ranges prevail, with spring ranges above 15 m; the highest in the world. These basins are largely intertidal with shallow tidal channels. There is some fine sediment-discharge from the rivers on the landward ends of these two basins but most of the intertidal sediment is reworked from the basins themselves or from erosion of the shorelines.

Like in the other examples, sediments show the typical trend of fine sediment at the

Figure 11.33 Extensive tidal flats with a mud-dominated surface in the Bay of St.-Malo.

Figure 11.34 Shallow trench in Bay of St.-Malo sediments showing both tidal bedding and flaser bedding.

distal end of the basins with an increase in grain size toward the lower or open portion of the basins. Some steep slopes along the basin have marsh fringes that are only tens of meters wide or less. Most of this tide-dominated basin is covered with various bedforms that are developed in sand. These range in scale from small ripples, through megaripples, to sand waves of different wave lengths (Figure 11.36).

The less energetic portions of the Bay of Fundy are places where tidal bedding is deposited (Figure 11.37). Here mud dominates the accumulations and their cohesiveness keeps them fairly stable.

11.8 Human Impact on Tidal Flats

As with all coastal environments, the dense coastal population in much of the world has had negative impact on tidal flats. Fortunately in many places where tidal flats are extensive

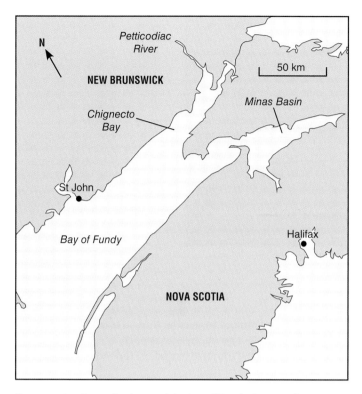

Figure 11.35 Generalized map of the Bay of Fundy showing the Minas Basin and Cobequid Bay.

Figure 11.36 Low tide photograph showing extensive and varied bedforms in Cobequid Bay in the Bay of Fundy.

Figure 11.37 Photograph showing tidal bedding in low-energy portion of the Bay of Fundy.

due to large tidal range the population is not so great. In addition, tidal flats are not very hospitable for humans. Large tidal flats also have some hazards. Flooding tidal cycles can cause problems for curious visitors. The coast along the Bay of St. Malo has sirens that go off as the tide floods to warn people to move to high ground. Turnigan Arm near Anchorage, Alaska has a tidal range in excess of 10 m and a researcher was drowned there because of failure to leave the area in time; the flooding tide came faster than the victim could escape the situation.

One of the most important interactions between humans and tidal flats has been along the North Sea coast of Europe. The extensive tidal flats there have been reclaimed for pasture land, providing farmers with additional land for sheep and cattle (Figure 11.38). Reclaiming is accomplished by digging a rectangular pattern of shallow ditches across the flats to enable sediment to be deposited. Eventually, the intertidal zone is reclaimed as uplands with vegetation. This not only destroys an important environment but it also causes problems on the adjacent barrier–inlet system (Figure 11.39). The new land causes a significant decrease in the tidal prism and therefore reduces flow through the tidal inlets. The inlet cross-sections are reduced and barriers are extended.

Some would say that such a result is a good thing but in time the inlets could be closed.

The above situation is a special case and the practice has been stopped in Europe. The other human impacts include significant dredging of tidal flats for a variety of purposes including ports and harbors. Many people view tidal flats as a worthless inconvenience and the result is their destruction. We are getting to the point where no significant construction is permitted on this environment and it is being protected almost everywhere.

11.9 Summary

The tidal flat environment appears at first to be one of little variation; just a flat surface over which the tide rises and falls with predictable regularity. In fact, there is great variation but it is gradual and subtle. Unlike many coastal environments that derive their sediment from land, tidal flats rely on waves and currents to provide the sediments that slowly accumulate. The cyclic nature of the processes, along with the relatively low-energy conditions, produce thin layers arranged in a predictable and recognizable fashion. Abundant burrowing organisms can destroy this layered record in many areas.

(a)

(b)

Figure 11.38 Examples of (a) shallow trenches to assist in reclamation of tidal flats, and (b) grassy area already reclaimed on the Wadden Sea intertidal flat lands.

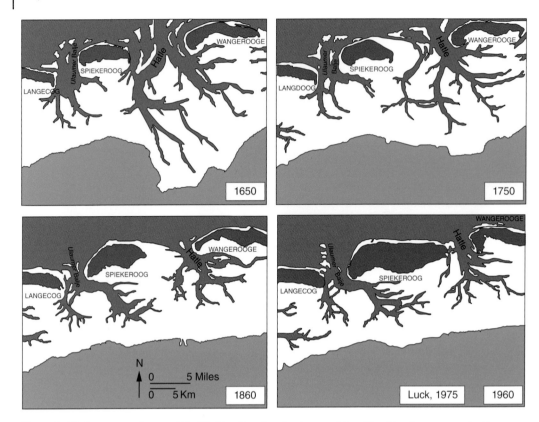

Figure 11.39 Sequence of maps of the Wadden Sea showing how reclaimed land has been converted to uplands (green) over centuries. (*Source:* FitzGerald et al. (1984)).

References

Dalrymple, R.W., Zaitlin, B.A., and Boyd, R. (1992). Estuarine Facies models: conceptual basis and stratigraphic implications. *J. Sedim. Petrol.* 62 (6): 1130–1146.

FitzGerald, D.M., Penland, S., and Nummedal, D. (1984). Control of inlet shape by sediment bypassing: East Frisian Islands, West Germany. *Mar. Geol.* 60: 355–376.

Klein, G.D. (1972). Determination of paleotidal range in clastic sedimentary rocks. Proc. 24th Intl. Geol. Congress, 6: 397–405.

Klein, G.D. (1977). *Clastic Tidal Facies.* Champaign, IL: Continuing Education Publishing.

Postma, H. (1961). Transportation and accumulation of suspended matter in the Dutch Wadden Sea. *Neth. J. Sea Res.* 1: 148–190.

Visser, M.J. (1980). Neap–spring cycles reflected in Holocene subtidal large-scale bedform deposits: preliminary note. *Geology* 8: 543–546.

Suggested Reading

Alexander, C.R., Davis, R.A., and Henry, V.J. (eds.) (1998). *Tidalites: Processes and Products*, Spec. Publication No. 61. Tulsa, OK: SEPM.

Amos, C.L. (1995). Siliciclastic tidal flats. In: *Geomorphology and Sedimentology of Estuaries*, Developments in Sedimentology No. 53 (ed. G.M.E. Perillo). Amsterdam: Elsevier.

Davis, R.A. and Dalrymple, R.W. (eds.) (2012). *Principles of Tidal Sedimentology*. New York: Springer.

DeBoer, P.L., van Gelder, A., and Nio, S.D. (eds.) (1988). *Tide-Influenced Sedimentary Environments and Facies*. Dordrecht: D. Reidel Publishing Company.

Eisma, D. (1998). *Intertidal Deposits; River Mouths, Tidal Flats and Coastal Lagoons*. Boca Raton, FL: CRC Press, (especially chapters 6–9).

Smith, D. G., Reinson, G. E., Zaitlin, B. A., and Rahmani, R. A., (eds.), 1991, Clastic tidal sedimentology. *Can. Soc. Petrol. Geol. Mem.* 16, Calgary, CSPG, 387 pp.

12

Coastal Wetlands

It is typical for a portion of the inner, protected margin of an estuary or a low-energy open coast to be covered by a vegetated intertidal environment. If covered with grasses or grass-like vegetation this environment is called a marsh. If covered with woody shrubs and trees, typically called mangroves, this environment is a swamp, or more properly, a *mangal*. These environments may be normal marine in salinity or they can range through brackish toward fresh water. This discussion will not include the freshwater marshes along the rivers that may grade into the estuary. The proportion of the estuary that supports the salt marsh environment ranges widely; from essentially all of the estuary except for tidal channels, to a border only a few meters wide. The proportion of the estuary that is covered by vegetation tends to be an indication of the maturity of the estuary or the degree to which it has been filled in with sediment. For example, some of the estuaries on the Georgia coast have little open water except near the inlet between the barrier islands. A similar situation exists in coastal southwest Florida, where mangroves dominate. Only tidal creeks dissect the extensive vegetated environment in these sedimentologically mature estuaries (Figure 12.1). By contrast, the German Wadden Sea is bordered by only a narrow marsh and the Bay of Fundy supports a narrow and discontinuous marsh environment where the surface gradients are steep (Figure 12.2).

Both salt marshes and mangrove mangals are special, vegetated intertidal environments and will be discussed in detail. Some comparisons will be made to demonstrate important differences between them.

12.1 Characteristics of a Coastal Marsh

A marsh is really the portion of the higher part of the intertidal environment that is covered by vascular plants. Above about neap high tide there is little energy to disturb the sediment substrate and the sediment that accumulates there tends to be relatively fine-grained with a fairly stable sediment surface. These factors provide the type of environment that supports vegetation; an undisturbed place of fine, organic-rich sediment. Various opportunistic and tolerant grasses thrive in this environment.

The marsh environment is commonly divided into the low marsh, which is approximately from neap high tide to mean high tide or slightly above, and the high marsh, which is from that level up to spring high tide.

12.1.1 Marsh Plants

There are two genera that are particularly prone to establish dense stands on such substrates; *Spartina* (Figure 12.3), which is present in two common species in North America, and *Juncus* (Figure 12.4), with one prominent species. Although not the only marsh taxa, these are the most widely distributed in North America.

Beaches and Coasts, Second Edition. Richard A. Davis, Jr. and Duncan M. FitzGerald.
© 2020 John Wiley & Sons Ltd. Published 2020 by John Wiley & Sons Ltd.

Figure 12.1 Aerial overview of a marsh showing only tidal creeks interrupting the marsh vegetation.

Figure 12.2 Narrow band of marsh vegetation along the margin of the Bay of Fundy in Canada.

Figure 12.3 Marsh on the Florida coast showing *Juncus*, the primary high-marsh genus in many North American marshes.

Figure 12.4 Channel margin along a tidal creek showing high growth form of *Spartina alterniflora*.

Figure 12.5 Outer margin of the low marsh (*Spartina*) with scattered oyster accumulations, a common situation in the southeastern United States.

The specific type of vegetation that develops marshes depends upon the elevation within the intertidal zone and the latitude, which in other words is a climatic control. In the middle and southern coasts of North America, *Spartina alterniflora* is the typical low-marsh grass (Figure 12.5), not because of the height of the plants but because of its substrate elevation. It is typically found between neap and spring high tide. In most estuaries this zone is a narrow range in elevation of a few tens of centimeters, but can be up to a meter or more in estuaries with very large tidal ranges. *S. alterniflora* is a coarse grass that grows in very dense populations. Individual plants are generally about knee-high but display great variability, reaching up to more than 2 m in height depending upon the specific location within the marsh and the availability of nutrients. Highest plants tend to be on the highest elevations; the levees of the channel margins and near spring high tide. The *S. alterniflora* plants at the lowest part of the marsh may be quite small and discontinuous.

The high marsh in some areas is dominated by *Spartina patens*. This species is generally fine and small in contrast to *S. alterniflora*. It grows best on the upper flat surface of the marsh environment. *Juncus* is the high-marsh grass in low to mid latitudes and is restricted to the elevation at about spring high tide. *Juncus roemerianus* is the species that is most common in southern North America; *Juncus gerardii is* most common north of Delaware and New Jersey. Commonly called the needle rush or black rush, it is as tall as a person, and it has a pointed end that has been known to penetrate shoes. This species attains its height throughout the extent of the spring tide position of the estuary margin. During the growing season it is a dark green color but attains a silvery hue during the fall and winter.

Other high-marsh plants include *Distichlis* and *Salicornia*. *Salicornia* (Figure 12.6), also called a salt wort, is a fleshy plant that rises only 10–30 cm above the substrate. It is the only common marsh plant that does not look like a grass, although *Juncus* is a rush not a grass. *Distichlis* looks very much like *S. patens* and they may occupy the same part of the marsh.

Relief on the marsh is typically low, but there are numerous subtle variations in elevation that cause distinct zonation of vegetation in the salt marsh environment because the plant species involved are quite susceptible to elevation differences. Quite subtle or local changes in relief or general morphology are reflected in the zonation of plant species and in their growth forms.

The low boundary of the marsh is the non-vegetated tidal flat or the margin of a tidal channel. The upper boundary can be a variety of environments but is typically characterized by some type of upland vegetation.

12.1.2 Global Distribution

The worldwide distribution of salt marshes can be organized into nine regions based on the vegetation communities.

Figure 12.6 Very high in the intertidal zone populated by *Salicornia*.

In the northern high latitudes is the Arctic Region that includes northern portions of North America and Russia along with Greenland and Iceland and northernmost Scandinavia. Here, marshes are fragmentary due to extreme weather conditions. Europe is divided into two regions: one along the north coast including the Baltic and the coast of Ireland and Great Britain and the other along the Mediterranean coast. Another region extends along the Atlantic coast of North America from northeastern Canada through the United States and including the Gulf Coast. It is this region that contains the most extensive marshes of North America and that will be given the most consideration in this chapter. The north and east coast of South America together comprise a region. The Pacific American region extends along the west coast of both North and South America. Australia and New Zealand comprise a region in the south Pacific. The eastern coast of Asia along with Japan completes the regions of the Pacific basin. The final region is really a special marsh environment that is restricted to high elevations in low-latitude areas that are otherwise dominated by mangroves. This situation exists in Florida and also parts of Baja California.

12.2 Marsh Characteristics

The physical environment of the marsh community is influenced by the degree to which it is protected from wave action, tidal regime, rate of sea-level rise, the topography of the coastal area, sediment supply and the nature of the substrate. The marsh environment is very similar to a river system: it is typically cut by meandering channels; the channels have point bars (Figure 12.7); there are natural levees along the channel banks (Figure 12.8); crevasse splays may form in breaches of the levees (Figure 12.9); and there may be meander cutoffs and oxbow lakes. In addition, the marsh surface tends to be extremely flat and horizontal, just like a floodplain.

Because the zonation of vegetation is so closely tied to the elevation within the intertidal zone, it is practical to zone the marsh in a similar fashion. The most commonly used approach is simply to subdivide the marsh into the low marsh and the high marsh. The low marsh is that part of the marsh from the beginning of vegetation up to at least mean high tide. This is generally dominated by *S. alterniflora*. The high marsh extends from about the mean high tide up to the limit of tidal activity. This portion of the marsh is

Figure 12.7 Photograph of a tidal creek meander showing a natural levee.

Figure 12.8 Aerial photo showing a tidal creek with multiple small point bars.

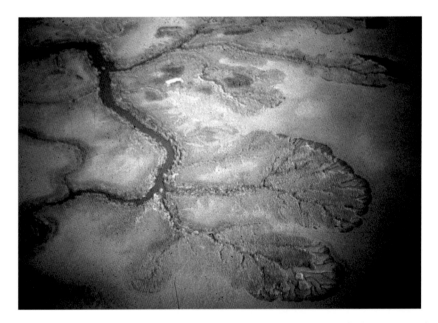

Figure 12.9 Aerial photo of a wetland on the Georgia coast showing multiple crevasse splays.

dominated by *Juncus roemerianus* and/or *Salicornia* depending on the overall setting.

There are, however, differences in marsh zonation and profiles depending upon the geographic location. For example, in New England (Figure 12.10) the lower marsh includes both *S. alterniflora* and *S. patens*, with the upper marsh being composed of *Salicornia*, *Distichlis* and a fringe of *Juncus*. An upland scrub forest typically borders the

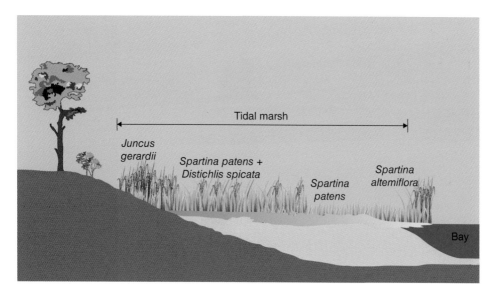

Figure 12.10 Diagram showing zonation of high-latitude marshes. (*Source:* Dawes (1998)).

marsh itself. Further to the south in Georgia and Florida, the typical zonation is a relatively narrow lower marsh of *S. alterniflora* and an extensive high marsh dominated by *J. roemerianus* (Figure 12.11).

12.2.1 Marsh Classification

A convenient way to consider marsh development is through their maturity. This can most easily be done by considering the relative distribution of the low- and high-marsh

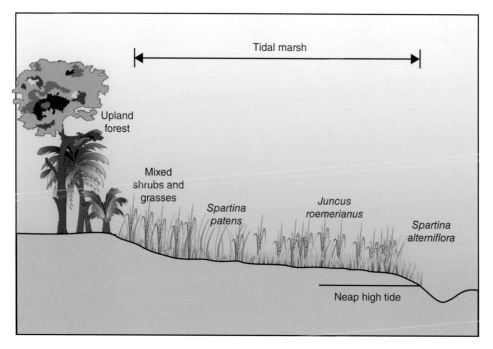

Figure 12.11 Diagram showing zonation of low- to medium-latitude marshes. (*Source:* Dawes (1998)).

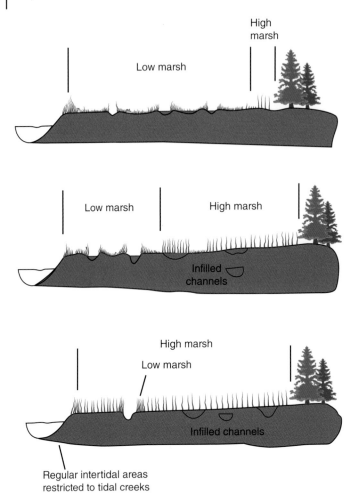

Figure 12.12 Diagrams showing the maturation of salt marshes from youthful to old age. (*Source:* Frey and Basan (1985)).

portions of the total vegetated environment (Figure 12.12). Without using absolute ages, we can consider young, mature and old marshes to reflect the progressive development of the marsh system assuming that sea level has not changed substantially.

A young marsh is one that has mostly low marsh vegetation, that is *S. alterniflora*, with perhaps only a fringe of high marsh, around the outer edge (Figure 12.12a). Tidal channels are abundant, providing good drainage, and sediment supply. This stage of marsh development lasts until sufficient sediment has been delivered to the upper intertidal area to support a significant upper marsh community.

The intermediate stage of marsh development (Figure 12.12b) has a near-equal distribution of high and low marsh. The tidal channels are fewer in number than in the young marsh. As the sediment continues to be delivered to the upper part of the intertidal zone, the marsh becomes more mature. Much of the intertidal zone is covered by marsh vegetation with only few large tidal creeks interrupting an otherwise continuous marsh environment. Continued sediment accumulation will cause encroachment of land plants into the marsh as the estuary is reduced in overall size.

The end product of this scheme of succession of marsh development is complete infilling of the intertidal zone up to the level of near spring high tide. The marsh is essentially all high marsh with only a fringe of low marsh, and tidal channels are widely spaced (Figure 12.12c). Because marshes are sediment sinks, this is their eventual fate unless sea level changes cause either enlargement of the estuary or abandonment at a high elevation. If this occurs, upland terrestrial vegetation will likely encroach into the highest part of the marsh.

12.3   Marsh Sedimentation

We have said that a marsh develops above the neap high tide level of the intertidal environment, and that as the estuary fills with sediment, the marsh increases its extent. In addition, there is an increase in the amount of high marsh as the marsh matures through sediment accumulation with time. How does the marsh grow in this manner and what are the mechanisms for delivering sediment to the marsh or potential marsh environment?

There are various ways for sediment to reach the marsh environment but two predominate. One is the settling lag–scour–lag mechanism (see Figure 11.19) for building up the tidal flat that was discussed in the previous chapter. In this manner the aggradation and progradation of the tidal flat will result in the sediment surface increasing in elevation thereby providing appropriate conditions for marsh vegetation to colonize the tidal flat. This building up of the tidal flat includes both sand being transported along the substrate as bedload, and mud transported in suspension. Each tidal cycle, especially those between mean tide range and spring tide range, brings sediment up to the level where marsh vegetation can become established. This type of accumulation encourages marsh expansion in response to sediment accumulation.

Once marsh vegetation has been established, the primary mode of sediment delivery is via suspended sediment. This sediment is typically mud and is provided both from normal high tide flooding of the marsh during near-spring conditions and also during storm conditions. Most estuaries have some mud in suspension during each tidal cycle. Each of the high tide phases of the tidal cycles provides a small to modest supply of mud to the marsh. The longer the slack-water period at high tide, the more sediment will settle out of suspension.

Storms provide the highest rate of sediment influx into the marsh environment. They do two primary things to help in this activity: the waves and currents generated during storms cause large amounts of fine sediment to be carried in suspension, and many storms create storm surge or storm tides in the estuaries where the marshes occur. As a consequence, there is a great deal of sediment made available to the marsh environment (Figure 12.13). This sediment is delivered in two primary ways: through the simple flooding of the marsh by sediment-laden water, and by breaching of the natural levees and deposition of a crevasse splay-type sediment deposit (see Figure 12.9). Both of these mechanisms provide considerable sediment to the marsh surface and both may produce enough sediment during a given storm to temporarily bury the marsh grass. These storm layers may be several tens of centimeters thick. Because marsh grass is very resilient, it will not die when buried but will grow up through the storm layer in weeks to months. This type of high sedimentation rate on the marsh surface results in the eventual elevation of the marsh above normal intertidal levels. The result is that the marsh environment disappears in favor of the upland environment.

Marsh vegetation tends to be quite dense and provides an excellent sediment trap in two ways. First, the grass slows the flow of tidal waters to assist settling out of fine suspended sediment particles to the floor of the marsh. Secondly, considerable amounts of

Figure 12.13 Infra-red aerial photograph of marsh area in South Carolina showing mud covering marshes after Hurricane Hugo in 1989.

fine sediment adhere to the marsh grasses as the sediment-laden water flows past. Both of these mechanisms provide for accumulation of generally muddy sediment on the marsh. Additional sediment accumulates on the marsh surface as the result of suspension feeders living within the marsh grass producing pellets that accumulate within the marsh and contribute to its aggradation. In high latitudes, ice can also be important in transporting sediment onto the marsh surface. This is very common in the New England area of the United States and along parts of the Wadden Sea on the North Sea coast of Europe.

12.3.1 Sediments

The general nature of salt marsh sediment is quite unlike that of other coastal environments except for the upper part of the intertidal flats. It is commonly an unequal mixture of mud and plant debris with small amounts of shell material, sand-size terrigenous particles, and large plant fragments. In general, marshes typically contain the finest sediments of all coastal environments. This is not always true, especially for those marshes

developed on washover deposits or flood tidal deltas associated with barriers; most of these are dominated by sand-size sediments. It is also possible for a particular area to have little mineral mud-sized sediments throughout, thereby making it impossible for mud to be a major component of marshes. The Florida peninsula falls into this category because marshes there form on sand-dominated substrates.

The coastal marsh accumulates a distinctive combination of sediment, structures, geometry and biogenic features. Although there is some nearly universal similarity among marsh deposits, there may be striking contrasts. Most marshes accumulate much plant debris and typically develop peat. Numerous benthic invertebrates may live within the marsh, particularly infaunal organisms such as various worms, burrowing crabs and snails.

As a consequence of all of these burrowing organisms, plus the effects of the roots of the marsh vegetation, many wetlands show considerable bioturbation in the substrate (Figure 12.14). There are, however, many that do accumulate well-bedded marsh sediments (Figure 12.15).

Figure 12.14 A highly bioturbated bank in a wetland tidal creek.

(a)

(b)

Figure 12.15 (a) and (b) Well-preserved bedding in marsh sediments on the German Wadden Sea coast.

12.3.2 Sea Level and Marsh Development

It should be apparent from the above discussions that the marsh environment is very delicately balanced with sea level. The entire marsh environment exists within much less than a meter of relief near high tide except in places with extremely high tidal ranges. The high-marsh environment lies within only about 10–15 cm of relief. As sediment accumulates on the marsh, the elevation can reach above spring high tide. But this situation is without considering sea level change; especially sea level rise.

In Chapter 4 we discussed the current situation regarding sea-level change and noted that globally, there is an annual rise of 3 mm. This is modest, but there are indications that this rate is increasing and there are many local and regional areas where it is much higher. If we consider the current eustatic rate, it means that a coastal salt marsh must accumulate 3 mm of sediment each year in order to maintain its current elevation relative to sea level. The desired situation is at least a balance between sea-level rise and sediment accumulation. In most coastal settings this is not a significant problem; such a balance exists. However, if predictions of increased rates of sea level rise come true, then we will have potential problems with marshes being drowned by the rise of sea level. There is considerable concern about this scenario becoming a major problem for marsh stability. Because marshes are among the most productive environments of all, this situation could cause major problems for the coastal ecosystem.

12.3.2.1 Mississippi Delta

Catastrophic conditions currently exist in the extensive marsh environment associated with the Mississippi River delta (Figures 12.16 and 12.17) on the coast of Louisiana. Recall from the chapters on sea level (Chapter 4) and deltas (Chapter 8) that this area is experiencing a relative sea level rise of almost 1 cm each year. While sea level rise is not a major problem along many coastal environments, it is a very big problem for a marsh. Remember, most of the marsh exists within a very small range in elevation. On the Mississippi Delta, an area of less than a meter spring tidal range, it is only about 10–15 cm.

As a consequence, a sea level rise of nearly a centimeter may cause much of the marsh to

Figure 12.16 Drowning wetlands on the Mississippi Delta.

Figure 12.17 Wetlands on the Mississippi Delta that are in danger of destruction from both sea-level rise and human activities.

be drowned (see Figures 12.14 and 12.17). If the rate of sediment influx amounts to a centimeter per year, then there is a balance between the rate of sea level rise and the rate of marsh accretion. In the case of the Mississippi Delta area, human interference with the discharge of the river coupled with the compaction of delta sediments and withdrawal of fluids under the delta as a result of drilling for oi and gas has contributed significantly to the high rate of relative sea level rise. As a consequence, the delta is subsiding and the amount of sediment supply from flooding of the river has been greatly reduced. The bottom line is that much of the coast of Louisiana is drowning. The state is currently losing about 65 km^2 each year of coastal salt marshes to drowning; a football field an hour.

12.4 Human Impact on the Marsh Environment

The fragile nature of the marsh makes it susceptible to negative impacts from human activities (see Box 12.1). In the case of the Mississippi Delta there are a few things that can be done to help to minimize the effects of previous human activities. One of the major aims of human intervention on the delta is to prevent flooding in order to protect industry, residences, and infrastructure. This is accomplished primarily by the construction of levees designed to eliminate flooding during high-discharge events. If gates were constructed at intervals along the major distributaries and opened during high-discharge events major splays would develop which would provide much needed sediment to the wetlands of the delta.

One practice that has helped to develop new wetlands in this area is the diversion of significant discharge into the Atchafalaya Basin just to the west of the active delta. Over the past several decades this practice has developed a new delta of significant size (Figure 12.18). The sediment complex now supports extensive wetlands.

12.5 Marsh Summary

Although marshes are very diverse in their characteristics and their dominant vegetation, they have many common factors. There

Box 12.1 Human Impact on Coastal Marshes

There are many ways in which we cause degradation of marshes, especially through development of population centers and also by exploration for oil and gas. The net result is deterioration and or loss of marshes; one of the most productive of all coastal environments.

The portion of the tidal range in which marshes can develop and continue to flourish is small; only from near-neap high tide to spring high tide. There are some exceptions to this generality but it applies to most marshes. In microtidal and mesotidal coasts this range is much less than a meter. Marshes are limited to a very narrow part of the intertidal environment. As sea level rises, especially if the rate increases, as seems to be the case, marshes fall into great jeopardy.

The only way that marshes can be sustained in this sea-level scenario is if the marsh surface can accumulate the same thickness annually as the amount of sea-level increase in that year. Currently, on a global basis that means the marsh surfaces need to accumulate 3 mm of mud to sustain their environment. If we think about conditions of suspended sediment supply around the world, this is not likely to happen in many places. Suspended sediment reaches coastal marshes via discharge from the mouths of rivers. The big rivers tend to be the most important because of their huge discharge of both water and sediment.

Presently, the amount of discharge, especially sediment-discharge has been significantly decreased, mostly by anthropogenic practices (Box Figure 12.1.1). Commercial navigation is

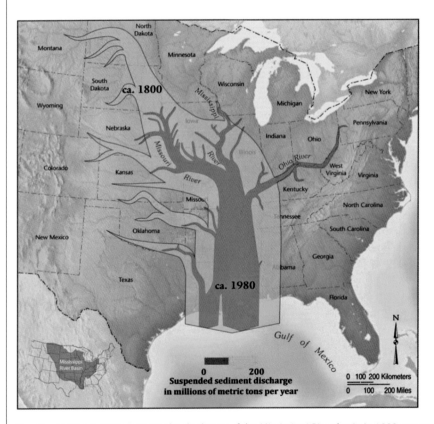

Box Figure 12.1.1 Map showing the discharge of the Mississippi River basin in 1800 as compared to 1980. (*Source:* USGS).

a major element in many river systems. As a result dams are important and very numerous. A dam prevents large volumes of sediment from moving down the river to the discharge point. Many river drain basins where agriculture is a huge part of the economy. A large portion of these basin require irrigation for crops. This decreases the discharge at the river mouth. A good example is the Colorado River that moves through much of the west and irrigates millions of acres and is also a water source for many municipalities. As a consequence there is zero discharge at its mouth in the Gulf of California.

Construction activities are also a major negative factor in the preservation of the marsh environment. Levees are common on all rivers that have problems with flooding (Box Figure 12.1.2). This is particularly important at locations where a river passes a population center (Box Figure 12.1.3). Levees are also built where canals are dredged for benefit of the

Box Figure 12.1.2 Natural levees on the lower Mississippi Delta.

Box Figure 12.1.3 Constructed walkway by the U. S. Army, Corps of Engineers.

Box Figure 12.1.4 Wetland surface of the Mississippi Delta with numerous canals and levees produced by the petroleum industry. (*Source:* Univ. Vermont).

petroleum industry (Box Figure 12.1.4). The job of a levee is to prevent flooding. Unfortunately the marsh cannot be sustained unless flooding takes place for it is floods that provide suspended sediment to nourish the marsh and keep it on pace with sea-level rise.

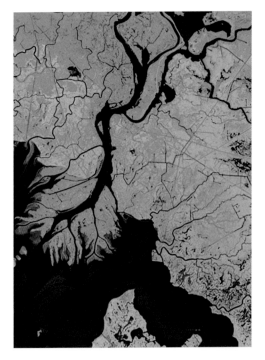

Figure 12.18 Infrared image of the Atchafalaya Basin showing the recently formed deltas that are prime locations for development of new wetlands. (*Source:* earthobservatory.nasa.gov).

are some generalizations that can be made about marshes. Most of these are related to their position along the intertidal zone. It cannot be stressed enough how important the elevation is within the marsh portion of the intertidal zone.

Marshes of all types are among the most important and most productive of all modern environments. They have high concentrations of photosynthetic organisms and they serve as a nursery ground for many finfish and shell fish. Because of their delicate position within the intertidal zone, their existence is threatened both by human activity and by sea level rise.

12.6 Mangrove Coasts

Stands of mangroves, called mangals, are tidal forest ecosystems that exist in protected marine through brackish water to freshwater conditions, as long as there is some tidal influence. Although there are various environmental conditions that influence the nature and extent of mangrove development, the most critical is air temperature; most mangroves cannot tolerate a hard freeze, which limits them to lower latitudes. Mangrove mangals are commonly considered as the low-latitude equivalent of coastal marshes. This comparison is not strictly correct in that there are two distinct differences between the two environments: marshes are populated by grasses whereas mangrove mangals are dominated by trees and shrubs, and mangroves occupy different positions within the intertidal zone than do marshes. As mentioned in the previous section, there are two areas in North America where mangroves and salt marsh vegetation occur together: much of the Gulf of Mexico and parts of the Baja California coast. In the Gulf, this is primarily because of temperature. The distribution of mangroves here has expanded greatly since the latter part of the twentieth century. In the 1960s mangroves were essentially absent on the northern coast of the Gulf; now they cover the entire area

with one species. The water temperature is now high enough along the shoreline for mangroves to survive. This extension of their distribution is attributed to climate change.

In this discussion we will consider how mangroves are distributed, both globally and within specific coastal systems. The zonation of mangroves and their influence on coastal processes, especially sediment transport and stability, will also be covered.

12.7 Mangrove Distribution

12.7.1 Global Distribution

More than 80 species of mangroves are recognized globally. The vast majority of these species flourish in Southeast Asia and Oceania. Their inability to tolerate hard freezes means that their global distribution is controlled by winter temperature. The Indo-Pacific Zone contains tremendous variety of mangrove taxa, whereas the Atlantic Zone includes only ten species. In the United States, only Florida, the Gulf Coast, and a small area of southern California are home to mangroves.

12.7.2 Local

Mangroves are restricted to protected waters where currents are sluggish and waves are small. This is typically associated with rather low-energy estuaries, lagoons and backbarrier environments. The primary factor in this distribution is the nature of mangrove propagation. Their seeds drop from the trees and float with the currents until they come to rest at the shoreline, where the propagules root and develop into seedlings (Figure 12.19). The long red mangrove seeds shown in the photo float in a vertical position until they run aground in about a decimeter of water. For germination to take place the seeds must maintain their position for some time; swift currents and wave action would prevent this.

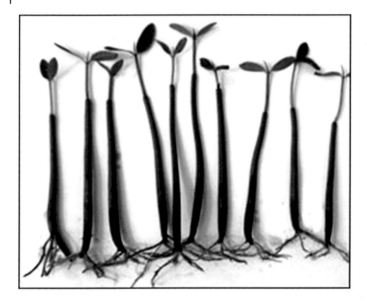

Figure 12.19 Photograph showing red mangrove seedlings that have germinated, with small leaves and rootlets.

(a)   (b)

Figure 12.20 (a) An example of the red mangrove in its natural environment on the coast of Mexico, and (b) a close-up photograph showing the nature of the prop roots of this species. (*Source:* (a) and (b) courtesy of J.W. Tunnell).

12.7.3 Zonation

In North America there are only four species of what are termed mangroves. There is a zonation of the prominent mangrove species that is related to their position within the intertidal zone, in a fashion similar to that of the grasses within the marsh environment.

The most seaward species is the red mangrove, *Rhizophora mangle* (Figure 12.20), identifiable by its large prop roots that anchor and stabilize the plant. This species commonly extends to below the low tide mark within the low part of the intertidal zone. Above this elevation but intermixed to some extent, is the black mangrove, *Avicennia*

(a)

(b)

Figure 12.21 (a) Example of the black mangrove showing its pneumatophores in its natural environment, and (b) a close-up of this species when it is in flower.

Figure 12.22 White mangrove along the coast of Florida.

germinans (Figure 12.21). This species inhabits the intertidal zone. The third typical mangrove of North American mangals is *Laguncularia racemosa* (Figure 12.22), the white mangrove, which inhabits the highest part of the intertidal zone and may extend up to the supratidal area. The fourth mangrove in North Americal is *Conocarpus erectus*, known as the buttonwood tree (Figure 12.23). Many people would not include this tree in

Figure 12.23 Photograph of a buttonwood tree on an upland environment in Florida.

the true mangrove community because it grows exclusively above the intertidal zone. The zonation across the intertidal zone in Florida is basically in this same order, with the red mangrove being lowest and the white the highest, just above spring tide (Figure 12.24). Unlike marshes, it is common for mangrove species to be somewhat intermixed; there is not a sharp boundary between species.

There are various types of mangrove environments, ranging from fringing to overwash locations in the intertidal zone. The fringing types occur along the protected open coasts of estuaries (Figure 12.25) and other coastal bays. Overwash mangals develop where the low-lying overwash deposits accumulate (Figure 12.26). Some mangroves even grow in bedrock (Figure 12.27).

12.8 Mangroves and Coastal Processes

Mangroves have some influence on coastal processes because of their prominent size and dense network of root structures.

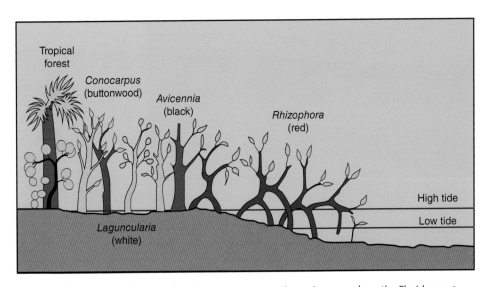

Figure 12.24 Schematic diagram showing mangrove zonation as it occurs along the Florida coast. (*Source:* Dawes (1998)).

Figure 12.25 Oblique aerial photograph of a mangrove mangal of reds on the southwest coast of Florida.

Figure 12.26 Photograph showing denuded mangroves on a washover environment on the Florida coast after Hurricane Andrew.

Although most of their influence centers on physical processes, some biological processes may also be involved. Some mangrove species, especially those of the genus *Rhizophora*, have significant influence on currents. The primary reason for this influence is the presence of the numerous, closely spaced, and resistant root structures that characterize nearly all mangrove species. In sites where open-water currents may be as high as

Figure 12.27 Photograph of red mangroves growing out of basalt bedrock in the Galapagos Islands.

100 cm s^{-1}, the currents within the dense mangrove root system may be as slow as 10 % of the open water.

As the tide floods and ebbs, the prop roots and pneumatophores (aerial roots) along with substantial burrowing structures at the sediment surface, produce major increases in roughness and friction. This generates a significant decrease in the flow velocity of the tidal currents and thereby greatly affects sediment transport and accumulation. Additional roughness is caused by the algae, barnacles, oysters and other organisms that may be growing on the root structures. All of these factors have an effect on waves as well. They tend to attenuate wave energy and thereby diminish the role of waves in erosion of the mangrove substrate.

Mangroves also have a significant influence on the effect of storms along coastal environments. Because of their location in low-latitude regions, tropical storms and hurricanes are likely to impinge on mangrove coasts. These storms bring intense wind, large waves, and storm surge. Mangroves are able to withstand these forces very well. The relatively low trees with very dense root systems are adapted to resist such intense conditions. They also

help to protect the sediment substrate in the mangal.

A good example of this situation is the passage of Hurricane Andrew across south Florida in August, 1992. The tremendous destruction that took place in the Miami area is well known, but few people are aware of what happened on the other side of Florida where mangrove mangals dominate the coastal zone. Here, along this very low-energy coast where mean wave height is only about 15 cm mangroves extend essentially to the open coast. Large waves combined with a storm surge of 1.5–2.0 m and wind of about 150 km h^{-1} would be expected to inflict major erosion on this undeveloped coast. Instead, the shoreline change was minimal; mangrove trees were broken off by the wind but their dense root systems prevented erosion (see Figure 12.27). This is an excellent demonstration of how mangroves are adapted to withstand intense storms and prevent erosion of the coastline (Figure 12.28).

Another aspect of the sediment trap effect of mangroves takes place primarily in the prop roots of the red mangroves. These structures are commonly at least a few centimeters in diameter. As such they physically block suspended sediment that adheres to

Figure 12.28 Photograph of an intertidal oyster reef in southwest Florida with red mangrove growing next to it.

their roots. This phenomenon can be important in muddy estuaries, especially those where tidal range is high.

12.9 Human Impact on Mangroves

Some people do not like mangroves. These are folks who live on the coast and have mangals between them and the water, obstructing their ability to see the shoreline and the water beyond. Development of these coastal areas has resulted in the complete destruction of mangroves locally. This is especially in places where dredge-and-fill construction has taken place; mostly in Florida. These finger canals and adjacent upland construction sites have destroyed very many mangroves (Figure 12.29).

Another detrimental anthropogenic activity has recently been permitted for mangroves: the State of Florida now allows mangroves to be trimmed to permit residents to see the shoreline and water. These trimming regulations cover a range of situations. Permits are required but the permission does not negate the problems for the trees. One of the regulations is that mangroves can be cut down to 6 ft

in height (Figure 12.30a). Another is that large and densely vegetated trees can be trimmed to as to provide windows for viewing the water (Figure 12.30b). We do not think such practices are proper environmental management.

12.10 Summary

The presence of dense vegetation on the intertidal zone is a distinctive feature of one of the most important of all coastal environments. These diverse and highly productive environments are tremendously productive in the form of photosynthesis and as a food supply to many types of herbivores. They also provide a home and a place for reproduction for many organisms.

Another key feature of these environments is their role in coastal protection, both in stabilizing sediment substrates and slowing erosion by waves and currents. This is especially the case for mangrove mangals, which can withstand direct attack from hurricanes and experience limited erosion.

These vegetated environments are also important sediment traps and substrate stabilizers.

Figure 12.29 Aerial photograph of what was the back-island wetland and is now dredge and fill construction for residences on the Florida coast.

(a)

(b)

Figure 12.30 Mangroves in Florida can now be trimmed a) to a 6-ft level or b) as "windows" in tall trees, all to let home-owners see the water. (*Source:* (a) and (b) courtesy of Florida Department of Environmental Protection).

References

Dawes, C.J. (1998). *Marine Botany*. New York: John Wiley and Son.

Frey, R.W. and Basan, P. (1985). Coastal salt marshes. In: *Coastal Sedimentary Environments*, 2e (ed. R.A. Davis), 225–302. New York: Springer.

Suggested Reading

Chapman, V.J. (ed.) (1977). *Wet Coastal Ecosystems*. Amsterdam: Elsevier.

Perillo, G.M.E., Wolanski, E., Cahoon, D.B., and Brinson, M.M. (2009). *Coastal Wetlands: An Integrated Ecosystem Approach*. Amsterdam: Elsevier.

Scott, D.B., Frail-Gauthier, J., and Mudie, P.J. (2014). *Coastal Wetlands of the World*. Cambridge: Cambridge University Press.

Tiner, R.W. (2013). *Tidal Wetland Primer*. Amherst: University of Massachusetts Press.

13

Beach and Nearshore Environment

The beach and nearshore environment as defined here will extend from the bar and trough topography that characterizes the surf zone across the dry beach to the vegetation line. It is probably the most active environment on the earth's surface. It is continuously in motion; sometimes under high-energy conditions and sometimes not. The beach is a thin strip of land that is typically only tens to a few hundreds of meters in width. It is a wave-dominated environment.

The discussion here will be divided into three parts: bar and trough topography, intertidal beach and dry beach. Each varies in space and time but within limits. Tropical storms and hurricanes cause extreme changes over short periods but, typically, recovery of the natural system takes place to a large extent.

13.1 Nearshore Environment

This is the area just offshore of the beach where the surf zone develops. It is characterized by a bar and trough topography. The gradient of the bottom, the level of wave energy and the sediment available control the number and position of the longshore sand bars that parallel the beach. Under most conditions there are two or three such bars (Figure 13.1). Note that the diagram also introduces additional terminology. The breaker zone is where waves break; typically over longshore bars (Figure 13.2). The surf zone is generally

from the outer bar to the shoreline unless quiet conditions prevail, in which case waves break only on the inner bar or not at all. The foreshore is essentially equivalent to the intertidal beach. The backbeach is the dry portion of the beach. It extends from the high tide line to the vegetation zone, where coppice mounds develop, and is typically essentially horizontal.

Where there is a steep gradient offshore of the shoreline a sand bar may not develop (Figure 13.3). Portions of the west coast of the United States down through most of South America, where there is essentially no continental shelf, display this feature. But in most coastal plain nearshore zones there are two or three longshore bars separated by shallow troughs (see Figure 13.2).

Longshore bars can take on multiple configurations; not just straight and shore-parallel (Figure 13.4). Shore-parallel bars commonly have saddles with a lower bar crest, as shown in box A. There is also a bar configuration where a part of the bar is attached to the shoreline, generally in a rhythmic pattern (B). There are also bars that are oblique and attached at one end to the shoreline (C). The fourth type of bar is one known as transverse, which extends essentially perpendicular to the shoreline. These bar types were recognized by the late W.F. Tanner at Florida State University while studying the coast of the Florida panhandle.

There is another important feature of shore-parallel sand bars: rip currents move offshore over paths of least resistance. The

Beaches and Coasts, Second Edition. Richard A. Davis, Jr. and Duncan M. FitzGerald.
© 2020 John Wiley & Sons Ltd. Published 2020 by John Wiley & Sons Ltd.

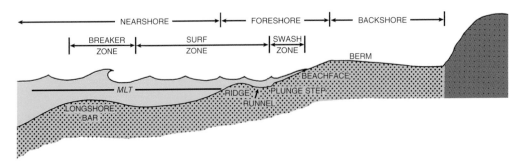

Figure 13.1 Diagram across the beach and nearshore environment to the first longshore bar, showing elements of this system.

Figure 13.2 Oblique aerial photograph of the beach and surf zone along Padre Island, Texas showing waves breaking over two longshore bars.

Figure 13.3 Aerial photograph of an area on the Pacific coast of South America where the nearshore zone has a steep gradient and longshore bars do not develop. (*Source:* Courtesy of O.H. Pilkey).

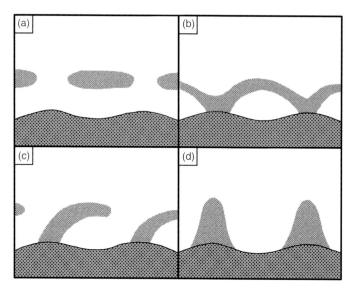

Figure 13.4 Diagrams of four different types of nearshore sand bars. After the work of W.F. Tanner.

Figure 13.5 Aerial photograph of the Oregon coast showing deeply incised rip channels in the intertidal zone. (*Source:* Courtesy of W.T. Fox).

question arises as to whether the saddles in longshore bars are eroded by rip currents or whether the rip currents flow there due to the saddle being the path of least resistance. A good example of currents creating saddles can be seen on the coast of Oregon. Here there are relatively deep and well-defined rip channels (Figure 13.5) that were probably mostly the result of erosion. Other rip channels can be formed at the shoreline and are oriented perpendicular to the coast (Figure 13.6). These can develop on low-energy coasts, such as Florida, or high-energy coasts with a steep offshore, such as Australia.

(a)

(b)

Figure 13.6 Rip channels that are essentially perpendicular from shore on (a) the Florida Gulf coast, and (b) the New South Wales coast of Australia.

Although rip currents excavate little or no sand there is considerable evidence that they do transport sediment seaward. This is clearly demonstrated in aerial photographs (Figure 13.7). The overall contribution to the sediment budget in the surf zone is small.

Substrate

The nearshore surf zone is virtually all well-sorted sand with some scattered shell debris.

Upper-flow regime conditions are generally present on the crests of the bars with plane beds but in the troughs. wave-generated combined-flow ripples prevail (Figure 13.8). The waves and combined-flow currents cause changes in the surface features across this shallow marine system (Figure 13.9). Few animals live on this surface or beneath it, though sand dollars are common just below the sediment interface of the sand bars in

Figure 13.7 Evidence of seaward sediment transport by a rip current on the coast of Lake Michigan.

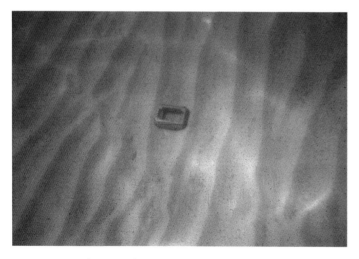

Figure 13.8 Underwater photograph of symmetrical wave-generated ripples in the nearshore zone.

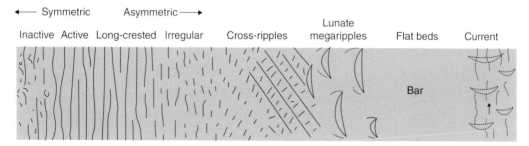

Figure 13.9 Diagrams showing the changes in bedforms as waves and currents move across the nearshore zone without longshore bars. As interpreted by H.E. Clifton and colleagues.

many areas, and sting rays may occur in the troughs when wave energy is low. Surf fishing in the troughs between sand bars is generally good.

13.2 Foreshore

The intertidal beach is commonly called the foreshore. There is a very wide range of width depending on tidal range and slope of the beach face. It can be a little as a few meters (Figure 13.10a) or a wide as hundreds of meters (Figure 13.10b). Part of this range is due to the magnitude of the tide at the location in question, part is due to erosion versus progradation and part is due to the availability of sand.

Ridge and Runnel
Beaches erode during storms. What happens to that sediment that is lost from the beach? There are four places to which eroded sand is transported: offshore beyond the surf zone, or what engineers call the depth of closure – the depth beyond that where waves can move sediment in a shoreward direction; alongshore – longshore currents can exceed a meter per second during storm conditions and considerable sediment is carried by these currents; the surf zone adjacent to the beach; and, actually, carried across the beach and deposited in a washover fan.

The shallow transport of sediment removed from the beach accumulates in intertidal depths to form what is called a ridge and runnel, sometimes referred to as a swash bar (Figure 13.11). The sediment accumulation, the ridge, is relatively wide and typically rises only a few tens of centimeters above the adjacent sand bottom. In some places, especially where tides are at least mesotidal, this sediment accumulation is completely in the intertidal zone (Figure 13.12). Regardless of the circumstances of the location of the sand ridge and adjacent runnel its shoreward migration is similar.

As the relatively small waves between storms move across the shallow nearshore zone, they generate shoreward-directed currents that move sediment. They cause the sand ridge to become asymmetrical, looking much like a large bedform. This large "bedform" eventually becomes intertidal. Under this circumstance the rising tide flows over the ridge causing shoreward migration (Figure 13.13). As the sand bar migrates landward it develops cross-stratification of the same scale as the thickness of the migrating sand body. The runnel surface is covered with ladderback ripples formed by a combination of wave-generated ripples and perpendicular current-generated ripples (Figure 13.14). Sediment is carried across the surface of the ridge and then cascades down the slipface to form cross-strata (Figure 13.15). As time passes, this ridge moves on to a storm beach to repair it by moving the previously eroded sand back to the beach (Figure 13.16). This is not the same sand but it is the same or nearly the same volume. These migrating ridges are much like the longshore sand bars except that they are intertidal. They commonly have rip channels to allow water that has made its way across the bar to return to a stable level (Figure 13.17).

This migration of the ridge and the welding to the shoreline will provide a sequence that will then be reworked by foreshore processes as the beach progrades. There is a stratigraphic sequence that we would expect to see if it is allowed to remain without reworking by foreshore processes. It begins at the base with a storm beach. This beach commonly displays relatively high seaward tips and lag deposits of heavy minerals (Figure 13.18). The runnel and ridge strata overlie the storm beach in this thin transgressive sequence. In reality, this complete sequence is rarely preserved due to reworking as the foreshore is molded by swash processes. We can, however, see cycles of erosion and accumulation in the stratigraphy of the foreshore beach with the heavy mineral storm beach strata at the base (Figure 13.19). This photograph shows two erosion periods with the heavy minerals and welded sand in between.

(a)

(b)

Figure 13.10 Photos of (a) a narrow foreshore on the Alabama coast and (b) a very wide foreshore (intertidal beach) on the Oregon coast.

(a)

(b)

Figure 13.11 (a) An intertidal ridge and runnel on the Georgia coast, and (b) a huge supratidal ridge on the Florida coast.

Foreshore

The foreshore is the wet beach where surging waves create back and forth swash. In fact such water motion creates an important sedimentary structure called swash marks (Figure 13.20). Another fairly common surface structure on the foreshore beach is the antidune (Figure 13.21). Antidunes tend to develop on gently sloping beaches with fine sand. They are formed by returning swash but the asymmetry of the bedform and the dip of the internal strata are in the upshore direction (Figure 13.22). They are an upper-flow regime structure.

Biota

There are two small animals that are common in the swash zone. Both are infaunal and

Figure 13.12 Photograph of a ridge and runnel complex where the ridge is actively migrating shoreward.

Figure 13.13 Close-up photograph of a migrating ridge, showing the shoreward-oriented cross-strata.

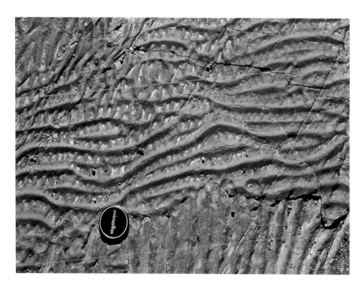

Figure 13.14 Ladderback ripples that characterize the runnel environment over which the ridge migrates.

Figure 13.15 Photograph of a ridge that has migrated up to the shoreline.

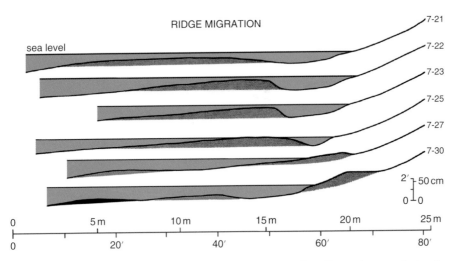

RIDGE MIGRATION

sea level

7-21
7-22
7-23
7-25
7-27
7-30

2′ ⊤ 50 cm
0 ⊥ 0

0   5 m   10 m   15 m   20 m   25 m

0   20′   40′   60′   80′

Figure 13.16 Diagram showing a time-series of ridge and runnel profiles as shoreward migration took place in only 10 days.

Figure 13.17 Oblique aerial photograph of a migrating ridge with a rip channel.

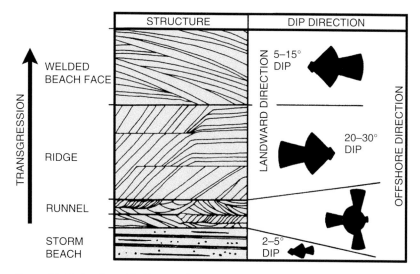

Figure 13.18 Stratigraphic diagram showing the theoretical sequence of stacked units of a ridge and runnel system that has migrated to an eroded beach.

Figure 13.19 Photograph of a shallow trench in beach deposits showing heavy mineral layers that represent the storm beach surface.

both are abundant in some locations. The small, multi-colored surf clam, *Donax variabilis* (Figure 13.23), is as dense as hundreds per square meter in some places. The other common creature is the ghost shrimp, *Callianassa major* (Figure 13.24). It burrows fairly deep and lines its burrows with aggregates of mud. It expels its fecal pellets to the surface. Another common beach organism that actually floats in from the open sea is the brown algae *Sargassum*. It can come to the coast in huge quantities (Figure 13.25) and cause problems for tourism. This algal complex contains a zoo of various small animals. Another floater that commonly makes its way to the beach is the Portuguese man o'

Figure 13.20 Photograph of a foreshore showing swash marks on the coast of Lake Michigan.

Figure 13.21 Photograph of a beach on Jekyll Island, Georgia showing a pavement of antidunes.

Figure 13.22 Close-up of a low trench showing the low angle cross-stratification of the antidunes.

Figure 13.23 Photograph of surf clams, *Donax variabilis*, that are very common on many foreshore beach environments.

(a)

(b)

Callianassa pellets

Figure 13.24 Close-up of (a) a ghost shrimp *Callianassa*, and (b) pellets produced by ghost shrimp.

Figure 13.25 Beach on the Texas coast onto which considerable *Sargassum* has come to rest.

Figure 13.26 Portuguese man o' war (*Physalia*) that has washed on to the foreshore beach.

war (*Physalia*) (Figure 13.26). It has dangerous tentacles that contain poison; even dead on the beach it can put stingers in your feet.

The sediments of the foreshore environment are varied in both grain size and composition. Grain size is related to both wave climate and availability, and ranges from fine sand (Figure 13.27) to very coarse gravel (Figure 13.28). Some beaches are a combination of shells and shell debris with sand, which produces a texture called bimodal; essentially a combination of two populations (Figure 13.29). Thanks to the generally high-energy wave action, beach sediment tends to be well-sorted and well-rounded (Figure 13.30). The composition of beach sand is mostly quartz because of its physical and chemical durability. Most beach sediment contains a small proportion of heavy mineral grains, generally only a percent or so. Erosion of the foreshore during storm conditions removes quartz and leaves the heavy minerals as a lag deposit. These lag deposits of heavy minerals accumulate as a thin layer across the erosional beach (Figure 13.31).

Figure 13.27 Fine sand, quartzitic sand beach on the Florida Gulf Coast.

Figure 13.28 Boulder beach on the west end of the Bay of Fundy, Nova Scotia, Canada.

(a)

(b)

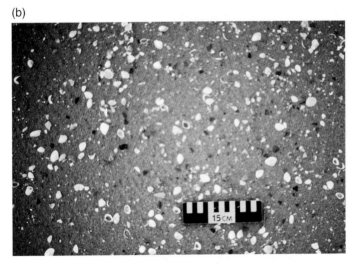

Figure 13.29 Photograph of a bimodal beach on the Atlantic coast of Florida with (a) shell debris on the sand, and (b) a close-up of shells on a fine sand beach.

There is a general relationship between the gradient of the foreshore and the grain size. In general, coarse sediment produces a steep foreshore (Figure 13.32). There are exceptions to this relationship, notably shell beaches and gravel beaches.

13.3 Backbeach

The backbeach, the generally dry part of the beach environment, is essentially horizontal. It ranges widely in width depending on whether the beach is erosional or prograding.

(a)

(b)

Figure 13.30 Close-up photograph of (a) well-sorted and well-rounded quartz beach sand, and (b) sorted and rounded gravel from a beach.

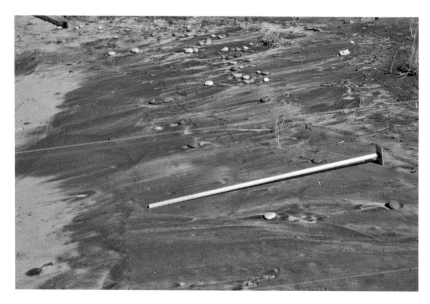

Figure 13.31 Storm beach surface covered with heavy minerals.

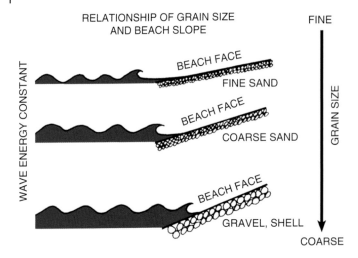

Figure 13.32 Diagram showing three different situations of foreshore slope relative to sediment grain size. (*Source:* courtesy of Joe Holemes, RPI).

Figure 13.33 Photograph of a dominantly carbonate beach on the north shore of Oahu, Hawaii.

Some erosional beaches typically have no backbeach, only a narrow and relatively steep foreshore.

Sediments

Like the foreshore, the backbeach has a wide range of sediments in terms of both composition and texture. The discussion here will be about the beach sediment above high tide even if it is not flat.

Beaches may be almost totally calcium carbonate, such as on the north coast of the island of Oahu in Hawaii (Figure 13.33). This sediment is eroded from the coral reefs just offshore to produce a carbonate beach in a generally terrigenous system of volcanic rocks. Another carbonate beach is at Acadia National Park in Maine (Figure 13.34). This beach is in a granite region so it is very unusual. It completely comprises mussel and other shell material broken up by waves and carried to this location by currents. Heavy mineral beaches are the natural beaches on the east coast of the North Island of New Zealand (Figure 13.35). The mineral is iron oxide derived from nearby volcanic sources;

Figure 13.34 Carbonate beach of shell debris near Acadia National Park on the coast of Maine.

Figure 13.35 Oblique aerial of an iron oxide sediment beach on the east coast of the North Island of New Zealand.

Figure 13.36 Shallow trench in a beach on the Lake Michigan coast where thick storm lag deposits of heavy minerals have accumulated.

on occasion Japan has purchased this sediment as raw material for its steel industry. Other beach compositions may be related to a process of erosion or to nearby mineral sources. Glacial sediments tend to be a virtually museum of textures and compositions. They can include heavy minerals that can cover a beach and may be in a layer several centimeters thick (Figure 13.36).

Surface Processes

The dry beach has processes and structures that originate primarily from wind. These structures tend to be directional due to their wind origin. Sand shadows are linear accumulations of sand on the lee side of a cobble or large shell (Figure 13.37). Their preservation potential is about zero but they can provide information on the recent few day's wind directions. Another indicator of direction is the current crescent structure (Figure 13.38). Current moving past a pebble or cobble will cause small U-shaped scour around the object.

Erosion is a very important and widespread beach process. Storms may reduce the beach width and sediment thickness in only a day or so. Chronic erosion can cause serious management problems. Many beaches experience a seasonal cycle of erosion and deposition with the former in the winter and the latter in summer. This is related to storms and wave climate. A good example of such a cycle occurs on the coast of Oregon (Figure 13.39).

Backbeach Biota

There is a variety of beach animals. Some are temporary visitors and others are full-time residents. One of the most curious beach animals is the ghost crab (*Ocypoda quadrata*) (Figure 13.40). It is rarely seen because it is a nocturnal creature. The beach at night can be full of them scurrying around. The main evidence of their existence is their burrow and excavation sediment (Figure 13.41).

Marine turtles are among the most interesting of all coastal animals. They only spend a few hours on the beach but it is a very important visit. Some species, such as loggerheads, come onshore during the night to build a nest and lay their eggs (Figure 13.42), others, like Kemp's ridley (Figure 13.43) do it during the day (Figure 13.44). The National Parks Service (NPS) marks and protects

Figure 13.37 Photograph of sand shadows where wind transport of sand has result in a small accumulation of sand in the lee of a cobble, showing wind direction.

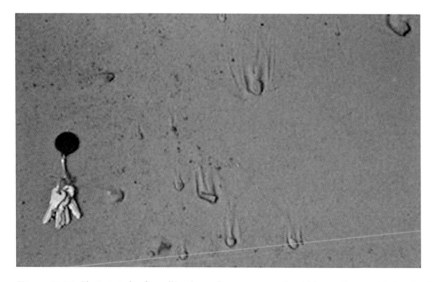

Figure 13.38 Photograph of small U-shaped scours around pebbles as the result of water transport in the direction of the open end.

loggerhead nests but leaves them on the beach (Figure 13.45) for a completely natural hatch of the dozens of 2–3 cm diameter eggs (Figure 13.46). Because of possible predation, the Kemp's ridley nests at the Padre Island National Seashore on the Texas coast are harvested and incubated by the NPS. After incubation the hatchlings, about 5–6 cm across, are released (Figure 13.47). The survival rate of all turtle hatchlings is very low.

(a)

(b)

Figure 13.39 Excellent examples of (a) a summer beach, and (b) a winter beach on the Oregon coast. (*Source:* (a) and (b) courtesy of W.T. Fox).

13.4 Human Impact on Beaches

The beach environment is probably the most heavily impacted by anthropogenic activities of the entire coast. The beach is fragile, heavily used and very often next to heavily populated areas. Most of the negative impact is the result of construction that is intended to protect the beach and prevent erosion. So called hard structures such as sea walls, groins and breakwaters are among the most widely used in this effort (Figure 13.48).

Figure 13.40 Close-up photograph of a ghost crab (*Ocypoda quadrata*) on a wet beach.

Figure 13.41 Photograph of a ghost crab burrow and excavation sediment.

The approach to shoreline protection has pretty much been replaced by beach nourishment, a process of dredging sand from a borrow area and depositing it on the beach to provide protection, recreational activities and esthetically pleasing conditions (Figure 13.49). Dredges remove sand and pump it to the construction site (Figure 13.50). Pipes are typically used to transport the slurry of water and sand to the desired location on the constructed beach (Figure 13.51). Large machines shape the beach to the design

Figure 13.42 A turtle track across the beach on Padre Island. (*Source:* Courtesy of U.S. National Park Service (NPS), https://www.nps.gov/bisc/learn/nature/images/DSC04315.JPG).

Figure 13.43 A Kemp's ridley turtle on the foreshore of a Texas beach. (*Source:* Courtesy of U.S. National Park Service (NPS), https://www.nps.gov/media/photo/gallery.htm?id=00B48C47-155D-451F-67B48D73D43C3928).

specifications (Figure 13.52). Although not common, some nourishment is carried out by trucking sand to the construction site (Figure 13.53). This causes problems with roads and the quality of the sediment used. Some of this type of nourishment is left for waves to mold the beach into a natural profile.

Figure 13.44 A Kemp's ridley turtle working on its nest. (*Source:* Courtesy of National Park Service (NPS) https://www.nps.gov/pais/learn/nature/images/IMG_1.JPG).

Figure 13.45 Markers designed to protect a turtle nest and its eggs on the Florida coast. (*Source:* Ianaré Sévi, https://commons.wikimedia.org/wiki/File:Protected_Sea_Turtle_Nest_(Boca_Raton_FL).jpg).

In some locations, especially on the European coasts, various plantings and other stabilizing features are installed in order to dissipate wind energy and capture sand to protect the beach (Figure 13.54).

13.5 Summary

The beach is one of the most dynamic environments on the Earth's surface. It is typically composed of sand and is wave-dominated.

Figure 13.46 Eggs laid by a Kemp's ridley turtle on the Texas coast. (*Source:* Courtesy of U.S. National Park Service).

Figure 13.47 Hatchlings of Kemp's ridley turtles being released on the Texas coast. (*Source:* Courtesy of U.S. National Park Service (NPS), https://www.nps.gov/pais/images/csthumbnail_large/20070210123753.jpg).

(a)

(b)

Figure 13.48 Examples of coastal construction include (a) a simple seawall on the Florida coast, (b) a combination of a riprap wall and groins, and (c) jetties that impact the beach.

Figure 13.49 Photograph of a completed beach nourishment project on the Florida coast.

Figure 13.50 A large suction dredge that pulls borrow sediment up and pumps it through large pipes to the shore or into large barges.

Figure 13.51 Oblique aerial photograph of a nourishment project being constructed on the coast of Alabama, with pipes to distribute the borrow material. (*Source:* U.S. Department of Transportation, https://www.fhwa. dot.gov/engineering/hydraulics/pubs/07096/images/image304.jpg).

Figure 13.52 Pipes and heavy equipment used to move borrow material to mold it to the design specifications.

Figure 13.53 Photograph of a nourishment project on the North Sea coast of Denmark. This project used borrow material that was transported by truck, each load forming a small pile.

Figure 13.54 Example of how various benign structures are used to stabilize the beach and attract more sediment, shown here on the coast of the Netherlands.

Box 13.1 Daytona Beach

Daytona Beach, Florida, USA has great historic significance for the sport of auto racing. First of all, it is one of the few beaches in Florida on which driving is easy. The beach is wide, gently sloping and generally smooth except after a bad storm. Racing on the beach began in 1902 and ended in 1958, peaking in the 1930s through the 1950s. Races with various vehicles were held, initially on the straightaway on the beach. Although the early races were popular (Box Figure 13.1.1) they generally lost money. Crashes were common, with traction being difficult.

Daytona Beach was the location of many land-speed record attempts. In 1927 Henry Seagrave

Box Figure 13.1.1 Old fashioned race cars racing on the beach in a straightaway track. (*Source:* State Archives of Florida/Coursen)

Box Figure 13.1.2 One of the early races on the oval track in the early 1950s. (*Source:* Florida Memory).

set the record of 207.79 mph in his 1000 hp. Sunbeam, and the culmination of record-setting came with Malcolm Campbell's1935 record of 276.82 mph.

In the early 1950s an oval track was conceived by combining the beach with State Highway 1A which parallels the beach about a block inland. The track included two one-mile straightaways, one on the highway and the other on the beach. With approaches and exits the course was 4.2 miles in the 1940s, after an earlier course of 3.2 miles. This provided a mechanism for having fairly long races using a conventional oval-shaped track (Box Figure 13.1.2). These beach races were the first of the stock car races and gave rise to NASCAR, the most popular spectator sport in the United States. Because the beach races had become so popular with large crowds, it was decided to construct a new track inland. The last race held on the beach was in 1958, the year before the current Daytona Speedway opened.

Storms can remove much of the beach sediment but much of it will return between storms. Waves can bring sediment to the beach as well as transport it along the shoreline by longshore currents. Some beaches experience a seasonal cyclicity with erosion in the winter and accretion in the summer.

Human impacts in this environment include structures such as seawalls, groins and breakwaters. This approach to managing and protecting the beach has been replaced by beach nourishment. This produces protection for the upland infrastructure and buildings and provides an excellent beach for recreation that attracts tourists.

Suggested Reading

Davis, R.A. (2014). *Beaches of the Gulf of Mexico.* College Station, TX: Texas A & M Press.

Davis, R.A. (2015). *Beaches in Space and Time.* Sarasota, FL: Pineapple Press.

Griggs, G. (2010). *Introduction to California's Beaches and Coast*. Berkeley, CA: University of California Press.

Hobbs, C.H. (2012). *The Beach Book*. New York: Columbia University Press.

Pilkey, O.H., Neal, W., Cooper, J.A.G., and Kelley, J.T. (2011). *The World's Beaches*. Berkeley: University of California Press.

Wang, P., Rosati, J., and Chung, J. (eds.) (2015). *Coastal Sediments 2015*. Singapore: World Scientific Press.

14

Coastal Dunes

Sand dunes are an important part of many, but not all, coastal areas. They are large piles of sand that accumulate as a result of similar processes and in generally similar shapes and patterns as dunes on inland deserts. The fundamental prerequisites in both cases are an abundant sediment supply and the wind to transport it. In most coastal areas, the wind is typically not a limiting factor but the sediment supply may be. Nevertheless, we have coasts where dunes may be more than 10 m in elevation (Figure 14.1).

Coastal dunes are not restricted to barrier islands, although nearly all barriers have at least small dunes. Some coasts without barriers have tremendous dune fields. Particularly good examples are the southern coast of Oregon, where the dunes extend a few kilometers inland from the coast, and the southeastern part of Lake Michigan, where dunes nearly 100 m high have developed. Among the largest dunes in the world are on the coast of Namibia (Figure 14.2) in southwest Africa, where barriers are absent (see Box 14.1).

14.1 Types of Coastal Dunes and Their Distribution

Any coast where sand accumulates in significant quantities has the potential for formation of dunes. Dunes are about the best protection we have against severe storms and their related large surges. Prevailing wind or diurnal sea breezes provide the typical transport mechanism for sand-sized sediment on most coasts. The prevailing wind typically has some onshore or shore-parallel component, and the sea breeze may be a major factor in many areas. It is, in fact, the dominant wind along the southwest coast of Australia near Perth. Basically any dry part of the beach or other coastal environment that is without or sparsely vegetated is subjected to eolian (wind) transport.

The initial development of coastal dunes is in the form of coppice mounds. These are small accumulations of sand (Figure 14.3) on a dry backbeach where opportunistic vegetation has initiated growth. The sand is trapped by the vegetation to form incipient dunes. They are somewhat fragile and are removed by any high water produced by storms. In the absence of storms they can eventually become true foredunes, linear dunes landward of the beach (Figure 14.4). Some coasts have several lines of dunes behind the beach (Figure 14.5), demonstrating their progradation due to abundant sediment and appropriate wind conditions. Each of these ridges was a foredune ridge at the time that it formed but was then fronted by a new one. Some barrier islands contain a complicated assortment of dune ridge arrangements that show sets of ridges at acute angles to one another (Figure 14.6). This condition indicates periods of erosion that separated periods of dune accumulation and barrier progradation. Some barriers have only a single foredune ridge, though it may be a high one.

Beaches and Coasts, Second Edition. Richard A. Davis, Jr. and Duncan M. FitzGerald.
© 2020 John Wiley & Sons Ltd. Published 2020 by John Wiley & Sons Ltd.

(a)

(b)

Figure 14.1 (a) Large dune, and (b) large blowout with a soil horizon along the west coast of Lake Michigan.

Dunes that form on the interior of barriers or on the coastal zone mainland, develop as the result of lack of vegetation to stabilize the substrate. This is generally due to an arid climate which gives rise to desert conditions or to removal of vegetation by grazing or other human-related activities. A good example of how a combination of these factors has contributed to an extensive, active dune complex is on Padre Island on the Texas Coast. Generally sub-arid conditions combined with abundant sand along the beaches have resulted in an extensive active dune complex in the Padre Island National Seashore, Texas on the southern part of the island (Figure 14.7). An additional assist has been provided by extensive cattle grazing on the island during the late nineteenth and early twentieth century. This portion of the Texas coast is one of considerable sediment accumulation and persistent onshore wind. As a consequence, the island is extremely wide and the mainland is dominated by an extensive dune complex. The barrier itself contains extensive active dunes that range from being only a meter or so high on the landward side near Laguna Madre, to several meters high in the central island. The limited

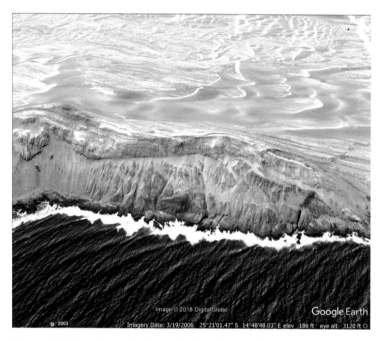

Figure 14.2 Huge dunes that have developed along the skeleton coast of Namibia. (*Source:* Courtesy of Know Namibia, http://knownamibia.com/locations/skeleton-coast).

Box 14.1 Huge Dunes from the Skeleton Coast of Namibia

Namibia is a country with a German colonial past in the southeastern portion of the African continent. The Skeleton Coast is mostly a national park and extends about 600 km along the northeastern portion of the country (Box Figure 14.1.1). Dunes rise a few hundred meters above the shoreline (Box Figure 14.1.2). The lack of rainfall results in an absence of vegetation so there is no stabilization of the dunes; sand is mobile and moves almost continuously. This is a true desert environment with annual rainfall typically less than 10 mm, about one-third of an inch. The name comes from a combination of the numerous boats and ships that have been wrecked on this coast (Box Figure 14.1.3), early mariners who washed ashore in the fifteenth century, and the skeletal remains of numerous species of animals. Fog is common and wind blows hard most of the time, producing big surf. Portuguese sailors referred to the coast as "the gates of Hell."

In spite of the absence of significant rainfall there are many living species of animals (Box Figure 14.1.4) including giraffes, lions, elephants, springboks, baboons and black rhinoceros. In fact there have been several research projects into how such a diverse population of animals has persevered. In part, the survival of these creatures is thanks to shallow wells dug in the low interdunes by baboons and elephants.

Box Figure 14.1.1 Map of Namibia showing the location of the "Skeleton Coast." (*Source:* karellafrica).

Box Figure 14.1.2 Aerial photo of the large and mobile dunes along the Namibia coast. (*Source:* the daily beast).

Box Figure 14.1.3 Photograph of the coast with large dunes and a wrecked ship on the beach. (*Source:* sararibookings).

Box Figure 14.1.4 Photograph of large active dunes with gemboks walking over the sand. (*Source:* theafricachannel).

development of the dunes on the landward part of the island is due to their destruction by storm surges associated with hurricanes. These small dunes are on the wind tidal flats where storm surges of a meter or so are fairly common and which can flood the dunes, destroying them by a combination of waves and currents. After the surge subsides, the wind and available sand must start again to construct the dunes.

Figure 14.3 Coppice mounds on the dry backbeach along the South Australia coast south of Adelaide.

Figure 14.4 A single foredune ridge along a barrier island.

A different situation is the coastal dune complex along the southern Brazilian mainland in the state of Santa Caterina. Here the strong wind off the Atlantic Ocean along with great amounts of sand have produced huge mobile dunes that extend 3–4 km inland. These coastal dunes have inundated buildings as they migrate in an offshore direction perpendicular to the coast (Figure 14.8).

Figure 14.5 A barrier island along the southeastern coast of Victoria, Australia displaying multiple dune ridges indicating the presence of a large sediment supply.

Figure 14.6 A barrier island along Padre Island on the Gulf Coast of Florida showing sets of low-lying dune ridges at angle to each other.

14.2 Dune Formation

Dry sand and appropriate wind conditions are common along the backbeach environment because it is rarely wet and generally void of vegetation. Wet sand has too much cohesion to respond to wind. The dry backbeach shows various evidences of wind transport including ripples (Figure 14.9), sand shadows (Figure 14.10), and heavy minerals

(a)

(b)

Figure 14.7 (a) Active sand dunes on the interior of Padre Island, Texas. (*Source:* Courtesy of Google Earth). (b) Active sand dunes on the interior of Padre Island, Texas. (*Source:* R. A. Davis).

or gravel lag concentrates of shells and shell debris. The sand shadows indicate a recent wind direction and may show scour around a shell or pebble. The gravel or shell lag deposit results from wind blowing the fine sand from the beach and leaving behind the larger particles that cannot be transported. After a while the large particles become concentrated and actually form almost a pavement. Such a pavement inhibits further wind erosion and is

called desert armor because of its importance in limiting wind erosion. Once the concentration becomes nearly continuous across an area then wind can no longer access the smaller particles that it can transport.

Much of the wind-blown beach sand tends to accumulate just landward of the active backbeach. It is stopped from further transport by any type of obstruction that may be present including bedrock cliffs, vegetation,

(a)

(b)

Figure 14.8 (a) Large coastal dunes migrating toward the shore and covering a small building in Santa Catarina, Brazil, and (b) the same situation with landward-migrating dunes on the North Sea coast of Denmark.

existing dunes or even human construction such as buildings or sea walls. Once the initiation of eolian sediment accumulation begins, it continues unless conditions change – for instance, loss of sediment supply, the destruction of the stabilizing factor or wave-induced erosion.

One of the best and most widespread aids in dune development is vegetation. Any type of plant serves as a focus for anchoring wind-blown sediment. Typically the relatively inactive backbeach is covered with opportunistic plants (Figure 14.11) such as the beach morning glory, beach *Spinifex* and

Figure 14.9 Wind-generated ripples on the sand surface of the backbeach environment.

Figure 14.10 Small wind shadows adjacent to cobbles on the beach that show the direction of wind.

Figure 14.11 Flowers on vines – opportunistic plants growing on the back beach.

marram grass. One of the most effective dune stabilizers on southern dunes up to the latitude of Virginia is sea oats (*Uniola*) while the American beach grass (*Ammophila*) extends from Virginia up to Nova Scotia in Canada. It is quite common to see small piles of sand around isolated plants on the back-beach area. After only months these piles increase in size and the plants spread, thus increasing their effectiveness. In fact, even pieces of wood or any other sizeable obstacle can act as a seed for dune development. As the sand becomes trapped by the vegetation this provides an enlarged area of stability. More sand becomes trapped and eventually a small dune develops. These small sand accumulations are typically a meter or two in diameter and about half that in height. Under the proper conditions they will eventually become larger and coalesce into a continuous foredune ridge.

Such small incipient dunes are quite vulnerable; even a modest storm can destroy them, requiring the building process to begin again. This is the reason that there is so much attention paid to preserving vegetation on the backbeach and at the foot of dunes. Absence of intense storms along with an abundant supply of sand and a regular mechanism for delivery of the sand eventually produces a dune. Dune size is dependent largely on the supply of sand-sized sediment.

Any major geomorphic or structural feature along the coast that presents some vertical component is also an effective trap for wind-blown sand. The base of a rocky cliff or sea wall can accumulate sediment and become vegetated. Assuming that sufficient sediment supply is available, this type of accumulation can eventually become a dune.

14.3 Dune Dynamics

The existence of dunes is testimony to the mobility of sand through wind transport on the coast, and attack by waves is an obvious factor in dune stability. Although vegetation is an effective stabilizer of these accumula-

tions, there are conditions when even vegetated dunes may become mobile or may be eroded.

The first and most obvious factor of these is the attack by waves. Even though dunes are out of the regular influence of waves, they are quite vulnerable to only modest surges produced by storms. In areas of generally erosive beach conditions, dune retreat is especially a problem because there is no backbeach to protect them (Figure 14.12). Elevated water level with storm waves superimposed produces swash and in some cases, direct wave attack at the toe of the dune. Sand is easily washed away and carried both offshore and along shore. Even though a dense dune grass cover is present, the sand is easily removed (Figure 14.13), commonly leaving a dense root system hanging over the scarp in the dune. Post-storm recovery may occur and return some, or even all, of the sand to the beach. Proper conditions can start the rebuilding process of the dune but it can take many years to restore the loss resulting from a single storm. It is generally rather easy to recognize dunes that have been eroded and then rebuilt by the change in profile and perhaps even in the type of vegetation. Rising sea level presents another scenario for dune erosion by providing continual increase in the accessibility of the dunes to wave attack.

The other major aspect of dune dynamics is concerned with the migration of part or all of the dune through eolian processes. The same mechanism that forms the dune, also can cause it to move, sometimes great distances. Generally, dune mobility is associated with an absence of vegetation. Climatic conditions may reduce or eliminate vegetative cover or over-grazing may remove much of the vegetation. Regardless of the reason the result is the same; sediment begins to move.

The most common process for dune migration is called blowover (Figure 14.14). The onshore wind component simply carries sand across the dune surface to the crest and permits it to move down the landward side by gravity. This creates a relatively steep slope whose angle with the

Figure 14.12 A huge scarp on a dune on the Lake Michigan coast that has recently been eroded by a storm.

Figure 14.13 Scarp on heavily vegetated dune on San Jose Island on the Texas coast.

horizontal, the angle of repose, is generally about 30°. In other words, the sand is able to maintain a slope of this gradient as it migrates landward. This is true for all dunes regardless of their location or direction of migration. The sediment may move as the result of individual grains rolling down the steep slope, the slipface or as the result of grain flow (Figure 14.15). This is a type of sediment gravity process whereby over-steepening of the slope causes an instability that results in large numbers of grains moving down the slope in an avalanche fashion. Anyone who has walked down a dune face

Figure 14.14 Dunes on the New South Wales coast of Australia that are being blown inland.

Figure 14.15 When the sand surface on a dune is oversteepened it fails and grains flow down the slope.

has seen this phenomenon take place as the dune is disturbed. Migration of large dunes pays little attention to trees, buildings or whatever is in its path. As long as the dune is larger than the obstruction it will move over it. Houses have been buried and then many years later exhumed as a result of migrating dunes.

14.3.1 Dune Structures

Because dunes produce bedforms of various scales and those bedforms move, they also produce various types of stratification. We can only see this internal stratification after dunes have been eroded and their internal nature exposed. There are instances when storms will decapitate a dune to expose the complex nature of cross-stratification. This can happen on foredunes (Figure 14.16) or on dunes in a barrier island (Figure 14.17).

14.4 Human Influence on Dunes

Various human activities have a major influence on coastal dunes. Some is beneficial and some detrimental. People everywhere recognize that dunes are fragile and most realize that dunes are beneficial. There are several ways by which people enhance the presence and protection of dunes. Probably the most important of these is the collection of sediment to make dunes, most commonly through the use of fences in the backbeach area to trap sediment as it is blown landward (Figure 14.18). As time passes, and in the absence of strong storms, considerable sand accumulates (Figure 14.19). In some locations Christmas trees are used for the same purpose (Figure 14.20) although they are much more vulnerable to the effects of wind and sand than is a fence.

There are multiple human activities that cause problems for dunes. Probably the most basic of these is the construction of walkovers. Some of these are placed directly on the sand (Figure 14.21a) and some are simply pathways (Figure 14.21b) through the dunes. In either case they lead to erosion by onshore wind. The worst of such practices is to completely remove dunes in order to view the water from residences (Figure 14.22).

Another poor activity for dune management is carried out by some communities on the Texas coast. Firstly dunes are prevented from prograding toward the shoreline as they attempt to do naturally. Daily scraping of the beach prevents this and keeps the beach very wide for tourism including parking vehicles. This coast experiences considerable accumulation of the floating brown algae *Sargassum* (Figure 14.23a). This accumulation is bad for tourism so it is scraped into large piles and the resulting blend of algae and sand is dumped in large notches that have been excavated in the foredunes (Figure 14.23b).

Figure 14.16 Stratification on an eroded surface on foredunes on the Netherlands coast.

Figure 14.17 Complex cross-stratification in a decapitated dune on Padre Island, Texas.

Figure 14.18 Fence placed on the backbeach that is designed to trap sediment and eventually build a dune.

14.5 Summary

Dunes serve as the best protection for depositional coasts. Granted, they are not as good of a protection from waves and erosion as cliffs or other bedrock, but they provide a barrier from wave attack during storms. The addition of vegetation, especially mature trees, results in what is almost like a dam against high water.

In order for dune development to proceed to a level where this protection is in place, it is

Figure 14.19 Dune that has buried the fence placed to collect sand.

Figure 14.20 Old Christmas trees placed on the backbeach at Surfside, Texas with the intent of trapping sand and helping to build dunes.

(a)

(b)

Figure 14.21 (a) Boardwalk over the dunes placed directly on the sand and (b) a pathway through the sand on the dune itself.

Figure 14.22 Location on Mustang Island, Texas where the entire dune has been removed so that people can see the water from their homes.

(a)

Figure 14.23 Management of *Sargassum* on beaches of Mustang Island, Texas include (a) scraping the algae into large piles and (b) placing this mixture of algae and sand into notches cut in the foredunes.

(b)

Figure 14.23 (Continued)

necessary to have a large supply of sediment and a well-developed beach. The dry beach is the immediate source of sediment to form the coppice mounds and then the foredunes.

Suggested Reading

Hesp, P. (1999). The beach backshore and beyond. In: *Handbook of Beach and Sshoreface Morphodynamics* (ed. A.D. Short), 145–169. Brisbane: John Wiley and Sons.

Houston, J.A. and Edmonson, S.F. (2001). *Coastal Dune Management*. Liverpool: Liverpool University Press.

Martinez, M.L. and Nordstrom, K.F. (eds.) (2007). *Coastal Dunes: Ecology and Conservation*. New York: Springer.

Nordstrom, K.F. (2004). *Beaches and Dunes of Developed Coasts*. Cambridge: Cambridge University Press.

U.S. Govt (2013). *Coastal Dunes: Dune Building Processes and a Primer for Dune Development and Management*. Washington, DC: U.S. Government Printing Office.

15

Barrier Systems

15.1   Introduction

Barriers are the sites of some of the world's most beautiful beaches. They form due to the combined action of wind, waves, and long-shore currents whereby thin strips of land are built a few to several tens of meters above sea level (Figure 15.1). They are called barriers because they protect the mainland coast from the forces of the sea, particularly during storms. They lessen the effects of storm waves, heightened tides and salt spray. The bays, lagoons, marshes and tidal creeks that form behind barriers provide harborages for pleasure boats and commercial craft, nursery grounds for fish and shellfish, and important sources of detritus and nutrients exported to the coastal ocean. Barriers occur through-out the world's coasts, but are most common along passive margins where sediment supplies are abundant and wave and tidal energy are conducive for onshore sand accumulation.

Barriers represent some of the most expen-sive real estate in many countries due to their development by the resort industry and because many people wish to live next to the sea. The lure of sandy beaches, salt air, water sports, and beautiful seascapes has always drawn vacationers and people wanting to erect summer cottages and permanent homes on barriers. Building lots on Figure Eight Island along the southern coast of North Carolina cost millions of dollars. Some of the qualities that make barriers so attrac-tive also make them vulnerable to storms,

rising sea level and erosion. Their sandy composition, general low elevation and grass and shrubbery cover allow waves and storm surges to overrun barriers during the passage of intense hurricanes and major extratropical storms (northeasters). The greatest natural disaster in terms of lives lost in the United States occurred in 1900 when a hurricane swept over Galveston Island off the Texas coast near Houston killing an estimated 8000 people.

In this chapter we explore the different types of barriers, the factors controlling their worldwide distribution, and how they become modified by coastal processes. Depending upon their evolution, barriers may have a variety of architectures. We dis-cuss the different components of barriers, what they look like in cross-section, and the numerous theories that have been put forth to explain their formation. One of the major forces impacting barriers today, as well as in the past, is rising sea level. The response of barriers to this inundation and the dynamic processes, which allow some of them to migrate onshore are also considered.

15.2   Physical Description

Barriers are wave-built accumulations of sediment that accrete vertically due to wave action and wind processes. Most are linear features that tend to parallel the coast, gener-ally occurring in groups or chains. Isolated barriers are common along glaciated coasts

Beaches and Coasts, Second Edition. Richard A. Davis, Jr. and Duncan M. FitzGerald.
© 2020 John Wiley & Sons Ltd. Published 2020 by John Wiley & Sons Ltd.

(a)　　　　　　　　　　　　　　　　　(b)

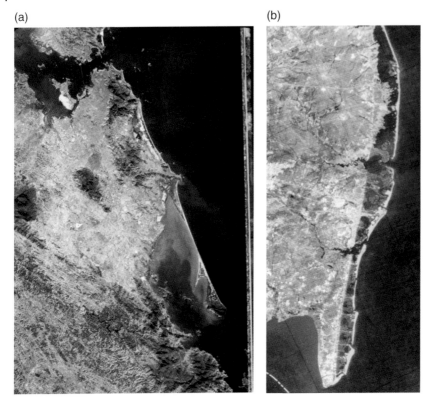

Figure 15.1 Barriers may form in a variety of coastal settings if the supply of sediment is adequate and wave and tidal conditions are conducive to sand accumulation onshore. (a) Barrier system pinned to bedrock islands and promontories along the Rio de Janeiro coast in Brazil, and (b) Barrier chain along the New Jersey coast. (*Source:* NASA).

such as in northern New England and eastern Canada, and along high-relief coasts such as those associated with collision coasts (see Chapter 2). Barriers are separated from the mainland by a region termed the backbarrier consisting of tidal flats, shallow bays, lagoons and/or marsh systems. An exception to this characterization occurs in instances where successive beach ridges have produced an out-building of the coast such as along the flanks of a river delta. Barriers may be less than 100 m wide or more than several kilometers in width. Likewise, they range in length from small pocket barriers of a few 100 m to those along open coasts that extend for more than 100 km. Generally, barriers are wide where the supply of sediment has been abundant and relatively narrow where erosion rates are high or where the source of sediment was scarce during their formation.

Barrier length is partly a function of sediment supply but is also strongly influenced by wave versus tidal energy of the region. This relationship will be discussed in more detail later in this chapter.

Barriers consist of many different types of sediment depending on their geological setting. Sand, which is the most common constituent of barriers, comes from a variety of sources including rivers, deltaic and glacial deposits, eroding cliffs and biogenic material. The major components of land-derived beach sand are primarily the minerals of quartz and feldspar. These durable grains are a product of physical and chemical decomposition of the bedrock from continents. In northern latitudes, where glaciers have shaped the landscape, gravel is a common constituent of barriers, whereas, in southern latitudes carbonate

(a)

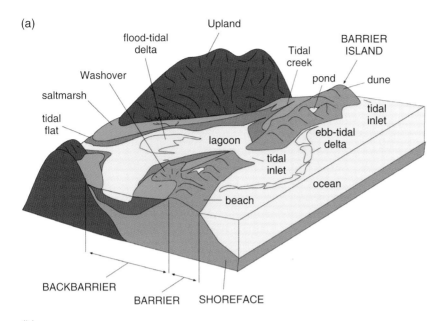

(b)

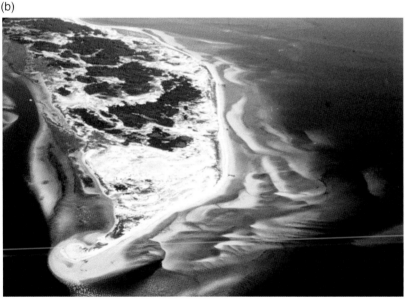

Figure 15.2 (a) Idealized barrier subenvironments. (b) Castle Neck, Massachusetts can be divided into three zones: 1. *Beach*, where sand bars are migrating onshore and weld to the lower shore; 2. *Barrier Interior*, consisting of dune and swale topography; 3. *Landward Margin*, where the barrier transitions to tidal flats and salt marsh. In tropical settings the margin may consist of mangroves.

material, including shells and coral debris, may comprise a major portion of the barrier sands. Along the southeast coasts of Iceland and Hawaii and along portions of the west coast of New Zealand, the barriers are composed of black volcanic sands derived from upland volcanic rocks.

Many environments make up a barrier and their arrangement differs from location to location, reflecting the type of barrier and the physical setting of the region. Generally, most barriers can be divided in three zones: the beach, the barrier interior, and the landward margin (Figure 15.2).

Beach

Due to continual sediment reworking by wind, waves and tides, the beach is the most dynamic part of the barrier. Beaches exhibit a wide range of morphologies depending upon a number of factors including the grain size of the beach, the abundance of sediment, and the influence of storms. Sediment is removed from the beach during storms and returned during more tranquil wave conditions. During the storm and post-storm period, the form of the beach evolves in a predictable fashion. The beach environment is discussed in detail in Chapter 13.

Barrier Interior

Along sandy barriers, the beach is backed by a frontal dune ridge (also called the foredune ridge) which may extend almost uninterruptedly along the length of the barrier provided the supply of sediment is adequate and storms have not incised the dune ridge. The frontal dune ridge is the first line of defense in protecting the interior of the barrier from the effects of storms. Landward of this region are the secondary dunes, which have a variety of forms depending on the historical development of the barrier and subsequent modification by wind processes. For example, devegetation of the Provincelands Spit at the tip of Cape Cod in Massachusetts by early settlers led to the formation of large parabolic dunes that are up to a half kilometer long and more than 35 m high. In other locations where barriers have built seaward through time, such as North Beach Peninsula in Washington or Kiawah Island in South Carolina, former shoreline positions are marked by a series of semi-parallel vegetated beach ridges (former foredune ridges), 3–7 m in height. The low areas between individual beach ridges, called swales, commonly extend below the water table and are the sites of fresh and brackish water ponds or salt marshes. In South Carolina, Georgia, and Florida alligators and water moccasins are known to inhabit these ponds. On developed barriers containing golf courses, these ponds often become the sites of water holes.

Landward Margin

Along the backside of many barriers the dunes diminish in stature and the low relief of the barrier gradually changes to an intertidal sand or mud flat or a salt marsh. In other instances, the margin of the barrier abuts an open-water area associated with a lagoon, bay or tidal creek. Along coasts where the barrier is migrating onshore, the landward margin may be dominated by aprons of sand that have formed by waves overwashing the barrier during periods of storms. In time, these sandy deposits may be colonized by salt grasses to produce arcuate-shaped marshes. Overwash deposits may also occur in the interior of the barrier. In still other regions, the backside of a barrier may be fronted by a sandy beach bordering a lagoon or bay. In this setting, wind-blown sand from the beach may form a rear-dune ridge outlining the landward margin of the barrier.

Barriers may terminate in an embayment or at a headland. In the case of barrier chains, individual barriers are separated by tidal inlets. These are the waterways that allow for the exchange of water between the ocean and the backbarrier environments. The end of the barrier abutting a tidal inlet is usually the most unstable portion of the barrier due to the effects of inlet migration and associated sediment transport patterns.

15.3 Distribution and Coastal Setting

General

Barriers comprise approximately 10% of the world's coastlines (Figure 15.3). They are found along every continent excep Antarctica, in every type of geological setting, and in every kind of climate. Tectonically, they are most common along *Amero-trailing Edge* coasts where low-gradient continental

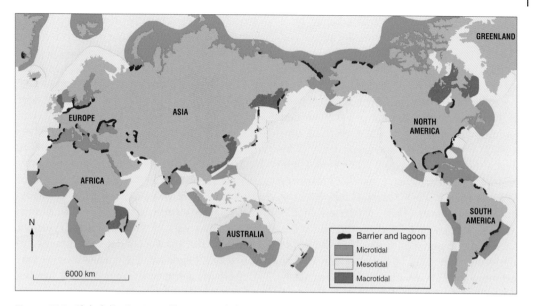

Figure 15.3 Global distribution of barriers and their tidal range setting. (*Source:* From Hayes (1979) using barrier data from Gierloff-Emden (1961) and tidal range data from Davies (1973)).

margins provide ideal settings for barrier formation. They are also best developed in areas of low to moderate tidal range[1] (microtidal to mesotidal range) and in mid to lower latitudes. Climatic conditions control the vegetation on the barriers, and, in backbarrier regions, the type of sediment on beaches, and in some regions such as the Arctic, the formation and modification of barriers themselves. The disappearance of barriers where tidal energy dominates, such as on the northwest (Big Bend) coast of Florida in the Gulf of Mexico and the German Bight in the North Sea attests to the requirement of wave energy in the formation of barriers.

Tectonic Controls

As discussed in Chapter 2, the tectonic setting of the coast dictates to a large extent the

sediment contribution to the coast, the width of the continental shelf, and the general topography of the coast. The tectonic coastal classification can be used to illustrate these relationships and explain the worldwide distribution of barriers. Amero-trailing edge coastlines tend to have abundant sediment supplies due to extensive continental drainage. Their low-relief coastal plains and continental shelves provide a platform upon which barriers can form and migrate landward during periods of eustatic sea level rise. The longest barrier chains in the world coincide with Amero-trailing edges and include the East Coast of the United States (3100 km) and the Gulf of Mexico coast (1600). There are also sizable barrier chains along southern coast of Brazil (960 km), East Coast of Indian (680 km), North Sea coast of Europe (560 km), Eastern Siberia (300 km), and the North Slope of Alaska (900 km).

The unparalleled concentration of barriers along the East Coast of the United States is undoubtedly a product of the erosion and denudation of the Appalachian Mountains,

---

1 Tidal range is the vertical difference between high and low tide. A classification of coasts based on tidal range consists of microtidal coasts (TR < 2.0 m), mesotidal coasts (2.0 ≤ TR ≤ 4.0 m), and macrotidal coasts (TR > 4.0 m).

which scientists have speculated may have once rivaled the Himalayas in elevation. The huge volume of sediment derived from the wearing-down of these mountains produced a wide, flat coastal plain and continental shelf region, much of which is veneered by layers of unconsolidated sediment or easily eroded sedimentary rocks. It is the reworking and redistribution of these surface sediments by rivers, tidal currents and waves that are responsible for extensive barrier construction along the East Coast. Likewise, the almost continuous chain of barriers in the Gulf of Mexico is related to the wide, flat coastal plain and continental shelf of this margin. It is thought by some scientists that sediment delivery to this coast, particularly along Texas, was much greater in the past when the climate of this region was wetter and rivers transported much larger sediment loads.

Marginal sea coasts contain a relatively small percentage of world's barriers even though these regions have some of the largest sediment-discharge rivers in the world, particularly along the Asian continent (e.g. Mekong, Yangtze, Yellow). While it would appear that many of these coasts have ample sediment to build barriers, the irregular topography of these margins leads to much of the sediment filling submarine valleys rather than forming barriers. In addition, the discharge from many Asian rivers has a very high suspended sediment component that may inhibit the concentration of sand-sized material. In fact, many of the marginal sea coastlines containing barriers, such as the chain found along the Nicaraguan and Honduran Caribbean coast, have no large sediment-discharge rivers associated with them. The origin of these barriers is probably tied to the reworking of previously deposited shelf sediments, similar to much of the East Coast of the United States and the Netherland and German North Sea coasts.

Most collision coasts contain few barrier island chains due to an overall lack of sediment. Most of the rivers discharging along these shores have small drainage areas and contribute little sediment for barrier construction. Additionally, the narrow, steep continental shelves of these margins result in high wave energy and rapid sediment dispersal along the coast. Along parts of the California coast some of the sediment that would ordinarily accumulate along the shore is drained from the beaches during storms through submarine canyons into the deep ocean basins. The proximity of canyons to the beaches is attributed to the narrow continental shelf. Barrier systems that do occur along the west coast are commonly isolated and related to specific sediment sources such as nearby rivers or eroding cliffs. Typically, they take the form of barrier spits attached to a bedrock headland. Two of the longest barrier chains along collision coasts are located adjacent to major river mouths, including the southern Washington barriers near the Columbia River and the Gulf of Alaska barrier chain fronting the Copper River Delta (Figure 15.4).

Only a small percentage of the world's barriers are found along Afro-trailing and neo-trailing edge coasts due to an overall lack of sediment along these margins. There is little sediment delivery to the Neo-trailing edge margins of the Red Sea, Gulf of Aden, and Gulf of California due to the immaturity in the development of river drainage along their coasts and the very low precipitation in these regions. Similarly, much of northern coast of Africa has little river discharge of sediment due to the arid conditions of the interior region. The extensive barrier system (300 km) that exists along the Ivory, Gold, and Slave Coasts on the west-central coast of Africa is related to the numerous moderately-sized rivers of this area.

Climatic Controls

By determining the amount of precipitation and evaporation of a continent, climate exerts a strong influence on the size and number of rivers as well as the overall volume of sediment delivered to the coast. A river's drainage area (defined as the continental area drained by the river) is another important factor governing sediment-discharge. Usually, the larger the drainage area (A_d) the greater is the sediment

(a)

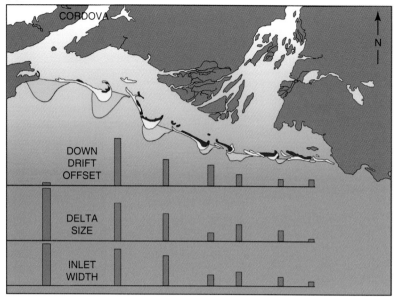

(b)

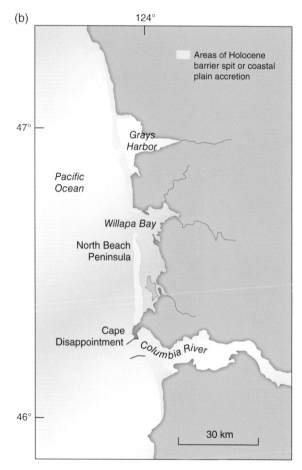

Figure 15.4 Barrier chains on collision coasts are generally found where a nearby river supplies abundant sediment. Examples include (a) Copper River Delta barrier chain (*Source:* From Hayes (1979)); and (b) Long Island–Grays Harbor barrier system along the southern Washington coast north of the Columbia River. (*Source:* From Dingler and Clifton (1994)).

delivery to the coast; however, there are exceptions. The Eel River in northern California releases almost twice the sediment load to the coast as does the much larger Columbia River to the north, despite having a drainage area two orders of magnitude smaller (Eel R. $A_d = 9000\,km^2$, Columbia R. $A_d = 661,000\,km^2$). The high sediment discharge of the Eel River is a product of the mountainous terrain it drains and a bedrock landscape that weathers easily producing abundant sediment.

Climate also dictates the kinds of plants, shrubbery, and trees colonizing barriers and the type of vegetation found in the backbarrier region. For example, salt grasses vegetate the backbarrier marshes of mid- to high latitudes, whereas, mangroves comprise the backbarrier region in low latitudes. Even the sediment making up the barrier is related to climatic controls and previous climatic conditions, such as the "Ice Age." Generally the shell and coral content of beaches is relatively high in tropical regions whereas terrigenous-derived gravel forms a significant component of northern beaches and barriers that have formed from these glacial deposits.

15.4 Summary

Barriers are best developed along continental margins where sand-sized sediment is abundant and the coastline is fronted by a moderately wide, gently sloping continental shelf that is backed by a coastal plain. Under these conditions, barriers are able to migrate onshore in a regime of slow sea-level rise. Direct contribution of sediment to the coast by rivers is not enough to ensure the presence of barriers as evidenced by the Asian marginal sea coasts where barriers are sparse. The fine-grained sediment discharged from the rivers of this region may overwhelm the ability of waves to concentrate sand and form barriers.

15.5 Barrier Types

Although barriers have many different forms, for ease of discussion they are grouped into three major classes based on their connection to the mainland (Figure 15.5). Barrier

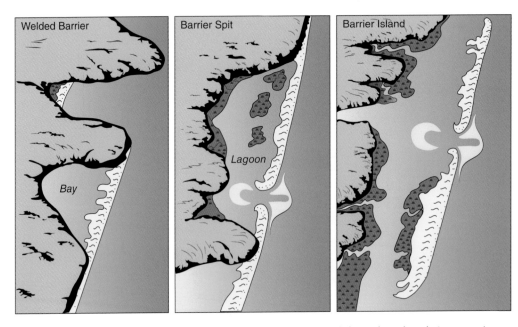

Figure 15.5 Barriers can be separated into three general morphological classes based on their connection with the mainland: welded barriers, barrier spits, and barrier islands.

spits are attached to the mainland at one end and the opposite end terminates in a bay or the open ocean. Welded barriers are attached to the mainland at both ends, and barrier islands are isolated from the mainland and surrounded by water.

15.5.1 Barrier Spits

Barrier spits are most common along irregular coasts where angular wave approach and an abundant supply of sediment result in high rates of longshore sediment transport. These conditions promote spit building across embayments and a general straightening of the coast. Barrier spits are the dominant barrier form along tectonically active coasts. In some instances, spit construction partially closes off a bay forming a tidal inlet between the spit end and the adjacent headland. In these instances, tidal currents flowing through the inlet prevent spit accretion from sealing off the bay. The Cape Cod shoreline in New England illustrates well this type of coastline evolution. Following deglaciation, the shoreline of Cape Cod appeared very different from what it looks like today, consisting of irregularly shaped, sandy glacial, and glacio-fluvial deposits (see Chapter 17). As eustatic sea level rose in response to melting glaciers and water being returned to the ocean basins, this landscape gave rise to large headland areas separated by broad embayments. Since 5000 years ago, waves associated with the rising ocean waters have attacked the unconsolidated headlands causing their retreat and the release of large quantities of sediment that have subsequently moved along the shore, leading to the construction of spits across the inundated lowland areas. In this way the shoreline of Cape Cod was smoothed into its present form.

Recurved Spits

Many spits have recurved ridges that conform to the general outline of the spit end. Such spits are called recurved spits. Each of the ridges corresponds to a former shoreline position and collectively they trace the evolution of the spit (Figure 15.6). Ridge formation is part of the spit extension process whereby sediment is added incrementally to the end of the barrier. Before the subaerial (above the water) portion of the spit can build into a bay or tidal inlet, a platform fronting the spit must first be created upon which sediment can accumulate and bars can form. This subaqueous portion of the spit commonly contains much more sand than the part of the spit above water. For example, Sandy Hook, which defines the northernmost end of the New Jersey coast, is accreting into Raritan Bay toward New York Harbor. The spit end of Sandy Hook has an average elevation of 5 m above mean sea level, whereas bay depths immediately offshore of the spit reach 12 m and more. Likewise, the northern spit end of Cape Cod is building into water depths of more than 50 m, contrasting to average land elevations of 10 m.

The spit platform is a shallow sloping sandy surface that extends from low tide to a depth of several meters below mean low water. The platform is fed with sediment through the longshore movement of sand along the updrift portion of the barrier, enabling the platform to build into deep water (Figure 15.7). Waves breaking over the spit platform aid in the development of swash bars, subaqueous bars having a length of 100 m or more and a height of one to several meters. These bars, which migrate onshore due to the action of breaking and shoaling waves, tend to wrap around the end of the spit becoming aligned with the waves that bend around the spit platform. As the bars move onshore, they first gain an intertidal exposure and eventually attach to the upper beach forming a ridge. Commonly, the welding process is incomplete and a low area (swale) is left between the newly accreted ridge and the dune line that defines the former end of the spit. Gradually, the ridge is colonized by dune grasses and builds vertically by wind processes.

Recurved spits exhibit many different beach ridge patterns characterizing different

(a)

(b)

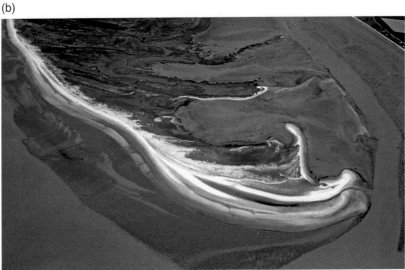

Figure 15.6 Recurved spit development. (a) Spit building across the mouth of the Santee River in South Carolina. Individual beach ridges represent pulses of sand that lengthen the spit in spurts. The sporadic supply of sand may be related to storm activity. (b) Similar episodic spit building defines Arcay Spit on the Vendée coast of France.

depositional styles. In some instances, the ridges indicate that there has been a simple extension of the spit, whereas in other cases they demonstrate that the barrier has widened as well as extended along the shore. In both examples, the ridge and swale morphology suggests that spit accretion occurs episodically and each ridge may represent an increase in the rate of sand delivery to the spit platform. In turn, the change in the flux of sand along the coast may be related to storm processes whereby sediment is excavated from updrift sandy cliffs, river deltas or beach deposits.

(a)

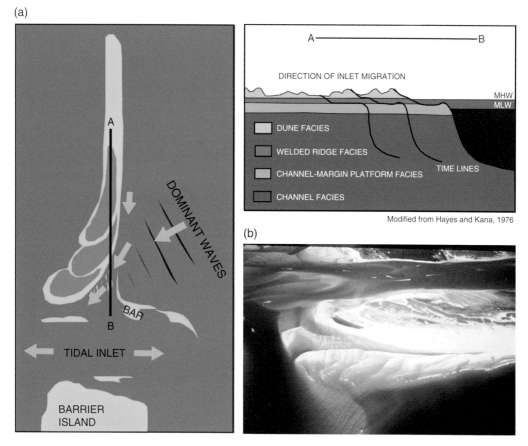

DIRECTION OF INLET MIGRATION

MHW
MLW

☐ DUNE FACIES

☐ WELDED RIDGE FACIES

☐ CHANNEL-MARGIN PLATFORM FACIES

TIME LINES

☐ CHANNEL FACIES

Modified from Hayes and Kana, 1976

(b)

DOMINANT WAVES

BAR

B

← TIDAL INLET →

BARRIER
ISLAND

Figure 15.7 Spit accretion across a bay or into a tidal inlet is preceded by extension of the spit platform. (a) The spit platform shown in cross-section. (*Source:* Modified from Hayes and Kana (1978).) (b) Aerial view of Chatham Inlet, Cape Cod, Massachusetts illustrating swash bars migrating onshore and building ridges, which lengthen the supratidal portion of the spit. The platform is constructed from sand derived via the longshore transport system.

Another common trend along recurved spits is the change in swale environment along the length of the spit, reflecting a gradual filling of a swale over time. At the spit end, a newly formed swale may be deep enough to permit tidal inundation from an opening along ridge on the lagoon side. Sedimentation in the swale resulting from washovers, wind-blown sand and, to a lesser extent, fine-grained material carried by tidal currents, transforms the ponded water region to a marsh environment and eventually to a sandy low inter-ridge area. Thus, the spit end tends to exhibit low ridges with water-filled swales, whereas the updrift, older portion of the spit contains vegetated beach ridges and sandy or grassy swales.

15.5.1.1 Spit Initiation

Spit formation is due to the deflection of sediment that is moving on and along a beach in the surf and breaker zones into a region of deeper water commonly associated with a bay or tidal inlet. This deflection may be initiated at small protrusions of the shoreline or at major headlands. Similarly, the process may occur at the edge of embayments along an otherwise smooth coastline. The explanation of why the longshore movement of sand forms spits rather than following a pathway

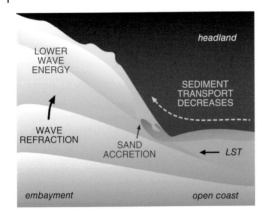

LOWER
WAVE
ENERGY

headland

SEDIMENT
TRANSPORT
DECREASES

WAVE
REFRACTION

SAND
ACCRETION

← LST

embayment

open coast

Figure 15.8 Spit development is due to sand deposition resulting from a decrease in the rate of longshore sediment transport. The rate of sand transport decreases downdrift of the headland due to a reduction in wave energy.

along the irregular shoreline lies in the fact that wave energy decreases in embayments and thus the capacity of waves to move sand is reduced (Figure 15.8).

Remember from Chapter 6 that the movement of sand on and along a coast is a function of wave energy and breaker angle among other parameters. Consider an idealized case in which a wave crest extends a long distance along the shore such that it can be followed from where it breaks at an angle along the exposed beach to where it breaks within the bay. The section of wave breaking along the exposed beach will be higher and will expend more energy in the nearshore zone than the wave crest that breaks inside the bay. This makes sense because as the wave crest bends into the bay, it travels further than the wave breaking along the exposed coast and thus uses up more energy interacting with the seabed. Because the rate of sand movement is a function of wave height, more sand is transported along the exposed beach than inside the bay. This means that as sediment transport decreases from a relatively high rate along the exposed beach to a lower rate inside the bay, some sediment must be deposited along the way. This sediment accumulates at the edge of the embayment because this is where the change

in the longshore transport rate is the greatest. It is this deposition of sediment that initiates spit growth.

15.5.1.2 Other Spit Forms

Spits exhibit many different forms in addition to recurved spits, including cuspate spits, flying spits, tombolos and cuspate forelands. Each of these accretionary landforms develops in specific geological settings and by particular coastal processes.

Cuspate spits are triangular accumulations of sand that extend from the shoreline into a semi-protected body of water, such as the lagoons and bays found behind barrier islands and spits (Figure 15.9). They range in length from a few tens of meters, such as those in the elongated bays along the southern coast of Martha's Vineyard, to spits which are kilometers in length including those on the lagoonal side of Santa Rosa Island in the western Panhandle of Florida or those in Patos Lagoon along the southern coast of Brazil. Their lengths reflect the size of the bay and the resulting wave energy produced inside the bay. Larger bays and lagoons have longer fetches which allow larger waves to develop and more sand to be transported along the bay shoreline. The semi-protective nature of the bays means that sand transport along the shoreline is governed by local wind-generated waves and not by ocean swell or storm waves. The multi-directional winds that are common to most coasts produce bi-directional sand transport along these protected shores. This back and forth movement of sand along bay shorelines is what forms cuspate spits. The tip of the spit may have a narrow intertidal to sub-tidal tail that changes its orientation as the wave climate changes. Growth and modification of the spit are often revealed in the pattern of beach ridges that are common to cuspate spits.

Flying Spits look similar to recurved spits and commonly exhibit cuspate-shaped beach ridges, indicating former shoreline positions. However, unlike recurved spits, they occur along straight to slightly irregular shorelines and extend into deep water at an acute angle

(a)

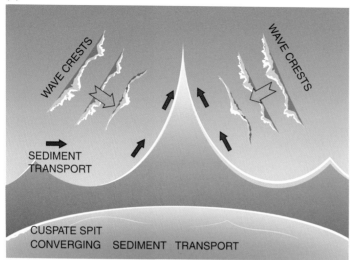

(b)

Figure 15.9 Cuspate spits occur in protected environments such as within bays and lagoons. (a) Model of cuspate spit formation due to bi-directional sand transport. (b) Example of a cuspate spit on the bay side of the Coatue barrier, Nantucket Island, Massachusetts. (*Source:* Courtesy of Miles Hayes).

to the beach (Figure 15.10). Like cuspate spits, they are found most commonly along semi-protected shorelines where wave energy is a product of local winds. Flying spits exist in the lagoons bordering the Texas coast; however, the best-known and studied flying spit is Presque Isle along the south-central shore of Lake Erie, near Erie, Pennsylvania. Presque Isle is 10 km in length and its spit end is approximately 2 km offshore of the beach. The reason why these spits build into deep-water is not well understood but it may be

(a)

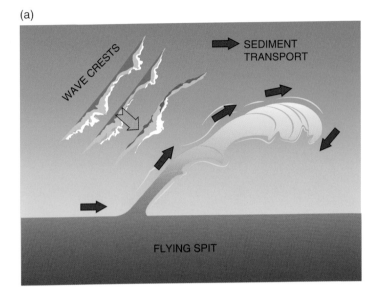

(b)

Figure 15.10 Flying spits are relatively uncommon features found in semi-protected waters. (a) Model of a flying spit. (*Source:* Courtesy of Miles Hayes). (b) Example at Presque Isle, Pennsylvania. (*Source:* U.S. Army Corps of Engineers, https://en.wikipedia.org/wiki/File:Presque_Isle_Pennsylvania_aerial_view.jpg).

related to some type of irregularity along the original shoreline or in the nearshore region. Because the trend of flying spits is normally almost parallel to the dominant wave crests of the region, some scientists have argued that spit construction is an attempt of the shoreline to minimize the longshore transport of sediment.

Tombolos are a type of spit that was first identified and named by Italians because they are common features along the Italian coast (Figure 15.11). In the strictest definition,

(a)

Time 1. Wave shadow

(c)

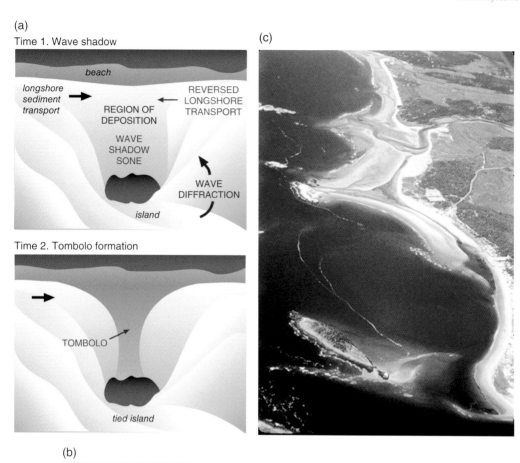

Time 2. Tombolo formation

(b)

Figure 15.11 A tombolo is an accretionary landform that connects the mainland to an island. (a) Model of the development of a tombolo. Examples of tomobolos in (b) Covachos, Cantabria in northern Spain, and (c) at Popham Beach, Maine.

tombolos are spits of land that build out from the shoreline eventually producing a sediment bridge from the mainland to an island. The largest of these occurs along the northern Tyrrhenian Sea coast of Italy at Orbetello where two tombolos, almost 8 km in length, have evolved due to the large size of the island (10 km long) and an abundant sediment supply along the coast. Moderate- and small-sized islands tend to have a single tombolo.

Tombolos form on the lee side of an island due to the obstruction and redistribution of wave energy caused by the island. The expenditure of wave energy on the island creates a "wave shadow zone" along the landward beach. The reduction in the rate of longshore sand transport in this region produces sediment accumulation and spit growth toward the island. Waves bending around the island also create a reversal in the direction of dominant sand movement along the shore, augmenting the sand-trapping processes in the wave-sheltered zone.

Cuspate Forelands are large triangular-shaped projections of the shoreline that may extend seaward more than 25 km. The apex of the foreland is commonly composed of beach ridges, which parallel the two converging shorelines (Figure 15.12). Alternatively, the ridges parallel a single shoreline and are truncated along the other, indicating a reorientation of the foreland (i.e. Cape Canaveral, Florida). Bays and lagoons or marshy areas generally occupy the region between the foreland and the mainland. Cuspate forelands may be isolated or can occur in a series such as those along the coast of North and South Carolina. Dungeness is a solitary foreland located along the southern coast of Britain where the English Channel is narrowed by the closeness of France. It is built of gravel ridges and is prograding seaward and to the northeast due to dominant wave energy from the southwest. The Carolina forelands, including Cape Hatteras, Cape Lookout, Cape Fear, and Cape Romain, form an evenly-spaced (110–165 km apart) scalloped shoreline with individual capes projecting 25–40 km beyond the adjacent embayments. A similar group of capes with the same approximate spacing (130 km) is found on the North Slope of Alaska along the Chukchi Sea (Icy Cape, Point Franklin, and Point Barrow). Formation of the Carolina capes has been related to giant eddies associated with the northward flowing Gulf Stream, wave refraction around offshore shoals, basement controls and the reworking of former delta deposits. Although rivers do not presently supply sand to the capes, excepting possibly the Cape Fear River, each of the capes is associated with one or more moderately-sized rivers, which may have delivered sediment to the coast during the Pleistocene, when sea level was lower. These deltaic deposits may have been the source of the cape sediments or may have preceded barrier development and influenced sand transport trends along the coast.

15.5.2 Welded Barriers

Welded barriers occur along irregular coasts where the supply of sediment is adequate for barrier construction (Figure 15.13). They are common features along rocky coasts such as parts of the west coast of the United States and along glaciated coasts including New England, eastern Canada, and Alaska. Most often, these barriers are backed by shallow water bays and lagoons; however, freshwater and brackish water marshes also exist. Welded barriers are most common along microtidal coasts where bay areas are diminutive in size. In this type of setting there is insufficient tidal energy to keep a tidal inlet open. If a welded barrier is breached during a storm, the inlet normally closes quickly, because the exchange of tidal waters between the bay and ocean cannot remove the sand dumped into the inlet by waves. Thus, environments with low tidal ranges and moderate to high wave energy promote the development of welded barriers.

The formation of a welded barrier can be a simple process whereby a spit builds across

(a)

(b)

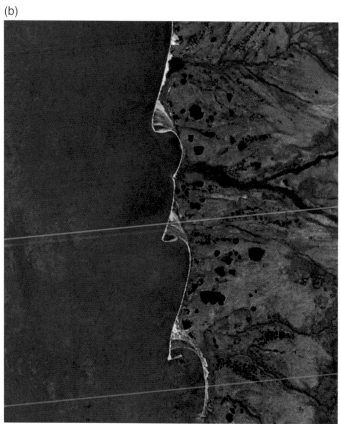

Figure 15.12 Cuspate forelands occur along many sandy coasts including (a) the North Carolina coast (*Source:* NASA), and (b) the Amur Sea (*Source:* Image © 2018 Google Earth).

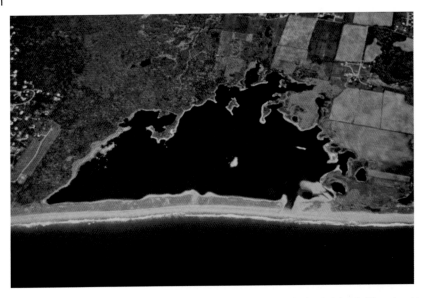

Figure 15.13 Welded barrier fronting a brackish Trustom Pond, Rhode Island. After the tidal inlet closed at the right side of the barrier, tidal exchange ceased and water in the pond became increasingly less saline.

a bay and attaches to the opposite shoreline. However, in many instances, their evolution represents a long-term response to rising sea level. This process is evident along the southern shore of Martha's Vineyard offshore of Cape Cod where low welded barriers front elongated bays. Historical evidence shows that some of these barriers were once cut by semi-permanent tidal inlets. Closure of the inlet occurred because the onshore migration of the barrier reduced bay area at a faster rate than rising sea level inundated the landward bay shoreline.

An interesting phenomenon of welded barriers occurs on the peninsula coast of Alaska and along other gravelly shorelines of the world. Here, welded barriers, which are composed chiefly of gravel and lesser amounts of sand, contain multiple gravel ridges constructed by storm waves. The barriers front ponded water areas and the permeable nature of their sediment permits the seepage of water from the pond through the barrier and into the ocean. Land drainage and the permeability of the barrier sediments control the level of water in the bays. During periods of high water influx, the level of the bay may overtop the crest of the barrier

cutting an ephemeral outlet as water drains into the ocean. After bay water levels have subsided, increased rates of sediment transport commonly accompanying storms and high wave events once again seal off the opening.

15.5.3 Barrier Islands

Barrier islands are relatively narrow strips of sand that parallel the mainland coast. They usually occur in chains and, excepting the tidal inlets that separate them, may extend uninterrupted for over a hundred kilometers. For example, Padre Island along the Texas coast is about 200 km long. Barrier chains may consist of a few islands or more than a dozen. The length and width of barriers and overall morphology of barrier coasts are related to several parameters including tidal range, wave energy, sediment supply, sea-level trends and basement controls. The fact they exist along most of the east and Gulf coasts of the United States with different geologic and oceanographic conditions suggests that barriers can form and be maintained in a variety of settings. In the next few sections of this chapter we will explore the formation of barrier islands, the different layers comprising barriers and what they look like in cross section, and the morphology of barrier island coasts and individual barrier islands.

15.5.3.1 Origin of Barrier Islands

The widespread distribution of barrier islands along the world's coastlines and their occurrence in many different environmental settings has led numerous scientists to speculate on their origin for more than 150 years (Figure 15.14). Any acceptable theory of their formation must explain the following:

1) Barrier chains are aligned parallel to the coast.
2) Most have formed in a regime of slow eustatic sea-level rise.
3) They are separated from the mainland by shallow lagoons, marshes and/or tidal flats.
4) Tidal inlets separate individual barriers along a chain.
5) They are composed of sand (some contain gravel).
6) They formed during periods of sand abundance (Question: where did the sediment come from?)

The different explanations of barrier island formation can be grouped into three major theories: Offshore bar theory (de Beaumont and Johnson); Spit accretion theory (Gilbert – Fisher); and Submergence theory (McGee – Hoyt). It will be shown that no one theory can explain the development of all barriers and moreover, it is sometimes difficult or impossible to prove how an individual barrier formed because many barriers have been drastically modified after they were formed due to rising sea-level, onshore migration and other related processes.

E. de Beaumont

One of the earliest ideas of barrier island formation was published in 1845 by a Frenchman, Elie de Beaumont, who studied coastal charts of barriers along the North Sea and Baltic Sea in Europe, those in the Gulf of Mexico, North Africa, and elsewhere. He believed that waves moving into shallow water churned up sand, which was deposited in the form of a submarine bar when the waves broke and lost much of their energy. As the bars accreted vertically, they gradually built above sea level forming barrier islands.

G.K. Gilbert

Some years later in 1885, de Beaumont's idea that barriers were formed from offshore sand sources was countered by G.K. Gilbert, who argued that the barrier sediments came from alongshore sources. Gilbert proposed that sediment moving in the breaker zone through agitation by waves would construct spits extending from headlands parallel to the coast. The subsequent breaching of spits by storm waves would form barrier islands. What is truly amazing about G.K. Gilbert is that this theory was conceived in western Utah while he was studying former lake deposits associated with the terraces along the Wasatch Mountains. The former lake was Lake Bonneville, which formed during the Pleistocene when the climate was cooler, reaching a size of 50,000 km^2 or about the size of present-day Lake Michigan. It was along this lake that spits had once developed. As the climate warmed in the Holocene, Lake Bonneville diminished to one tenth its size to what is now Great Salt Lake.

W.D. McGee

A third barrier island theory was published by W.D. McGee in 1890. He reasoned that the east and Gulf coasts of the United States were undergoing submergence, as evidenced by the many drowned river valleys that occur along these coasts including Raritan, Delaware and Chesapeake Bays. He believed that during submergence, coastal ridges were separated from the mainland, forming lagoons behind the ridges. He called the barriers "keys" and used Dolphin, Petit-Bois, Massacre, Horn, Dog, Ship, and Cat Islands along the coast of Mississippi as examples where coastal submergence had formed barrier islands.

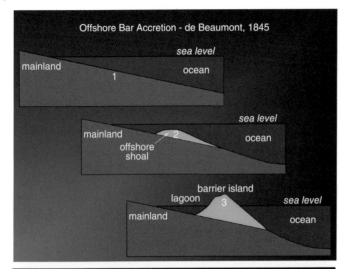

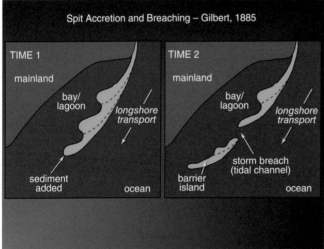

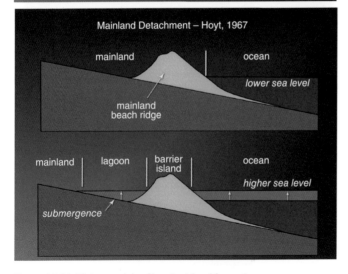

Figure 15.14 Major models of barrier island formation.

D.W. Johnson

At the turn of the century there were three viable explanations of barrier island formation, however, none of these theories had been tested using laboratory experiments or through the collection of field data. In 1919, D.W. Johnson re-investigated the various theories and studied the shore-normal profile of barrier coasts. He reasoned that if barriers had formed from spits, then the offshore profile should inter-sect the mainland at the edge of the lagoon. Additionally, the profile should appear as though sand had simply been deposited onto a uniform sloping nearshore ramp. However, Johnson discovered that most barrier coasts do not exhibit this type of profile, rather they seem to indicate that sand had been scooped from the nearshore and moved onshore to build the barrier. Thus, Johnson became a champion of the offshore bar theory.

John Hoyt

During the next 50 years the formation of barrier islands was a subject of much debate, but it was not until sediment cores began to be gathered through barriers and in lagoonal regions that progress was made in under-standing their development. John Hoyt worked along the coast of Georgia and many of his conclusions concerning barrier island formation were based on coring studies there. In an article published in 1967, he cor-rectly argued that if barriers had developed from offshore bars or through the breaching of spits, then open-ocean conditions would have existed along the mainland prior to bar-rier formation. Before becoming sheltered, breaking waves and onshore winds would have formed beaches and dune systems at these locations. Hoyt was able to show that for most barrier systems an open-ocean coast never existed along the present mainland coast. In situations where rising sea level and marsh development has led to an encroach-ment of lagoonal deposits onto the mainland, sediment cores from these regions have showed no evidence of beach and nearshore sediments or the shells of organisms that commonly inhabit the nearshore region.

Hoyt maintained that if coastal dunes or beach ridges were gradually submerged by rising sea level, then the mainland shoreline would never have been exposed to waves and thus nearshore deposits would never have developed. Hoyt went on to show that width of the lagoon was a function of the level of submergence and slope of the land surface. Steeper slopes and lower levels of submer-gence produced narrow lagoons; whereas flat land gradients and greater levels of submer-gence led to wider lagoons. In his theory the low areas along the beach ridges became the sites of tidal inlets. He also noted that once the barrier was formed it could then be mod-ified by waves, sediment supply and sea-level changes.

John Fisher

Armed with sedimentologic data, Hoyt had made a compelling case for barrier islands forming by submergence of coastal dunes or beach ridges due to rising sea level. However, in 1968 John Fisher produced a critical review of the submergence theory pointing out that long, straight, and continuous dune ridges would not occur along a coast being inundated by rising sea level. Moreover, in his own research along the Outer Banks of North Carolina he found that sediments beneath the lagoonal deposits were more like those occurring behind a spit than in a coastal upland as suggested by Hoyt. In light of these observations, Fisher became a strong propo-nent of Gilbert's spit accretion theory.

Recent Barrier Studies

Since the late-1960s there have been many studies of barrier islands aimed at determin-ing the sedimentary layers making up the barrier and deciphering the manner in which these layers were deposited. This research has been aided by the radiometric dating of organic material contained within the barri-ers' sediments, such as shells, peat and wood, thereby providing a chronology (timing) of barrier construction. In addition to coring the barrier sediments, ground-penetrating radar (GPR) has also been employed to study barriers. This device sends electromagnetic energy into the ground and where the electrical conductivity of the sediment changes, such as at the interface of two sediment layers having different grain sizes, mineral compositions or organic content, some of the energy is reflected back to the surface. These signals are received by an antenna and after processing provide an X-ray view of the sediment layers comprising the barrier. Sediment cores are taken in conjunction with the GPR surveys to determine the composition of the sediment layers and the environment in which they were deposited. Additional information concerning former barriers and their associated tidal inlets and lagoons has been gathered from the inner continental shelf using high-resolution shallow seismic surveys, a technology similar in principle to GPR. These advancements have provided new insights about the formation of coastal barrier systems.

Scientists now accept the idea that barriers can form by a number of different mechanisms. For example, along the west coast of Florida there is historical documentation and direct observations that barriers have formed from subtidal bars migrating onshore, primarily during storms. Other barrier systems have undoubtedly developed from spits such as the barriers along the outer coast of Cape Cod or those found along the southern coast of Washington. In both cases, the spits are fed by abundant sand sources derived from eroding glacial cliffs in Cape Cod and the Columbia River in Washington. Along the coast of Louisiana former lobes of the Mississippi River delta have been reworked by wave action forming beach ridge complexes. Prolonged sinking of the marshes (subsidence) behind the barriers has converted these former vegetated wetlands to open-water areas leading to barrier detachment from the mainland (Figure 15.15).

In considering the formation of barrier islands, it is important to recognize that almost all the world's barriers are less than 6500 years old and most are younger than 4000 years old. Most barriers formed in a regime of rising sea level but during a time when the rate of rise began to slow. Sedimentological data from the inner continental shelves off the East Coast of the United States, in the North Sea, and in southeast Australia suggest that barriers once existed offshore and have migrated to their present positions. When we core the landward side of barrier islands, which in most cases represents the oldest part of the barrier, we discover that these sediments consists of overwash deposits, often overlying lagoonal units. This sequence of sediment layers indicates that the barrier was migrating onshore during its initial development. Many barriers eventually stabilized and then prograded seaward when the supply of sediment became more plentiful and the rate of sea-level rise continued to slow.

Final Observation

If the initial barriers migrated onshore to their present position and most of the structure of a barrier developed after it stabilized onshore, then it is a bit ironic that so much attention has been given to how barriers formed when little or none of this signature has been preserved in the present barrier system.

15.6 Prograding, Retrograding, and Aggrading Barriers

The overall form of barriers, their stability, and future erosional or depositional trends are related to the supply of sediment, the rate of sea-level rise, storm cycles, and the topography of the mainland. When a barrier builds in a seaward direction it is said to prograde and is called a prograding barrier. The opposite of this condition occurs when a barrier retreats landward, called a retrograding barrier. If a barrier builds vertically and maintains its form as sea level rises, it is labeled an aggrading barrier. These different types are described below.

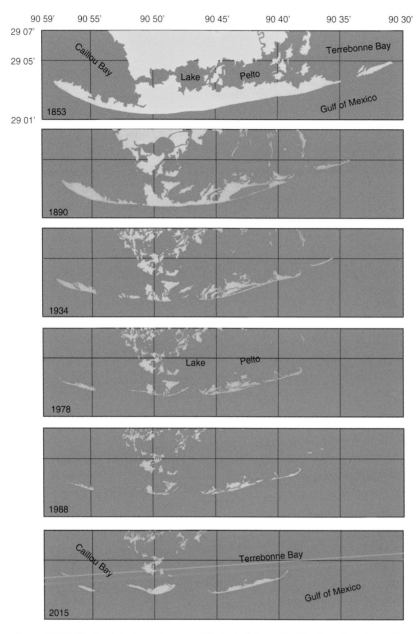

Figure 15.15 Barrier detachment process illustrated along the Isles Dernières located on the Mississippi River Delta. As the delta plain subsides, marshlands are converted to bays. In a period of only 125 years, semi-protected Pelto and Little Pelto Bays were transformed to large open-water environments (*Source:* From Boyd et al. (1987)).

15.6.1   Prograding Barriers

Prograding barriers form in a regime of abundant sand supply during a period of stable or slowly rising sea level (Figure 15.16). The sand to build these barriers may come from along the shore or from offshore sources. These conditions were met along much of the east coast of the United States and many other regions of the world about 4000–5000 years ago when the rate of sea-level rise

(a)   Prograding Barrier

Time 1

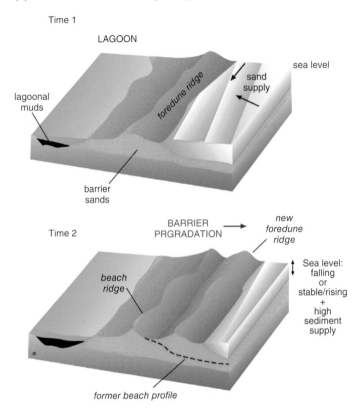

(b)

Figure 15.16 Prograding barriers develop in regions of abundant sediment supply. (a) Model of a prograding barrier. (b) Bull Island is a beach-ridge barrier located north of Charleston, South Carolina, which formed through the addition of prograding ridges. Periods of erosion are indicated by truncated beach ridges.

slowed and sand was contributed to the shore from eroding headlands (examples: Provinceland Spit in northern Cape Cod, Massachusetts; Lawrencetown barrier along the northeast shore of Nova Scotia), from the inner continental shelf (examples: the barrier system along Bogue Banks, North Carolina; the Algarve barrier chain in southern Portugal; the Tuncarry barrier in southeast Australia; the East Frisian Islands along the German North Sea coast), and directly from rivers (examples: most barrier systems in northern New England; the barrier chain situated north of the Columbia River along the southern coast of Washington).

Prograding barriers commonly build in spurts when the supply of sand is plentiful. Increases in the rate of sediment contribution to the downdrift coast can occur during storms when unconsolidated cliffs are eroded, releasing large quantities of sand to the littoral system, or during floods when rivers transport high sediment loads to the coast. In other instances, sediment is moved onshore from the inner shelf during long-term accretionary wave conditions. These processes cause beaches to build in a seaward direction, widening the berm and separating the foredune ridge from the ocean by an enlarging expanse of sand.

A common end-product of shoreline progradation is the development of beach ridges. As described earlier, beach ridges are dune systems that usually become vegetated first by grasses, then shrubbery, and finally by trees. This change in the maturity of the vegetative cover occurs as the ridges are displaced further and further from the shoreline and become more protected from salt spray and the effects of storms. Each beach ridge marks a former shoreline position and their overall pattern indicates how the barrier grew and evolved through time. The interval of time between the formation of successive beach ridges may be tens to hundreds of years.

Beach ridges form by a number of different mechanisms involving both erosional and depositional processes (Figure 15.17):

- *Scarp development* When a beach builds out gradually, the wind molds the upper berm into random hummocks (small incipient dunes). During moderate to large storms, the berm erodes back, producing a continuous scarp along the beach that cuts across these hummocks. This scarp provides a locus against which wind-blown sand is deposited. In this way a dune ridge is established along the entire length of the barrier. Repetition of these processes forms a beach-ridge barrier such as those along the central South Carolina and Georgia coasts.

- *Wrack lines* During the winter and early spring, storms coupled with spring high tides float organic material, largely composed of dead marsh grass, out of the back-barrier dispersing it along adjacent beaches. The high water levels and large waves accompanying storms cause the debris to be deposited along the upper portion of the beach. Depending upon the region, these wrack or drift lines, as they are called, may also consist of eel grass, seaweed, or driftwood. The wrack line traps wind–blown sand causing dune development. In this way, a dune ridge and future beach ridge may form along the entire length of the barrier.

- *Bar migration* A barrier can prograde through the addition of sand bars that migrate onshore and attach to the upper beach. Landward-migrating bars are particularly common in the vicinity of tidal inlets and are discussed in much greater detail in Chapter 16 If one of these bars incompletely welds to the upper berm, then a low area or swale develops between the bar and the landward beach. The bar itself commonly builds vertically from wind-blown sand, forming a dune ridge. When a series of bars migrates onshore and attaches to the beach, the resulting ridge and swale topography often becomes a beach-ridge barrier.

In fact, beach ridges may develop by a combination of two or more of the above

(a)

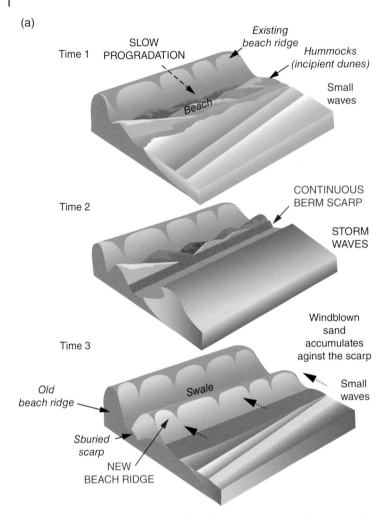

Figure 15.17 Various mechanisms of beach ridge formation. (a) Dune scarping. (b) Accumulation of sand around a wrack line. (c) Bar welding.

processes. A barrier composed of beach ridges is easily identified as one having evolved through progradation; however, this morphology does not necessarily indicate that the barrier is still prograding. Conversely, sedimentation conditions along the barrier may have changed such that now it has become erosional.

15.6.2 Retrograding Barriers

This type of barrier forms when the supply of sand is inadequate to keep pace with relative sea-level rise and/or with sand losses. Stated in terms of a sediment budget, a barrier becomes retrogradational when the amount of sand contributed to the barrier is less than the volume transported away from the barrier (Figure 15.18). The sand may be lost offshore during storms, moved along shore in the littoral system, or transported across the barrier by overwash. The end result is erosion to the front of the barrier causing a decrease in width of the beach and ultimately a destruction of the foredune ridge. Eventually, a narrowing and lowering of the barrier profile produce a retreat of the barrier across the adjacent bay, lagoon, or marsh system. This landward migration of the barrier, termed barrier rollover, is accomplished

(b)

Wrack-Line Stabilization

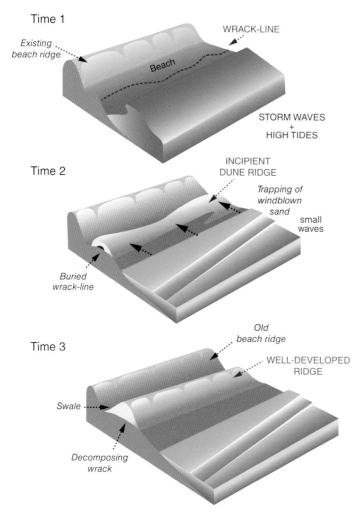

Figure 15.17 (Continued)

primarily during storms by overwash. Overwash is a cannibalistic process whereby storm waves transport sand from the beach through the dunes, depositing it along the landward margin of the barrier. In this way the barrier is preserved by retreating landward (Figure 15.19).

Historical records demonstrate that some barriers, such as the Isles Dernières off the Louisiana coast, have migrated onshore from one to several times their widths during the past 100 years. The major cause of barrier retreat in Louisiana and much of the Gulf Coast region is attributed to relative sea-level rise and exhausted sediment supplies. Retreat of the Isles Dernières is coincident with storm surges and large waves associated with the passage of hurricanes. A somewhat different scenario of barrier retreat has occurred along the central Maryland coast. Here in 1933 a severe hurricane breached the northern end of Assateague Island, creating a tidal inlet (Ocean City Inlet). In 1935, jetties built to stabilize the entrance to the inlet blocked the southerly transport of sand that had

(c) Bar Migration and Attachment

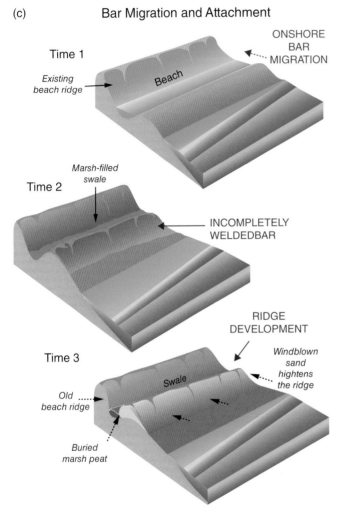

Figure 15.17 (Continued)

once nourished Assateague Island. In addition, strong ebb tidal currents produced in the jettied channel transported large quantities of sediment offshore, forming a massive sand shoal known as an ebb-tidal delta. Due to these sand-trapping mechanisms, the sediment-starved shoreline immediately downdrift of the inlet began to erode. Eventually, erosion along northern Assateague Island reached a critical width (~200 m) and low overall elevation such that overwash activity produced barrier rollover. In less than 50 years after the island was breached, the barrier south of the inlet had retreated one island width across the adjacent lagoon.

Retrograding barriers are identified by their overall narrow width, single or nonexistent foredune ridge, and washover aprons. Because these barriers have been migrating over various types of backbarrier settings, including lagoons and marshes, the sedimentary components comprising these environments are commonly exposed along the front side of the barrier, usually in the intertidal zone. This explains the appearance of stiff muds and the remnants of peat deposits along the lower beach of many

(a)

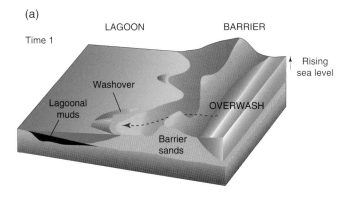

LAGOON   BARRIER

Time 1

Rising
sea level

Washover
Lagoonal
muds   OVERWASH

Barrier
sands

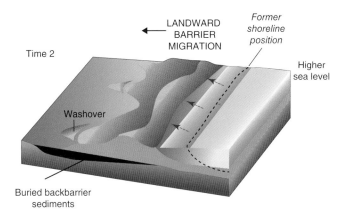

*Former
shoreline
position*

LANDWARD
BARRIER
MIGRATION

Time 2

Higher
sea level

Washover

Buried backbarrier
sediments

(b)

Figure 15.18 Retrogradational barriers develop in a regime of rising sea level and a depleted sediment supply. (a) Model of a retrogradational barrier. (b) Barrier rollover is occurring along the Magdalen Islands by storm overwash.

(a)

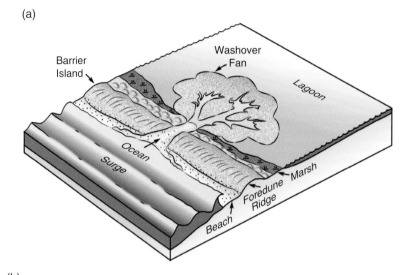

(b)

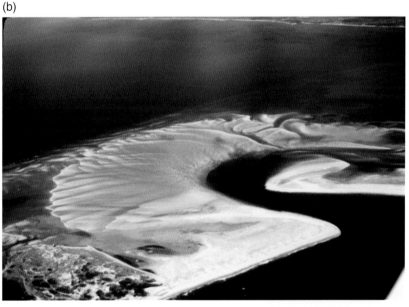

Figure 15.19 (a) Washovers are the primary mechanism whereby a barrier migrates onshore (*Source:* From Davis (1994)). (b) In some instances a washover is a precursor to barrier breaching and tidal inlet formation. This aerial view of Monomoy Island along the southeast coast of Cape Cod, Massachusetts depicts a case in which a overwash activity progressed into a breaching of the island and transformation of the washover fan into an expansive flood-tidal delta more than 1.5 km in width.

retrograding barriers. In some instances, tree stumps may extend through the sands of the intertidal zone, indicating that the barrier has migrated onshore so far that trees once growing on the mainland have resurfaced as stumps on the seaward side of the barrier.

15.6.3 Aggrading Barriers

If a barrier has built vertically during a regime of rising sea level and occupies approximately the same footprint as it did when it first formed or stabilized, it is termed an aggrading barrier (Figure 15.20).

(a)

Aggrading Barrier

Time 1

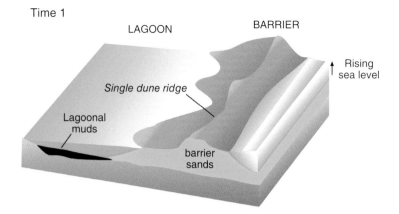

LAGOON

BARRIER

Single dune ridge

Lagoonal muds

barrier sands

Rising sea level

Time 2

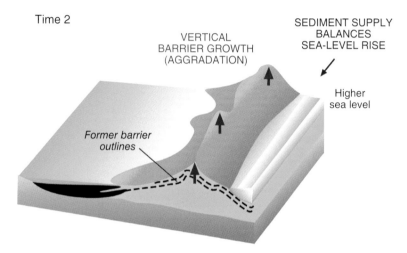

VERTICAL BARRIER GROWTH (AGGRADATION)

SEDIMENT SUPPLY BALANCES SEA-LEVEL RISE

Former barrier outlines

Higher sea level

(b)

Figure 15.20 Aggrading barriers remain approximately in the same position, building vertically in a regime of slow sea level rise. (a) Model of an aggrading barrier. (b) Good Harbor Beach, in northern New England, is a pocket barrier positioned between bedrock promontories and backed by a salt marsh and tidal creek system. Coring studies have revealed that the barrier has accreted vertically during the past 2500 years and is more than 7 m in thickness.

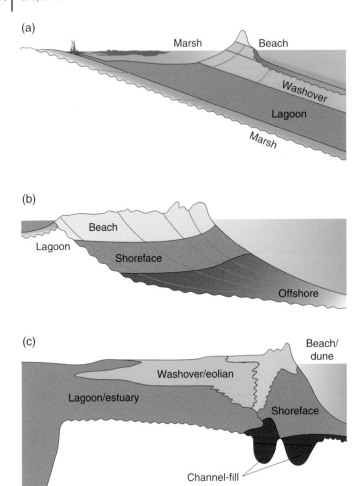

Figure 15.21 Stratigraphic models of different types of barrier sequences (*Source:* From Galloway and Hobday (1980)), (a) In the transgressive model, barrier sands are underlain by washover and lagoonal facies. (b) Regressive barriers tend to be relatively wide and commonly contain prograding beach ridges. Barrier sands overlie shoreface units. (c) Aggradational barriers represent steady-state conditions whereby the supply of new sand just compensates for rising sea level. Under these conditions the barrier neither migrates onshore or offshore, but rather builds vertically.

These barriers are rare because it requires that sediment is supplied (to the barrier) at a rate that exactly compensates for rising sea level. Too little sand and the barrier migrates onshore (retrograding), too large a supply and the barrier builds seaward. Part of Padre Island along the Texas coast is an example of an aggrading barrier. Without subsurface information aggrading barriers are difficult to recognize, because morphologically they may appear similar to non-beach ridge, prograding barrier or even a retrograding barrier that has stopped moving onshore.

15.7 Barrier Stratigraphy

Barriers exhibit a variety of architectures consisting of many different types of sedimentary deposits depending upon their evolutionary development (Figure 15.21).

Figure 15.22 Coke Island in southern North Carolina is a relatively thin retrogradational barrier consisting largely of coalescing overwash fans deposited on top of the marsh.

The sequence and composition of the layers making up the barrier, termed its stratigraphy, are defined by a set of grain size, mineralogical, and other characteristics of the layers. Factors such as sediment supply, rate of sea-level rise, wave and tidal energy, climate and topography of the land dictate how a barrier develops and its resulting stratigraphy. For example, barriers that have formed in the vicinity of the Mississippi River delta consist of fine to very fine sand because this is the most abundant sand size delivered by the river. In contrast, barriers along glaciated coasts tend to consist of coarser-grained sediment, including gravel, due to the coarse-grained, often heterogeneous, nature of glacial deposits found along these coasts. The rivers of these regions also tend to deliver fine to coarse sand to the coast. The gravelly sand barriers along the northeast shore of Nova Scotia exemplify this condition, having been formed from the erosion of glacial features called drumlins (see Chapter 17). In some regions the sediment comprising the barriers may have come from more than one source. For example, on the Gulf Coast of Florida, barrier sediments are composed of two populations: carbonate sand derived from shells and other carbonate material, and terrigenous sand originally sourced from the Appalachian Mountains and transported south along the Florida coast.

Barriers exhibit highly variable thicknesses from the thin deposits (~2 m) such as the Chandeleur Islands off the Louisiana coast or (1.2 m) Coke Island in North Carolina (Figure 15.22) to thick sand deposits, including the Tuncurry barrier in southeast Australia, where barrier sands extend to depths of more than 20 m or Plum Island in northern Massachusetts which ranges from 15 to 20 m in thickness. Along the west coast of Florida there is little new sand being added to this coast due to the lack of any riverine supply and the dearth of sediment on the inner continental shelf. Thus, the barrier deposits here are mostly less than 5 m thick. Generally, there is a direct correspondence between barrier thickness and sediment abundance. Plentiful sand supplies cause barriers to build seaward and aid in dune construction. Both processes contribute to thick barrier sequences. Another important factor affecting barrier thickness is accommodation space, which defines how much room is available for the accumulation of barrier sands. Steeper-gradient coasts produce more accommodation space as a barrier progrades than do flat-lying coasts. Likewise as spit builds into a deep bay, the thickness of the barrier sands will increase as the accommodation space increases. Rising sea level can produce a similar effect. For example, in Cape Cod Bay the Sandy Neck barrier has been building across Barnstable Bay for the past 3500 years. During that time, sea level has risen approximately 3 m and thus if the barrier maintains nearly the same elevation above mean high water through time, then the spit will be about 3 m thicker at its end than where spit growth was initiated.

Barrier sequences often contain tidal inlet deposits, especially along barrier coasts where tidal inlets open and close and/or where tidal inlet migration is an active process. A tidal inlet migrates by eroding the downdrift side of its channel while at the same time sand is added to the updrift side of its channel. In this way, the updrift barrier elongates, the downdrift barrier becomes shorter, and the migrating inlet leaves behind channel-fill deposits underlying the updrift barrier (Figure 15.23). Independent studies along New Jersey and the Delmarva Peninsula, North Carolina, and South Carolina indicate that 20–40% of these barrier coasts are underlain by tidal inlet fill deposits. Long-term tidal inlet migration along Shackleford Banks in North Carolina has produced inlet fills 10–20 m thick beneath 90% of the island.

In terms of prograding, retrograding, and aggrading barriers, each of these systems has a diagnostic stratigraphy that reflects the manner in which it developed (Figure 15.21). If we were able to cut a deep trench through the barrier, the layers of the sediments comprising the barrier would be revealed. In actuality, the stratigraphy of barriers is determined from numerous sediment cores drilled throughout the barrier, and in some instances augmented with ground-penetrating radar and other geophysical information. In each

case discussed below, the stratigraphy is described from the base of Holocene (post Pleistocene, 10,000 years ago to present) barrier contact to the surface.

- *Prograding barrier* Because this type of barrier builds in a seaward direction, the barrier sequence is commonly thick (10–20 m) and overlies offshore deposits, usually composed of fine-grained sands and silts. The barrier sequence consists of nearshore sands, overlain by beach deposits, and topped by dune sands. The contacts between the units are gradational and for the most part the sedimentary sequence coarsens upward except for the uppermost fined-grained dune sands.
- *Retrograding barrier* This barrier type migrates in a landward direction over the marsh and lagoon by overwash processes. The Holocene sequence typically bottoms in backbarrier sediments, however, if the barrier has retreated far enough landward, mainland deposits may be preserved, forming the base of the sequence. In this instance, we may find tree stumps, soils, and other deposits. The mainland units are overlain by a variety backbarrier deposits including lagoonal silt and clay and marsh peat, which had formed in intertidal areas. In the vicinity of tidal inlets, backbarrier deposits consist of channel sands and large sand shoals called flood-tidal deltas (see Chapter 16 on tidal inlets). Overlying the backbarrier deposits is the thin barrier sequence (<3–4 m) consisting of washovers, beach deposits, and dune sediments if they are present.
- *Aggrading barrier* These barriers build upward in a regime of rising sea level and in an ideal case, the deposits from the same environmental setting are stacked vertically. In most cases, however, the barrier has shifted slightly landward and seaward through time due to changes in sediment supply and rates of sea-level rise. Therefore, most aggrading barriers exhibit some interstacking of various units. For example,

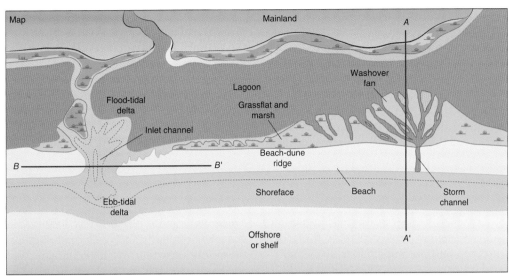

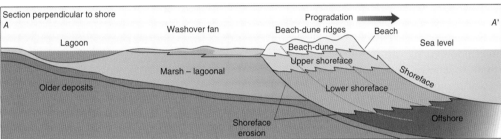

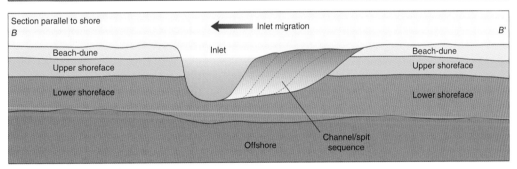

Figure 15.23 The stratigraphy of a barrier is dependent of its evolutionary history (*Source:* From McCubbin (1982). Reproduced with permission of American Association of Petroleum Geologists). It may contain elements of storm washovers (section *A–A'*), beach ridge progradation (section *A–A'*), or tidal inlet migration and spit accretion (section *B–B'*).

in the rear of the barrier the sequence may consist of washover and dune units inter-layered with marsh and lagoonal deposits. Aggrading barriers tend to be thick (10–20 m) and for reasons stated previously, are uncommon.

15.8 Barrier Coast Morphology

There are many factors that determine the location and size of individual features along a coast, such as the slope of the land dictating the size of a lagoon or a former stream valley controlling the position of a tidal inlet. Despite the many factors influencing coastal morphology in coastal plain settings, the overall distribution of barriers, tidal inlets and various backbarrier environments is primarily related to the relative magnitude of wave and tidal energy. In a simplification of their respective roles, waves are responsible for transporting sediment alongshore, which tends to elongate barriers. The rise and fall of the tides cause a filling and emptying of backbarrier areas. Tidal inlets through which this exchange of water occurs are the sites of strong tidal flow. These currents transport sand in an onshore and offshore direction.

15.8.1 Hayes Models

In a scheme conceived by Miles Hayes and later modified by him and others, depositional coastlines are separated into three classes based on the wave height and tidal range of the region. The three major divisions are wave-dominated, mixed-energy and tide-dominated settings (Figure 15.24). Barriers are found almost exclusively in the

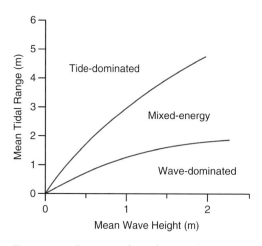

Figure 15.24 Depositional coastlines can be separated into three major types based on their wave and tidal energy. Barrier coasts occur almost exclusively in the wave-dominated and mixed-energy environments. Tide-dominated coasts are generally funnel-shaped and associated with a river. (*Source:* From Davis and Hayes (1984) after Hayes (1979)).

wave-dominated and mixed-energy environments. It should be noted that it is the ratio of wave height and tidal range that dictate the presence and distribution of barriers. For example, in the Ten Thousand Island region along the southern Florida Gulf Coast the tidal range is only about 1 m, but this section of coast, consisting primarily of mangrove islands, is clearly tide-dominated. The lack of barriers along this coast is the result of very low wave energy. Remember that barriers are wave-built accumulations of sand and therefore, where wave energy is insufficient to concentrate sand, beaches and barriers do not form. Similarly, in the German Bight of the North Sea extensive barrier development disappears toward the apex of the bight at the entrance to the Elbe River where the tidal range increases to almost 3.5 m. The large tidal range and expansive tidal flats of this region diminish wave energy, precluding the formation of beaches and barriers. Moreover, the importance of wave energy is illustrated along the west coast of the United States, where barriers exist despite spring tidal ranges approaching 4 m. Here an abundant sand supply and strong wave energy overcome the effects of tides to produce some very long progradational barriers (e.g. North Beach, and barriers fronting Grays Harbor, Washington).

Given the same ratios of wave height and tidal range, depositional coasts throughout the world exhibit similar morphologies as described below:

15.8.1.1 Wave-dominated Coast

These coasts are dominated by wave-generated longshore sediment transport; tides play a secondary role. They are characterized by long linear barrier islands and few tidal inlets (Figure 15.25). The backbarriers are composed of mostly open-water lagoons or bays. Marshes occur along the backside of the barriers, commonly on former washover deposits, and at the edge of the mainland. Large sand shoals are usually found on the landward side of the tidal inlets. Sand shoals on the seaward side are diminutive in size.

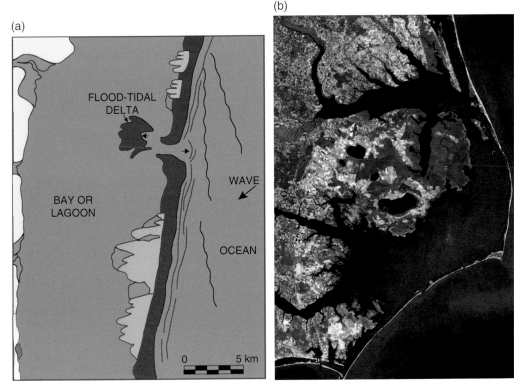

Figure 15.25 (a) Model of a wave-dominated barrier coast. (*Source:* From Hayes (1975, 1979)). These coasts are characterized by long barrier islands backed by lagoons and bays with few tidal inlets. (b) The Outer Banks of North Carolina are an example of a wave-dominated coast. The cuspate forelands of Cape Hatteras (north) and Cape Lookout (south) are joined by long, linear barrier islands that are interrupted by few tidal inlets.

Coastlines fitting into this class include the coasts of Texas, the Panhandle of Florida, the Outer Banks of North Carolina, Maryland, northern New Jersey, the barrier coast along the Nile River delta, and Southeast Iceland.

15.8.1.2 Mixed-Energy Coast

In this model both wave and tidal processes are important in shaping coastal morphology. Barriers on these coasts tend to be short and stubby (drumstick-shaped, see Chapter 16, Section 16.8.5) and tidal inlets are more numerous than along wave-dominated coasts, reflecting the greater role of the tides (Figure 15.26). The backbarriers of these regions are mostly filled with sediment and covered by expansive marshes incised by tidal creeks. Open-water areas commonly increase in extent near the inlets. In at least two locations of the world, the backbarriers of this coastal type (East Frisian Islands along the Germany North Sea coast and the Copper River Delta barriers in Alaska) consist of extensive tidal flats. It is likely that the change from open-water lagoons of wave-dominated coasts to the intertidal environments of mixed-energy coast is the result of more sediment being transported into the backbarrier by tidal currents. In addition, an increase in tidal range produces larger intertidal areas, which promote the development of marshes and stabilization of sediment. Mixed-energy coasts include the barrier coasts of northern New England, southern New Jersey, Virginia, southern North Carolina, South Carolina, Georgia, much of

(a)

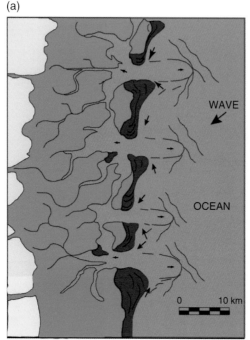

(b)

Figure 15.26 (a) Model of a mixed-energy barrier coast. (*Source:* From Hayes (1975, 1979)). Typically, these coasts contain short stubby barrier islands and numerous tidal inlets that connect to marsh and tidal creek backbarrier systems. (b) The beach ridge barriers in Georgia are an example of a mixed-energy coast. Many of the barriers along the Georgia Bight have formed around pre-existing Pleistocene-age barrier islands.

the west coast of Florida, the Frisian Islands in the North Sea (see Box 15.1), the Algarve in southern Portugal, and the Copper River delta barrier system in Alaska.

15.8.1.3 Tide-Dominated Coasts

Many tide-dominated coasts coincide with very large funnel-shaped embayments (>100 km across). This coastline configuration enhances the tidal wave, which produces large tidal ranges, strong tidal currents, and a sedimentation regime that is dominated by onshore–offshore transport. Due to the low wave energy along these coasts, barriers and tidal inlets are absent. Rivers commonly discharge sediment at the head of the embayments. The coarse-grained component of this sediment is deposited within the embayment and transported to the inner shelf forming large subtidal sand ridges that parallel the tidal flow. The fine-grained sediment tends to accumulate onshore forming expansive tidal

flats. Landward of the tidal flats are wide marshes. In equatorial regions, marshes are replaced by mangroves. The Bay of Fundy in Nova Scotia, the Gulf of Cambay, India, the head of the Bay of Bengal, Bangladesh, the mouth of the Amazon River and Bristol Bay, Alaska are examples of tide-dominated coasts.

15.8.2 Georgia Bight

A particularly interesting section of coast, demonstrating the relative influence of wave versus tidal energy, is seen in the Georgia Bight encompassing the region from North Carolina to Florida. The arcuate shape of this coast produces relatively narrow and steep continental shelves along the flanks of the bight and a wide, shallow shelf at the apex of the bight. From the Outer Banks of North Carolina to the Georgia coast the continental shelf widens, amplifying the tidal wave as it moves across the shelf resulting in larger

Box 15.1 Three Hundred Year History of the East Frisian Islands

The East Frisian Islands provide a unique opportunity to assess the long-term historical development of a barrier island chain. Hahns Homeier and Gunter Luck of the Forschungsstelle Norderney, Germany assembled a series of detailed maps that depict shoreline and bathymetric changes of this barrier system that extend back to 1650. There are few barrier coasts in the world in which the historical database goes back this far. The sequential maps demonstrate patterns of inlet migration, growth of individual barriers, reduction in size of the backbarrier drainage areas, and decrease in size of the ebb-tidal deltas. Interestingly, some of the historical changes are a product of a strong wind regime but others, including some large-scale morphological changes, are solely related to the actions by humans.

The East Frisian Islands are composed of seven barriers and five tidal inlets spanning the southeast coast of the North Sea between Ems River on the west and Weser River to the east (Box Figure 15.1.1). The barrier chain is 90 km long and separated by from the mainland by 4–12 km wide tidal flat, which is incised by a network of tidal channels. The winds in this region blow from the westerly quadrant during the entire year, producing a net easterly longshore transport rate of 270,000 m^3 $year^{-1}$ of sand. Spring tidal ranges vary systematically from 2.5 m at the western end of the chain to 2.9 m at the eastern end.

Much of the sand moving eastward in the littoral system is eventually delivered to the *seegats* (tidal inlets) where it bypasses the inlets through a complex pattern of wave and tidal processes. Sand enters the inlets from the west and is transported across the ebb-tidal delta, ultimately reaching the downdrift inlet shoreline in the form of large landward migrating swash bars. These bars are elongate parallel to shore and are 1–1.5 km in length, containing more than 100,000 m^3 of sand. As discussed in greater detail in Chapter 16 on tidal inlets, the shape of the barriers is dictated by where the swash bars weld to the beach. Drumstick-shaped barriers develop where the bars attach to the updrift end of the island, hump-backed barriers form when bars weld to the middle of the island, and a downdrift bulbous barrier shape is a product of bars moving onshore at the distal eastern end of the island.

The East Frisian islands have evolved in a regime of rising sea level like most other barrier sytems in the world. The fact that the

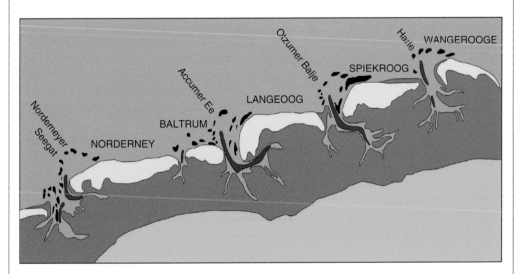

Box Figure 15.1.1 The East Frisian Islands are located along the coast of Germany on the North Sea.

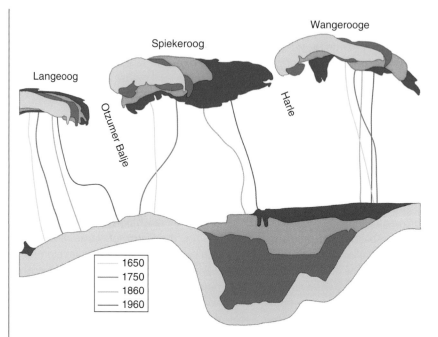

Box Figure 15.1.2 Changes in areal extent of the drainage system of Harle Inlet. Land reclamation of tidal flats has decreased the size of the backbarrier, resulting in smaller inlet tidal prism. (From FitzGerald, 1988).

Frisian barriers have grown in size during the past 300 years is evidence that the supply of sand to the island chain has more than compensated for land loss due to continued sea-level rise. The increase in dimensions of the barriers is a product of new sand being added to islands as well as due to human modification of the backside of the barriers through poldering (Box Figure 15.1.2). Poldering is the process whereby land is reclaimed from the sea. The practice involves building dikes across tidal flats and allowing sediment-laden tidal waters to enter regions that have been diked. After the suspended sediment from the seawater is deposited, the clear water is discharged from the dikes at low tide and new muddy sea water is allowed to flood through the dikes during the next high tide. After this procedure is repeated many hundreds of times, the tidal flat region accretes to an elevation where it can become productive farmland. Poldering explains how the backsides of the barrier have grown in size since 1650.

During the same period of time, individual barriers along the East Frisian chain have lengthened and some have prograded seaward. Between 1650 and 1960 the barriers increased in aerial extent by almost 42 km^2, an increase of 80%. Of this new land 56% was attributed to poldering and 44% was the result of barrier accretion. The source of new sand was puzzling to scientists until an historical analysis was made of the backbarrier region. It was found that from 1650 to 1960 the area drained by the tidal inlets (drainage area) decreased by 30%, amounting to 149 km^2 loss in tidal flats and open-water areas. The decrease in drainage area was due in part to spit accretion at the eastern end of the barriers but primarily a result of poldering, not only along the landward side of the barriers but also along the mainland shore. As the size of the backbarrier areas decreased, so too did tidal prism of the inlets. Because inlet tidal prism controls the volume of sand contained in the ebb-tidal deltas, extensive poldering in the backbarrier during the 1650–1960 period

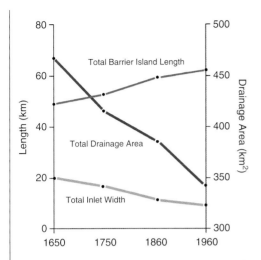

Box Figure 15.1.3 Graph showing morphological changes of the East Frisian inlets between 1650 and 1960. During this 310-year period the drainage areas decreased in size, resulting in smaller tidal inlets and longer barrier islands. (*Source:* From FitzGerald (1988)).

ultimately led to smaller equilibrium-sized ebb deltas. As the ebb deltas reduced in volume, wave energy transported the deltaic sand back onshore, increasing the sand supply to the adjacent beaches.

The end result of backbarrier poldering has been highly beneficial to the stability and long-term depositional trend along most of East Frisian Island chain. The striking growth of the barriers between 1650 and 1960 is reflected in an increase in the total length of the barriers by more than 14 km. Most of this increase has been at the expense of the tidal inlets, which collectively have narrowed by over 10 km (Box Figure 15.1.3). Thus, the history of the East Frisian Island chain demonstrates that changes to the backbarrier can affect the sediment supply to the fronting barriers and more importantly, human alterations can have a pronounced impact on the erosional–depositional trends of a barrier coast.

tidal ranges. Conversely, as the inner continental shelf flattens along this same stretch of coast, a greater proportion of the deepwater wave energy is attenuated. Thus, all other factors being approximately equal, tidal energy increases and wave energy decreases from North Carolina to Georgia. The coastal response to this change in physical setting is dramatic (Figure 15.27). The wave-dominated Outer Banks consists of long linear barriers interrupted by few tidal inlets and separated from the mainland by broad shallow sounds. Contrastingly, the mixed-energy Georgia coast is composed of relatively short beach ridge barriers separated by large tidal inlets with well-developed sand shoals. The backbarrier consists of marsh and tidal creeks. Whereas other factors control the width of the backbarrier, location of inlets, sediment supply and longshore transport directions, the overall morphology of the coast is a function of wave versus tidal energy, which, in turn, is strongly influenced by continental shelf width.

15.9 Barrier Coasts: Morphology and Evolution

In some areas of the world, the historical records are long enough to record the physical processes that produce major changes to the coast, thereby revealing how these barriers evolve. In other locations, detailed stratigraphic studies have led to a good understanding of barrier development. Using these studies, examples of different barrier settings, their morphology, and evolution are presented below.

15.9.1 Eastern Shore of Nova Scotia

Along glaciated coasts, local glacial deposits are commonly the major source of sediment to form barriers. These deposits are usually composed of till (a poorly sorted mixture of gravel, sand, silt and clay) or somewhat better-sorted outwash sediments consisting

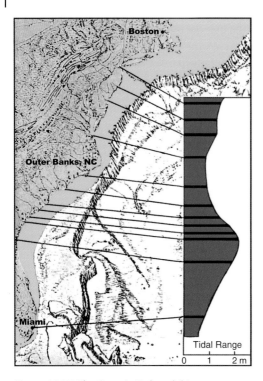

Figure 15.27 The Georgia Bight exhibits a systematic alteration in coastal morphology that is largely a product of changes in wave energy versus tidal energy which in turn is a function of continental shelf width. (*Source:* From FitzGerald (1996) using data from Nummedal et al. 1977 and Hayes (1979)).

of sand and fine gravel layers laid down by glacial meltwater streams. Due to the mixed nature of these deposits, the barriers that are derived from this sediment are composed of sand and gravel. In 1987, Ron Boyd and colleagues at Dalhousie University proposed a six-stage model to explain the evolution of the eastern shore of Nova Scotia (Figure 15.28). According to their model, when the last continental glaciers retreated from a position on the continental shelf, a variety of glacial deposits were left behind (Stage 1). As sea level rose, the large valleys excavated by the glaciers were drowned and transformed into embayments bordered by glacial headlands (Stage 2). As these headlands eroded due to wave attack, sand, and gravel were released., building spits into the adjacent bays. Where two spit systems growing from opposite sides

of the embayment joined, a baymouth barrier (a type of welded barrier) was formed (Stage 3) (Figure 15.29). As long as sediment supplies are adequate the barriers may prograde, resulting in a number of beach ridges and swales. As the headlands continue to erode, boulder retreat pavements define their former extent. Eventually, continuing sea-level rise and diminishing sediment supplies lead to erosion and breaching of the barriers (Stage 4). By Stage 5, the barriers are mostly destroyed and some of the sediment once contained in them is transported onshore by flood-tidal currents and storm wave action, forming intertidal and subtidal shoals. Continued sea level rise causes portions of the subaqueous shoals to migrate onshore and attach to new headland areas. Sediment contributed from the shoals and from the erosion of the new headlands helps to re-establish the barriers at the landward sites (Stage 6). At this point, the growth of the new barrier proceeds to Stage 3. With time, there may be several cycles of barrier growth, retreat, destruction, and re-establishment. Along glaciated coasts such as the eastern shore of Nova Scotia it is possible to find examples of different stages of this evolutionary model.

15.9.2 Mississippi River Delta Barriers

The Mississippi River delta in the Gulf of Mexico is composed of seven major delta lobes that were deposited approximately during the past 7000 years. Each lobe represents a period when the Mississippi River was debouching its sediment load essentially in one primary location. Thus, deltaic sedimentation has been characterized by periods of delta progradation followed by abandonment and delta building at a different nearby site. This process of delta lobe switching has produced five different barrier shorelines along the Holocene delta plain.

Shea Penland, John Suter, and Ron Boyd presented a model of barrier evolution for the Mississippi River delta in 1985 (Figure 15.30).

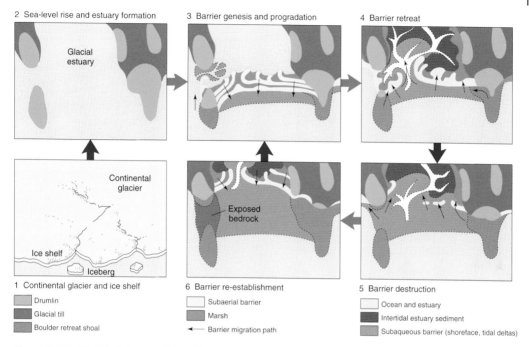

2 Sea-level rise and estuary formation

Glacial estuary

3 Barrier genesis and progradation

4 Barrier retreat

1 Continental glacier and ice shelf

Continental glacier

Ice shelf

Iceberg

6 Barrier re-establishment

Exposed bedrock

5 Barrier destruction

Drumlin

Glacial till

Boulder retreat shoal

Subaerial barrier

Marsh

← Barrier migration path

Ocean and estuary

Intertidal estuary sediment

Subaqueous barrier (shoreface, tidal deltas)

Figure 15.28 Model of shore evolution for the drumlin coast of the Eastern Shore of Nova Scotia. The pathway of barrier evolution (progradational versus retrogradational) is controlled by sediment supply and the rate of sea-level rise. (*Source:* From Boyd et al. (1987)).

Figure 15.29 The welded barrier and spits of central Nova Scotia portray Stage #3 of the drumlin coast model of Boyd et al. (1987).

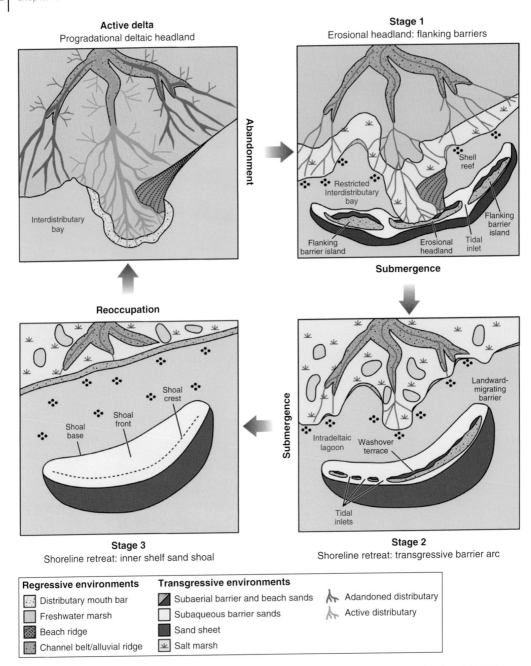

Figure 15.30 Model of barrier evolution for the Mississippi River Delta. (*Source:* From Penland et al. (1988) *Source:* Penland, Boyd, and Suter, 1988. Reproduced with permission of Society for Sedimentary Geology). The driving forces behind this model are the subsiding delta plain, the rising sea level and the changes in the locus of sediment-discharge by the Mississippi River.

In the first stage of their model, abandonment of the active delta leads to a subsidence of the delta plain and a high rate of relative sea-level rise. The active delta is transformed into an eroding headland with retreat of the shoreface.

Sediment eroded from the headland forms flanking barrier spits, which are subsequently breached during storms. A barrier island arc replaces the flanking barrier spit stage as the headland region continues to subside and sea level continues to rise rapidly (Stage 2). In time, the barrier island arc is separated from the mainland by a large expanse of shallow open water. The barrier island arc migrates onshore through overwash activity and the construction of flood shoals on the landward side of the inlets. For a time, the arc maintains a position above sea level by migrating on top of a stacked sequence of flood shoals and washover deposits. Eventually, the arc is transformed into a subtidal shoal as sea-level rise outpaces the ability of the barrier to build vertically (Stage 3). At this point, the active delta may reoccupy this region and process can begin anew. Each stage in the evolution of the barrier coast can be seen today along the Mississippi River delta.

15.10   Summary

Barriers occur on a worldwide basis but are predominantly found along Amero-trailing edge continental margins in coastal plain settings where the supply of sand is abundant. The coastal plain setting appears to be an important requirement for the wide distribution of barriers in that it provides a sediment source and a platform upon which sediment can accumulate. Barriers have many different forms but can be grouped into three major classes dependent on their connection of the mainland (barrier islands, barrier spits, and welded barriers). Barriers have formed by different mechanisms and most have migrated onshore to their present position. Retrograding barrier continuously move onshore through rollover processes, whereas, prograding barriers build seaward due to abundant sediment supplies and/or stable sea levels to slow rates in sea-level rise. Aggrading barriers are accrete vertically, keeping pace with rising sea level. The stratigraphy of barriers is dependent on their evolution and factors such as sediment supply, rate of sea-level rise, wave and tidal energy, climate and topography of the land.

References

Boyd, R., Bowen, A.J., and Hall, R.K. (1987). An evolutionary model for transgressive sedimentation on the eastern shore of nova scotia. In: *Glaciated Coasts* (ed. D.M. FitzGerald and P.S. Rosen), 87–114. New York: Academic Press.

Davies, D.K. (1980). *Geographic Variation of Coastal Development*. Harlow: Longman Group.

Davies, J.L. (1973). *Geographical Variation in Coastal Development*, 435. Edinburgh: Oliver and Boyd.

Davis, R.A. Jr. *The Evolving Coast*, 232. Scientific America Library.

Davis, R.A. Jr. and Hayes, M.O. (1984). What is a wave-dominated coast? *Mar. Geol.* 60: 313–329.

Dingler, J.R. and Clifton, H.E. (1994). Barrier systems of California, Oregon, and Washington. In: *Geology of Holocene Barrier Island Systems* (ed. R.A. Davis), 115–166. Berlin: Springer.

FitzGerald, D.M. (1988). Shoreline erosional–depositional processes associated with tidal inlets, Hydrodynamics and sediment dynamics of tidal inlets. In: *Hydrodynamics and Sediment Dynamics of Tidal Inlets*, (Lecture Notes on Coastal and Estuarine Studies Volume 29) (ed. D.G. Aubrey and L. Weishar), 186–225. Berlin: Springer.

FitzGerald, D.M. (1996). Geomorphic variability and morphologic and sedimentologic controls on tidal inlets. *Journal of Coastal Research*, Spec Issue # (23): 47–71.

Galloway, W.E. and Hobday, D.K. (1980). *Terrigenous Clastic Depositional Systems*. New York: Springer.

Gierloff-Emden, H.G. (1961). Nehrungen und Lagunen. *Petermanns Geogr. Mitt.* 105: 81–92; 161–176.

Hayes, M.O. (1975). Morphology of sand accumulations in estuaries. In: *Estuarine Research*, vol. 2 (ed. L.E. Cronin), 3–22. New York: Academic Press.

Hayes, M.O. (1979). Barrier island morphology as a function of tidal and wave regime.

In: *Barrier Islands* (ed. S.P. Leatherman), 1–28. New York: Academic Press.

Hayes, M.O., and T. Kana (1978). Terrigenous Clastic depositional Environments, Dept. of Geology, University of South Carolina, Columbia, South Carolina.

McCubbin, D.G. (1982). Barrier-island strand plain facies. In: *Sandstone Depositional Environments* (ed. P.A. Scholle and D. Spearing), 247–280. Tulsa, OK: A.A.P.G. Publishers.

Nummedal, D., Oertel, G., Hubbard, D.K., and Hine, A. (1977). Tidal inlet variability - Cape Hatteras to Cape Canaveral. Proc. of Coastal Sediments '77. ASCE, Charleston, SC, pP. 543–562.

Oertel, G. (1975). Ebb-tidal deltas of Georgia estuaries. In: *Estuarine Research*, vol. 2 (ed. L.E. Cronin), 267–276. New York: Academic Press.

Penland, S., Boyd, R., and Suter, J.R. (1988). Transgressive depositional systems of the Mississippi Delta plain: model for barrier shoreline and shelf sand development. *Jour. of Sed. Pet.* 58: 932–949.

Suggested Reading

Carter, R.W.G. and Woodroffe, C.D. (eds.) (1994). *Coastal Evolution*. Cambridge: Cambridge University Press.

Coates, R. (ed.) (1973). *Coastal Geomorphology*. Binghamton, NY: State University of New York Press.

Cronin, L.E. (ed.) (1978). *Estuarine Research*, vol. 2. New York: Academic Press.

Davis, R.A. (ed.) (1994). *Geology of Holocene Barrier Island Systems*. Berlin: Springer.

Fletcher, C.H., and J.F. Wehmiller, eds. (1992). Quaternary Coasts of the United States: Marine and Lacustrine Systems. SEPM Spec. Pub. #48.

Hayes, M.O., and T. Kana (1978). Terrigenous Clastic depositional Environments, Dept. of Geology, University of South Carolina, Columbia, South Carolina.

King, C.A.M. (1972). *Beaches and Coasts*. New York: St. Martin's Press.

Leatherman, S.P. (1983). *Barrier Island: From the Gulf of St. Lawrence to the Gulf of Mexico*. New York: Academic Press.

Oertel, G.F., and S.P. Leatherman (1985). *Marine Geology*, 63. Special Issue: Barrier Islands.

Nummedal, D., Pilkey, O.H., and Howard, J.D. (eds.) (1987). Sea Level Fluctuation and Coastal Evolution, SEPM Spec. Pub. #41.

Schwartz, M.L. (ed.) (1973). Barrier Islands. Stroudsburg, PA, Dowden, Hutchinson, and Ross.

16

Tidal Inlets

16.1 Introduction

Tidal inlets are found along barrier coastlines throughout the world. They provide a passageway for ships and small boats to travel from the open ocean to sheltered waters (Figure 16.1). Along many coasts of the world, including much of the east and Gulf Coasts of the United States, the only safe harborages, including some major ports, are found behind barrier islands. The importance of inlets in providing navigation routes to these harbors is demonstrated by the large number of improvements that are performed at the entrance to inlets such as stabilization by the construction of jetties and breakwaters, dredging of channels and the operation of sand-bypassing facilities.

Tidal inlets are also conduits through which nutrients are exchanged between backbarrier lagoons and estuaries and the open coast. Numerous species of finfish and shellfish rely upon tidal inlets for access to backbarrier regions for feeding, breeding, and as nursery grounds of their young. The fact that many fish travel through inlets in search of food makes tidal inlets prize locations for saltwater sport-fishing. In many lagoons, tidal inlets maintain the salinities, temperatures, and nutrient levels that are necessary for the reproduction and growth of valuable shellfish. For example, along the coast of Massachusetts, there are several sites in which the saltwater passageways (tidal inlets) to the bay or lagoon periodically

close. When this occurs, the freshwater influx gradually reduces the salinity in the lagoon, making the environment inhospitable to many saltwater shellfish. The spits fronting these lagoons are then artificially opened to maintain the proper habitat for various types of clams that are harvested by local fishermen.

Understanding tidal inlet processes is not only important for the maintenance of navigable waterways, but also for the management of adjacent barrier shorelines. Tidal inlets interrupt the longshore transport of sediment affecting both the supply of sand to the downdrift beaches and erosional–depositional processes along the inlet shoreline. As will be shown, the largest scale of shoreline changes along barriers occurs in the vicinity of inlets and these are a direct consequence of tidal inlet processes. The changes may be due to inlet migration, concentrated wave energy, large bars migrating onshore, sand losses to the backbarrier, and other processes that will be treated in this chapter. This information may become particularly important when considering the purchase of real estate on barrier islands in the vicinity of inlets (Figure 16.2).

16.2 What Is a Tidal Inlet

A tidal inlet is defined as an opening in the shoreline through which water penetrates the land, thereby proving a connection

Beaches and Coasts, Second Edition. Richard A. Davis, Jr. and Duncan M. FitzGerald.
© 2020 John Wiley & Sons Ltd. Published 2020 by John Wiley & Sons Ltd.

Figure 16.1 Tidal inlets serve as passageways to harbors and conduits through which nutrients are exported to coastal waters.

Figure 16.2 The apartment building at the northeastern end of Wrightsville Beach, North Carolina is endangered by the southerly migration of Masons Inlet.

between the ocean and bays, lagoons, and marsh and tidal creek systems. The main channel of a tidal inlet is maintained by tidal currents (Figure 16.3).

The second half of this definition distinguishes tidal inlets from large, open embayments, or passageways along rocky coasts. Tidal currents at inlets are responsible

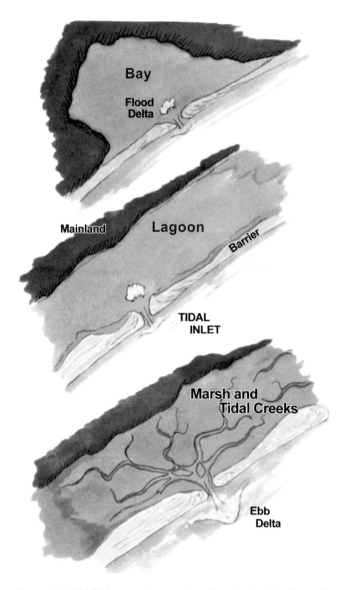

Figure 16.3 Tidal inlets are the openings along barrier shorelines. They allow the exchange of tidal waters between the ocean and backbarrier, which consists of bays, lagoons and marsh and tidal creeks. They are fronted by ebb-tidal deltas and backed by flood-tidal deltas.

for the continual removal of sediment dumped into the main channel by wave action. Thus, according to this definition, tidal inlets occur along sandy or sand and gravel barrier coastlines, although one side may abut a bedrock headland. For example, along the coast of Maine the entrance to York Harbor is bordered on both sides by bedrock and there is very little mobile sediment found

in the channel or seaward of the harbor. In this case, the tidal currents generated by a 2.7 m tidal range remove little or no sediment from the entrance channel. Because tidal currents are not required to sustain the dimensions of the channel, the entrance to York Harbor is not a tidal inlet (Figure 16.4).

Some tidal inlets coincide with the mouths of rivers (estuaries) but in these cases inlet

Figure 16.4 Entrance to York Harbor, Maine. This is a bedrock passageway. It is not a tidal inlet because there is little sediment deposited in the channel by wave action and tidal currents are not needed to keep the channel open.

dimensions and sediment transport trends are still governed, to a large extent, by the volume of water exchanged at the inlet mouth and the reversing tidal currents, respectively.

Tidal currents are produced at inlets due to the rise and fall of the tides (Figure 16.5). During the rising tide, the water level of the ocean rises at a faster rate than that inside the inlet. The water surface slope created by this condition causes the sea to flow into the inlet. This landward flowing water is called a flood-tidal current. During the falling tide, the water level of the ocean drops ahead of that of the bay inside the inlet. The seaward-sloping water surface produces ebb-tidal currents.

At most inlets over the long term, the volume of water entering the inlet during the flooding tide equals the volume of water leaving the inlet during the ebbing cycle. This volume is referred to as the tidal prism. The tidal prism is a function of the open-water area in the backbarrier and the tidal range. For example, a rough estimate of the tidal prism going into and out of

Bay-Ocean Tidal Lag

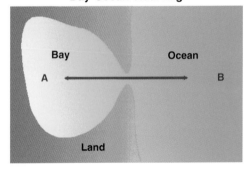

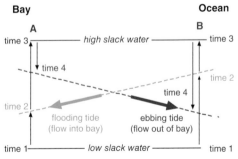

Figure 16.5 Tidal currents at inlets commonly reach speeds of 1–2 m s^{-1}. They are produced by a constriction of the tidal wave whereby the changing water level in the ocean precedes the tide level inside the inlet.

Mobile Bay, Alabama is determined by multiplying the area of Mobile Bay by the tidal range inside the bay. For backbarriers containing large intertidal areas such as marsh and tidal creeks or tidal flats, calculating the tidal prism is more difficult and it must be determined from tidal current and channel cross-section measurements. Frictional effects imparted by the inlet channel and backbarrier system also affect the tidal prism.

16.3 Inlet Morphology

A tidal inlet is specifically the area between the two barriers or between the barrier and the adjacent bedrock or glacial headland. Commonly, the sides of the inlet are formed by the recurved ridges of spits, consisting of sand that was transported toward the backbarrier by refracted waves and flood-tidal currents. The deepest part of an inlet, which is termed the inlet throat, is normally located where spit accretion of one or both of the bordering barriers constricts the inlet channel to a minimum width (Figure 16.6). This constriction is similar to placing your thumb over the nozzle of a hose to increase the velocity of the water flowing from the hose. Likewise, the minimum cross section of the inlet throat is the site where tidal currents reach their maximum velocity. Commonly, the strength of the currents at the throat causes sand to be removed from the channel floor leaving behind a lag deposit consisting of gravel or shells or in some locations exposed bedrock.

16.3.1 Tidal Deltas

Closely associated with tidal inlets are sand shoals and tidal channels located on the landward and seaward sides of the inlets. These sand deposits develop in response to tidal inlet and backbarrier processes. Waves breaking along adjacent beaches deliver sand to the inlet, dumping some of it into the main channel. Depending upon the tidal cycle, this sand is transported seaward by the ebb currents or landward by the flood currents. As the tidal waters flow beyond the constriction of the barriers, the currents expand laterally losing their velocity and their ability to transport sand. The sand that is deposited landward of the inlet forms a flood-tidal delta and the sand deposited on the seaward side forms an ebb-tidal delta.

16.3.1.1 Flood-tidal Deltas

The presence or absence, size, and development of flood-tide deltas are related to a region's tidal range, wave energy, sediment supply and backbarrier setting. Tidal inlets that are connected to one broad backbarrier

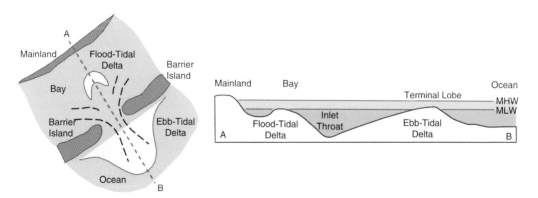

Figure 16.6 Longitudinal and cross-sections of a tidal inlet. Note that the inlet throat is the narrowest and deepest region of the tidal inlet.

channel and tidal marsh system (mixed-energy coast) usually contain a single relatively large flood-tidal delta lobe (i.e. Essex River Inlet, Massachusetts; Figure 16.7a). Contrastingly, inlets such as Drum Inlet along the Outer Banks of North Carolina (a wave-dominated coast), that are backed by large shallow bays, may contain flood-tidal deltas with numerous lobes (Figure 16.7b). Along some microtidal coasts, such as Rhode Island, flood deltas form at the end of narrow inlet channels cut through the barrier. Temporal and spatial changes in the locus of deposition at these deltas produce a multi-lobate morphology resembling a lobate river delta. (See Chapter 8 on deltas.) The small tidal range of this region prevents their reworking by ebb-tidal currents as occurs on mesotidal coasts.

To some extent, delta size is related to the amount of open water area in the backbarrier and the size of the tidal inlet. Along the mixed-energy coast of Maine, where tidal inlets are comparatively small (width < 100 m), flood-tidal deltas are correspondingly small and stacked in an alternating pattern along the main tidal creek. Tidal inlets along the barrier coast of central South Carolina have no flood-tidal deltas because the backbarrier has almost completely filled with fine-grained sediment and marsh deposits, resulting in tidal channels that are too narrow and deep for delta development. In some cases, deltas may have become colonized and altered by marsh growth, and are no longer recognizable as former flood-tidal deltas. This may be the case for some inlets in central South Carolina. At other sites, portions of flood-tidal deltas are dredged to provide navigable waterways and thus they become highly modified.

Flood-tidal deltas are best revealed in areas with moderate to large tidal ranges (1.5–3.0 m) because in these regions they are well exposed at low tide. As tidal range decreases, flood deltas become largely subtidal shoals. Mixed-energy flood-tidal deltas have similar morphologies consisting of the following components (Figure 16.8):

- *Flood ramp* – This is a landward-shallowing channel that slopes upward toward the intertidal portion of the delta. The ramp is dominated by strong flood-tidal currents and landward sand transport in the form of landward-oriented sandwaves.[1]
- *Flood channels* – The flood ramp splits into two shallow flood channels. Like the flood ramp, these channels are dominated by flood-tidal currents and flood-oriented sand waves. Sand is delivered through these channels onto the flood delta.
- *Ebb shield* – This defines the highest and landwardmost part of the flood delta and may be partly covered by marsh vegetation. When the ebb currents reach their strongest velocity in the backbarrier the tide has fallen such that the ebb shield is out of the water. Thus the ebb shields protects the rest of the delta from the effects of the ebb-tidal currents.
- *Ebb spits* – These spits extend from the ebb shield toward the inlet. They form from sand that is eroded from the ebb shield and transported back toward the inlet by ebb-tidal currents.
- *Spillover lobes* – These are lobes of sand that form where the ebb currents have breached through the ebb spits or ebb shield depositing sand in the interior of the delta.

Through time, some flood-tidal deltas accrete vertically and/or grow in size. This is evidenced by an increase in areal extent of marsh grasses which require a certain elevation above mean low water to exist. At migrating inlets new flood-tidal deltas are

---

1 **Bedforms**- Sediment moved along the channel bottom by tidal currents is commonly organized into repetitive, elongated packets of sand called bedforms. Like any waveform, each bedform has a crest and trough. A group of bedforms covering any one area tend to parallel one another and their crests are aligned perpendicular to the flow. Sandwaves are a type of bedform usually ranging in height (vertical distance from crest to trough) from 0.5 to 3 m with a spacing (distance from crest to crest) of 10 m to more than 100 m. Sandwaves commonly floor channels and extend across the width of the channel.

(a)

(b)

Figure 16.7 The morphology of flood-tidal deltas is a function of inlet size, open-water area of the bay, tidal range, and other factors. (a) View of single flood-tidal delta in New Zealand. (b) Multiple delta lobes at Chatham Inlet, Massachusetts. (*Source:* Photo taken by Albert Hine in early 1970s.)

formed as the inlet moves along the coast and encounters new open water areas in the backbarrier. At most stable inlets, however, sand comprising the flood delta is simply recirculated. The transport of sand on flood deltas is controlled by the elevation of the tide and the strength and direction of the tidal currents. During the rising tide, flood currents reach their strongest velocities near high tide when the entire flood-tidal delta is

(a)

(b)

Flood Tidal Delta

Ebb Tidal Delta

0   500
meters

N

(c)

Flood Tidal Delta

1. Flood Ramp
2. Flood Channel
3. Ebb Shield
4. Ebb Spit
5. Spillover Lobe

(d)

Flood Tidal Delta

1. Main Ebb Channel
2. Marginal Flood Channel
3. Swash Platform
4. Terminal Lobe
5. Swash Bars
6. Channel Margin Linear Bars

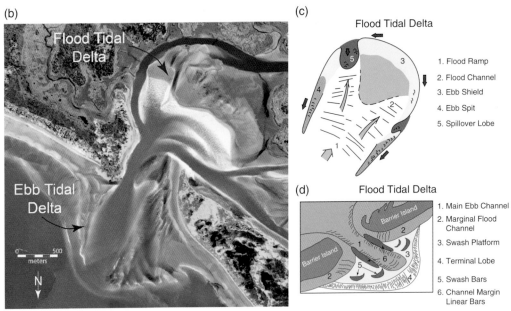

Figure 16.8 Miles Hayes conceived mixed-energy models of tidal deltas in the early 1970s working along the coast of New England. (a) Oblique aerial photograph of Essex River Inlet. (b) Overhead aerial photograph of Essex River Inlet. (c) Model of flood-tidal delta. (*Source:* After Hayes (1975).) (d) Model of ebb-tidal delta. (*Source:* After Hayes (1975).)

covered by water. Hence, there is a net transport of sand up the flood ramp, through the flood channels and onto the ebb shield. Some of the sand is moved across the ebb shield and into the surrounding tidal channel. During the falling tide, the strongest ebb currents occur near mid to low water. At this time, the ebb shield is out of the water and diverts the currents around the delta. The ebb currents erode sand from the landward face of the ebb shield and transport it along the ebb spits and eventually into the inlet channel, where once again it will be moved onto the flood ramp, thus completing the sand gyre.

In some locations, such as Shinnecock Inlet on Long Island, New York and Ogunquit River Inlet, Maine flood-tidal deltas have been mined for their sand which is pumped onto eroding beaches. However, this practice may actually create a sediment sink in the backbarrier which, in turn, may contribute to the erosion of beaches along the adjacent inlet shoreline.

16.3.1.2 Ebb-tidal Deltas

These are an accumulation of sand that has been deposited by the ebb-tidal currents and which has been subsequently modified by waves and tidal currents. Ebb deltas exhibit a variety of forms dependent on the relative magnitude of wave and tidal energy of the region as well as geological controls. Despite this variability, most ebb-tidal deltas contain the same general features including (Figure 16.8):

- *Main ebb channel* – This is a seaward-shallowing channel that is scoured in the ebb-tidal delta sands. It is dominated by ebb-tidal currents.
- *Terminal lobe* – Sediment transported out the main ebb channel is deposited in a lobe of sand forming the terminal lobe. The deposit slopes relatively steeply on its seaward side. The outline of the terminal lobe is well defined by breaking waves during storms or periods of large wave swell at low tide.
- *Swash platform* – This is a broad shallow sand platform located on both sides of the main ebb channel, defining the general extent of the ebb delta.
- *Channel margin linear bars* – These are bars that border the main ebb channel and sit atop the swash platform. These bars tend to confine the ebb flow and are exposed at low tide.

- *Swash bars* – Waves breaking over the terminal lobe and across the swash platform form arcuate-shaped swash bars that migrate onshore. The bars are usually 50–150 m long, 50 m wide, and 1–2 m in height.
- *Marginal-flood channels* – These are shallow channels (0.2–2.0 m deep at mean low water) located between the channel margin linear bars and the onshore beaches. The channels are dominated by flood-tidal currents.

As stated previously, the deepest section of an inlet occurs at the inlet throat, where depths exceeding 8 m are common. Moving out the inlet channel, depths gradually shallow to the point one or two kilometers seaward of the inlet throat where water depths may be less than 2 m. Waves breaking over the terminal lobe lead to numerous boating accidents each year including the loss of lives. Boaters may be caught unaware of the breaking wave conditions because in the deeper, landward portions of the main ebb channel the waters may be relatively calm. Breaking waves along the periphery of the ebb delta are usually due to a combination of near-low-tide conditions which produce shallow water depths, large waves and ebb-tidal currents. The ebb currents cause a shortening of the distance between the incoming waves. This stacking phenomenon produces steep waves, leading to breaking waves.

16.3.2 Ebb-Tidal Delta Morphology

The general shape of an ebb-tidal delta and the distribution of its sand bodies tell us about the relative magnitude of different sand transport processes operating at a tidal inlet (Figure 16.9). Ebb-tidal deltas that elongate with a main ebb channel and channel margin linear bars that extend far offshore are tide-dominated inlets. Wave-generated sand transport plays a secondary role in modifying delta shape at

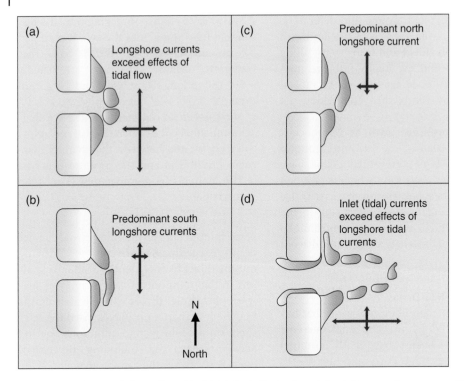

Figure 16.9 The morphology of an ebb-tidal delta indicates the relative influence of wave versus tidal energy as well as the dominant direction of longshore sediment transport. (*Source:* From Oertel (1975).)

these inlets. Because most sand movement in the inlet is in an onshore–offshore direction, the ebb-tidal delta overlaps a relatively small length of inlet shoreline. As will be demonstrated, this has important implications concerning the extent to which the inlet shoreline undergoes erosional and depositional changes.

Wave-dominated inlets tend to be small relative to tide-dominated inlets. Their ebb-tidal deltas are pushed onshore, close to the inlet mouth by the dominant wave processes. Commonly, the terminal lobe and/or swash bars form a small arc outlying the periphery of the delta. In many cases the ebb-tidal delta of these inlets is entirely subtidal. In other instances, sand bodies clog the entrance to the inlet, leading to the formation of several major and minor tidal channels.

At mixed-energy tidal inlets the shape of the delta is the result of tidal and wave processes. These deltas have a well-formed main ebb channel, which is a product of ebb-tidal currents. Their swash platform and sand bodies substantially overlap the inlet shoreline many times the width of the inlet throat due to wave processes and flood-tidal currents.

Ebb-tidal deltas may also be highly asymmetric such that the main ebb channel and its associated sand bodies are positioned primarily along one of the inlet shorelines. This configuration normally occurs when the major backbarrier channel approaches the inlet at an oblique angle or when a preferential accumulation of sand on the updrift side of the ebb delta causes a deflection of the main ebb channel along the downdrift barrier shoreline. Both conditions occur at Parker River Inlet along the North Shore of Massachusetts and thus its ebb-tidal delta significantly overlaps the downdrift shoreline of Castle Neck whereas very little of the ebb delta overlaps the updrift shoreline of Plum Island.

16.4   Tidal Inlet Formation

The formation of a tidal inlet requires the presence of an embayment and the development of barriers. In coastal plain settings, the embayment or backbarrier was often created through the construction of the barriers themselves, like much of the east coast of the United States or the Frisian Island coast along the North Sea. In other instances, the embayment was formed due to rising sea level inundating an irregular shoreline during the late Holocene. The embayed or indented shoreline may have been a rocky coast such as that of northern New England and California or it may have been an irregular unconsolidated sediment coast such as that of Cape Cod in Massachusetts or parts of the Oregon coast. The flooding of former river valleys has also produced embayments associated with tidal inlet development. The coastal processes responsible for the formation of tidal inlets are described below.

16.4.1   Breaching of a Barrier

Rising sea level, exhausted sediment supplies, and human influences have led to erosion along much of the world's coastlines, including its barrier island chains and barrier spit systems. This condition has caused a thinning of many barriers such that they are vulnerable to breaching during storms. Breaching occurs when a barrier is cut forming a channel (Figure 16.10). It is by far the most common mechanism by which tidal inlets form today. The breaching process normally occurs during storms after waves have destroyed the foredune ridge and storm waves have overwashed the barrier depositing sand aprons (washovers) along the backside of the barrier. Even though this process may produce a shallow overwash channel, seldom are barriers cut from their seaward side. In most instances, the breaching of a barrier is the result of the storm surge heightening waters in the backbarrier bay. When the level of the ocean tide falls, the elevated bay waters flow across the barrier toward the ocean, gradually incising the barrier and cutting a channel. If subsequent tidal exchange between the ocean and bay is able to maintain the channel, a tidal inlet is established. The breaching process is enhanced when offshore winds accompany the falling tide and if an overwash channel is present to facilitate drainage across the barrier. Along the Gulf Coast of the United States hurricanes have been responsible for the development of numerous tidal inlets (e.g. Hurricane Pass, Florida). Many of the tidal inlets that are formed through breaching are ephemeral and may exist for less than a year, especially if stable inlets are located nearby. Barriers most susceptible to breaching are long and thin and wave-dominated. For example, although there are only four stable inlets along the Outer Banks of North Carolina today, historical records indicate that at least 26 former inlets have opened and closed at various locations in the past. The reason why inlets close will be discussed later in this chapter.

16.4.2   Spit Building across a Bay

The development of a tidal inlet by spit construction across an embayment usually occurs early in the evolution of a coast. The sediment to form these spits may have come from erosion of the nearby headlands, discharge from rivers, or from the landward movement of sand from inner shelf deposits. As discussed in Chapter 15, Barrier Systems, most of barriers along the coast of the United States and elsewhere in the world are less than 5000 years old, coinciding with a deceleration of rising sea level. It was then that spits began enclosing portions of the irregular rocky coast of New England, the west coast, parts of Australia, and many other regions of the world.

As a spit builds across a bay, the opening to the bay gradually decreases in width and in cross sectional area (Figure 16.11). It may also deepen. Coincident with the

Time 1: ~20 years prior to breaching

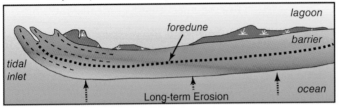

Time 2: ~2 years prior to breaching

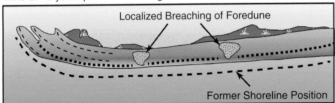

Time 3: Storm overwash

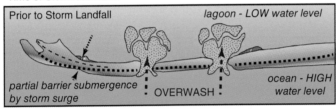

Time 4: Storm ebb-surge and breaching

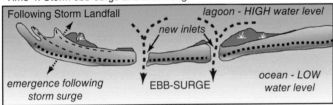

Figure 16.10 Generally the formation of a new tidal inlet is associated with the breaching of a barrier during a storm. The stages in this process involve a thinning of the barrier through long-term erosion, destruction of the fore-dune ridge, storm overwash, and finally a deepening of a channel through the barrier.

decrease in size of the opening is a corresponding increase in tidal flow. The tidal prism of the bay remains constant, so as the opening gets smaller, the current velocities must increase. Again, this is similar to gradually placing your thumb over an increasingly larger portion of the nozzle to a hose. For the flow out the hose to remain constant, the velocity has to increase. The tidal inlet is formed as the bay reaches a stable configuration.

The equilibrium size of a tidal inlet can also be explained in terms of sediment transport. Waves and flood-tidal currents are responsible for delivering sediment to the inlet and dumping a large portion of the sand into the inlet channel. The inlet responds to this deposition and decrease in cross sectional area by increasing the tidal flow, thereby increasing the transport capacity of the tidal currents. Thus, the tidal inlet reaches an equilibrium state when the

(a)

Time 1. Spit growth across a bay   Time 2. Spit extension and inlet formation

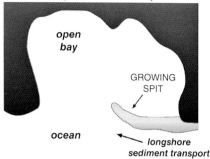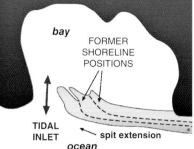

(b)

Figure 16.11 Spit construction across an embayment can create a tidal inlet. (a) Model of inlet formation due to spit accretion. (b) Aerial photograph of spit building and inlet development in Slocum Embayment, Buzzards Bay, Massachusetts.

amount of sand dumped into the inlet equals the volume removed by the tidal currents.

In fact, there are some tidal inlets, such as Barnstable Harbor Inlet located in Cape Cod Bay, Massachusetts, that are still developing because Sandy Neck continues to build across the Barnstable Bay. As the width of the inlet decreases, the equilibrium throat cross-section is maintained by the inlet channel deepening. This spit and inlet system has been evolving over the past 3500 years.

16.4.3 Drowned River Valleys

In many locations tidal inlets are located at the sites of drowned river valleys. A drowned river valley is valley that was enlarged when sea level was lower and rivers extended their pathways across the continental shelf to shorelines that were many miles seaward of where they are today. Sea level lowering was in response to the growth of continental glaciers during the Pleistocene Epoch. When the ice sheets retreated northward and water from the melting ice was returned back to ocean basins, rising sea level flooded the enlarged valleys, forming drowned river valleys. Due to the freshwater discharge and saltwater mixing at these locations, most drowned river valleys are estuaries.

Tidal inlets have formed at the entrance to drowned river valleys due to the growth of spits and the development of barrier islands which have served to narrow the mouths of the estuaries (Figure 16.12). They are delineated as tidal inlets when the dimensions of the inlet throat and overall sediment transport trends are a consequence of the saltwater tidal prism and the reversing tidal currents. Thus, the entrance to Chesapeake Bay is not a tidal inlet because its mouth has not been constricted through barrier construction, whereas the entrances to Mobile Bay in Alabama and Grays Harbor in Washington are tidal inlets due to barrier development.

It has been shown through stratigraphic studies, particularly along the east coast of the United States, that in addition to drowned river valleys, many tidal inlets are positioned in paleo-river valleys in which there is no river leading to this site today. These are old river courses that were active during the Pleistocene when sea level was lower and they were migrating across the exposed continental shelf. Tidal inlets become situated in these valleys because the sediment filling the valleys is easily removed by tidal currents. Once

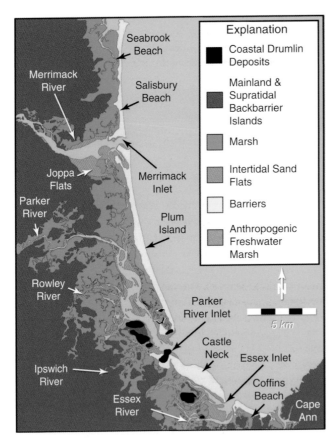

Figure 16.12 The location of tidal inlets commonly coincides with former river valleys. This situation is exemplified by the inlets that occupy drowned river valleys along the Merrimack barrier system in northern New England.

a tidal inlet migrates to one of these former valleys the inlet channel scours vertically, excavating the former riverine sediments and becoming anchored. Commonly, the sediment layers on either side of the paleo-valley are more resistant to erosion than are the valley-fill sediments. Therefore after a tidal inlet occupies a paleo-valley further migration of the inlet is impeded. Drowned river valleys and paleo-river channels comprise at least 25% of tidal inlet locations today, especially deep inlets (depth > 8 m).

16.4.4 Ephemeral Inlets

The most common type of ephemeral inlet is the one that is a product of hurricanes (Figure 16.13). During Hurricane Alicia in 1983, 185 km h^{-1} (115 mph) winds and 3-m storm surge were responsible for cutting 80 tidal inlets along the Texas coast. None of these hurricane passes, as they are called, lasted for more than a month. The ephemeral inlets were filled with sediment that was transported onshore and along the coast by wave action.

A special case of tidal inlet formation occurs at welded barriers along glaciated coasts such as sections of Alaska, New England, and Canada. Welded barriers at these sites are usually short in length (<1 km), composed of sand and gravel, and are backed by small fresh- to brackish-water ponds and lakes. Freshwater inflow to the backbarrier is derived from small streams and precipitation. Under normal conditions the freshwater influx is insufficient to cause overtopping of the barriers because the sand and gravel comprising the barrier permits water to percolate through the barrier sediment and drain into the ocean. However, during intense rain storms and/or melting snow, stream discharge may increase substantially until water in the pond flows across the barrier, cutting a channel and forming a tidal inlet. These inlets are usually short-lived and last only a few months because spit accretion and the formation of flood tidal deltas seal off the channel (Figure 16.14).

16.5 Tidal Inlet Migration

Some tidal inlets have been stable since their formation whereas others have migrated long distances along the shore. In New England and along other glaciated coasts, stable inlets are commonly anchored next to bedrock outcrops or resistant glacial deposits. Along the California coast most tidal inlets have formed by spit construction across an embayment, with the inlet becoming stabilized adjacent to a bedrock headland. As previously discussed, in coastal plain settings stable inlets are commonly positioned in former river valleys. One factor that appears to separate migrating inlets from stable inlets is the depth to which the inlet throat has eroded. For example, along the South Carolina coast tidal inlets deeper than 8 m are stable, whereas inlets shallower than 3–4 m have histories of migration. Deeper inlets are often entrenched in consolidated sediments that resist erosion. The channels of shallow migrating inlets are eroded into sand.

Tidal inlets migrate when the longshore transport of sand is added predominantly to one side of the inlet causing a constriction of the flow area (Figure 16.15a). As the tidal currents scour the channel to remove this sand, the downdrift side of the inlet channel is eroded preferentially and the inlet migrates in that direction. Generally, the rate of inlet migration tends to be high along wave-dominated coasts, where the inlet channel is scoured into sand and there is abundant sediment supply.

Although the vast majority of tidal inlets migrate in the direction of dominant longshore transport, there are some inlets that migrate updrift. In these cases the drainage of backbarrier tidal creeks control flow through the inlet. When a major backbarrier tidal channel approaches the inlet at an oblique angle, the ebb-tidal currents coming from this channel are directed toward the margin of the inlet throat. If this is the updrift side of the main channel, then the inlet will migrate in that direction. This is similar to a

(a)

(b)

Figure 16.13 Tidal inlets may form during the passage of major storm when waves batter and dismantle dunes and washover from incipient channels across the barrier. When floodwaters exit the bays, water is often funneled through these channels, deepening them and sometimes forming a permanent inlet. In most cases, the inlet channel closes after a few months. (a) Several ephemeral inlets were opened along the northeast coast of Dauphine Island, Mississippi in 1979 as a result of Hurricane Frederick. Hurricane passes, as they are called in Gulf Coast region, usually close shortly after they are formed because they are unable to capture a signification portion of the bay tidal prism. (b) In 2011, Hurricane Irene cut temporary hurricane passes through the Outer Banks of North Carolina.

(a)

(b)

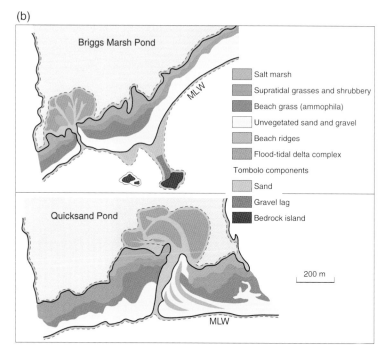

Figure 16.14 Ephemeral tidal inlets form along sand and gravel welded barriers on glaciated coasts. Inlets develop when the inflow of freshwater causes lake levels to overtop the barrier. As water drains across the barrier a channel is cut, forming an ephemeral tidal inlet. (a) Aerial photograph of welded barriers along the southern coast of New England. (b) Map of two ephemeral inlets illustrating how they close due spit accretion and the deposition of flood tidal deltas.

(a)

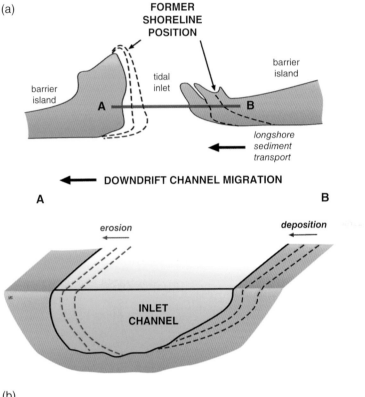

FORMER
SHORELINE
POSITION

barrier
island

tidal
inlet

barrier
island

A

B

longshore
sediment
transport

DOWNDRIFT CHANNEL MIGRATION

A

B

erosion

deposition

INLET
CHANNEL

(b)

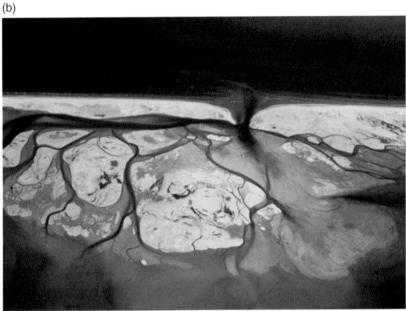

Figure 16.15 Migrating tidal inlets generally occur along coasts having a dominant longshore sediment transport direction such that sand is added preferentially to one side of the inlet channel. This type of inlet is usually shallow. The inlet channel is not eroded into resistant sediments, which would impede the process of migration. (a) Model of a migrating inlet. (b) This inlet is migrating left to right. Flood-tidal deltas formed at former inlet positions are vegetated, whereas the flood-tidal delta immediately landward of the inlet is relatively immature and mostly intertidal to subtidal.

river where strong currents are focused along the outside of a meander bend, causing erosion and channel migration. Inlets that migrate updrift are usually small to moderately sized and occur along coasts with small to moderate net sand longshore transport rates.

16.6 Tidal Inlet Relationships

Tidal inlets throughout the world exhibit several consistent relationships that have allowed coastal engineers and marine geologists to formulate predictive models. These models are effective tools when undertaking tidal inlet projects and are used by engineers and coastal managers to plan jetty construction, channel dredging, and the use of ebb-tidal deltas for beach nourishment material. The models are based on field data collected at many different tidal inlet locations. Through statistical analysis (regression analysis), whereby inlet parameters are plotted against one another, two important correlations have been discovered: Inlet throat cross-sectional area is closely related to tidal prism; and ebb-tidal delta volume is a function of tidal prism.

16.6.1 Inlet Throat Area–Tidal Prism Relationship

It has long been recognized that the size of a tidal inlet is tied closely to the volume of water going through it. In 1931 Morrough O'Brien quantified this relationship for inlets on the west coast (hence it is called the O'Brien Relationship) by plotting the cross-sectional area of the inlet throat (measured at mean sea level) versus its tidal prism during spring tide conditions (Figure 16.16). Because the data plot approximately along a line, he was able to derive a simple equation to represent this correlation. Later, in 1969 he showed

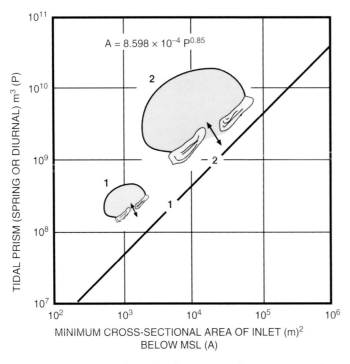

$$A = 8.598 \times 10^{-4}\, P^{0.85}$$

Figure 16.16 O'Brien's Relationship demonstrates that a strong correlation exists between an inlet's spring tidal prism and its throat cross-sectional area (O'Brien 1931, 1969).

that the relationship could be extended to inlets along the East and Gulf Coasts and since that time other scientists have revealed that, with slight modification, the relationship exists for inlets all over the world. Although it seems very reasonable that the opening of an inlet should be controlled by its tidal prism, the reason why this correspondence exists globally is because the filling and emptying of the backbarrier are governed by the rise and fall of the ocean tides. Most barrier coasts experience semidiurnal tides (two tidal cycles daily) and therefore the ocean tidal forcing of the filling and emptying of backbarrier areas worldwide has the same duration, approximately 6 hours and 13 minutes. This concept is illustrated well by comparing two inlets along the Gulf Coast – Midnight Pass in Florida, with a tidal prism of 7.4 million cubic meters, and the entrance to Mobile Bay in Mississippi which has a much larger tidal prism of 960 million cubic meters. If both inlets are to discharge their tidal prisms over the same time interval, it is easily understood that the opening to Mobile Bay (29,280 m^2) has to be much larger (100 times) than Midnight Pass Inlet (300 m^2).

Although it has been stressed that inlet size is primarily a function of tidal prism, to a lesser degree inlet cross sectional area is also affected by the delivery of sand to the inlet channel. For example, tidal inlets with jetties, which are stone or concrete structures built perpendicular to the entrance of an inlet, prevent the wave-generated transport of sand into the inlet. At these sites tidal currents can more effectively scour sand from the inlet channel and therefore they maintain a larger throat cross section than would be predicted by the O'Brien Relationship for inlets with no jetties. Similarly, for a given tidal prism, Gulf Coast inlets have larger throat cross-sections than Pacific coast inlets. This is explained by the fact that wave energy is greater along the West Coast and therefore the delivery of sand to these inlets is higher than at Gulf Coast inlets.

16.6.1.1 Variability

It is important to understand that the dimensions of the inlet channel are not static but rather the inlet channel enlarges and contracts slightly over relatively short time periods (<1 year) in response to changes in tidal prism, variations in wave energy, effects of storms, and other factors. For instance, the inlet tidal prism can vary by more than 30% from neap to spring tides due increasing tidal ranges. Consequently, the size of the inlet varies as a function of tidal phases. Along the southern Atlantic coast of the United States water temperatures may fluctuate seasonally by 16 °C (30 °F). This causes the surface coastal waters to expand, raising mean sea level by 30 cm or more. In the summer and fall when mean sea level reaches its highest seasonal elevation, spring tides may flood backbarrier surfaces that normally are above tidal inundation. This produces larger tidal prisms, stronger tidal currents, increased channel scour, and larger inlet cross-sectional areas. At some Virginia inlets this condition increases the inlet throat by 5–15%. Longer-term (>1 year) changes in the cross-section of inlets are related to inlet migration, sedimentation in the backbarrier, morphological changes of the ebb-tidal delta, and human influences.

16.6.1.2 Application

The O'Brien Relationship is a very useful concept when designing modification projects for inlets. For example, when an inlet is to be jettied and dredged to provide a navigable waterway for large ships entering and leaving a port, the dimensions of the channel have to be planned (Figure 16.17). If the channel is dredged to dimensions larger than what is in balance with the existing tidal prism, the channel will fill with sediment until the cross-section decreases to the equilibrium area. Remember that tidal prism is primarily a function of the open-water area in the backbarrier and the tidal range in the backbarrier and under most conditions will not change if the size of the inlet is enlarged. Therefore, if the improved inlet has an equilibrium cross sectional area of 12,000 m^2 and

1. Natural Channel Configuration

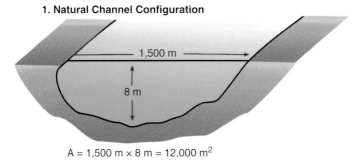

A = 1,500 m × 8 m = 12,000 m²

2. Stabilized Channel Configuration

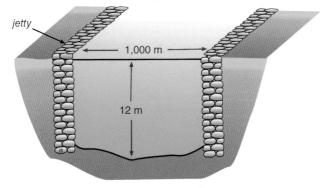

A = 1,000 m × 12 m = 12,000 m²

Figure 16.17 Application of O'Brien's Relationship. Disregarding the effects of friction, if the jetties are positioned closer together, then tidal currents will scour the channel deeper. Inlet dimensions will increase until the equilibrium cross sectional area that is indicated by O'Brien's Relationship is achieved.

navigational constraints require a 12 m deep channel, then the jetties should be positioned approximately 1000 m apart.

16.6.2 Ebb-Tidal Delta Volume–Tidal Prism Relationship

In the mid-1970s Todd Walton and graduate assistant, William Adams, did further statistical analysis (regression analysis) of various inlet parameters and discovered that, like inlet cross sectional area, the volume of sand contained in the ebb-tidal delta was closely related to the tidal prism. This relationship has come to be known as the Walton and Adams Relationship (Figure 16.18). As we have already discussed, the ebb-tidal delta comprises the sand that is diverted from the longshore transport system and transported seaward by ebb-tidal currents. The greater the ebb discharge, the more sand that is contained in the ebb-tidal delta. Walton and Adams also showed that the relationship was improved slightly when wave-energy was taken into account. This was accomplished by separating the data set into three inlet classes based on their wave energy: high wave-energy coasts such as inlets along Oregon and Washington; moderate wave-energy coasts including New Jersey, Outer Banks of North Carolina, and Delaware; and low wave-energy coasts such as the Gulf Coast. Waves are responsible for transporting sand back onshore, thereby reducing the volume of the ebb-tidal delta. Therefore, for a given tidal prism, ebb-tidal deltas along the West Coast contain less sand than do equal sized inlets along the Gulf or east coast.

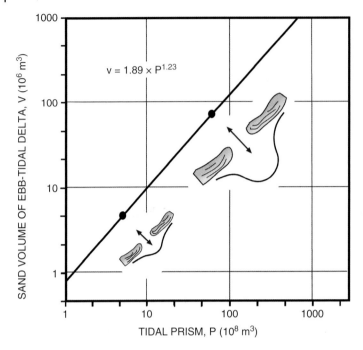

Figure 16.18 Walton and Adams Relationship indicates a strong correspondence exists between an inlet's tidal and the volume of its ebb-tidal delta. (*Source:* From Walton and Adams (1976).)

16.6.2.1 Variability

The Walton and Adams Relationship works well for inlets all over the world. Field studies have shown, however, that the volume of sand comprising ebb-tidal deltas changes through time due to the effects of storms, changes in tidal prism, or processes of inlet sediment bypassing. When sand is moved past a tidal inlet, it is commonly achieved by large bar complexes migrating from the ebb delta and attaching to the landward inlet shoreline. These large bars may contain more than 300,000 m^3 of sand and represent more than 10% of sediment volume of the ebb-tidal delta.

16.6.2.2 Application

Due to pervasive shoreline erosion, many barrier systems in the United States and elsewhere are being nourished with sand obtained from offshore sites, backbarrier and inlet dredging, and land sources. As these borrow sites become depleted, ebb-tidal deltas are also being mined for their sand.

The Walton and Adams Relationship helps engineers compute ebb-tidal delta volumes and the effects to adjacent beaches. The relationship is also used to determine how nearby beaches will respond when a tidal inlet is formed due to storm breaching or if an artificial cut is made through a barrier. Immediately following inlet formation, the ebb-tidal delta grows until it reaches an equilibrium volume as predicted by the Walton and Adams Relationship. The sediment that builds the ebb delta is sand that is removed from the longshore transport system, thereby causing erosion. The relationship helps coastal engineers calculate rates of change.

16.7 Sand Transport Patterns

The movement of sand at a tidal inlet is complex due to reversing tidal currents, effects of storms, and interaction with the longshore transport system. The inlet contains short-term and long-term reservoirs of sand, varying from the relatively small sandwaves flooring the inlet channel that migrate meters each tidal cycle to the large flood-tidal delta shoals where some sand is recirculated but the entire deposit may remain stable for hundreds or even thousands of years. Sand dispersal at tidal inlets is complicated because in addition to the onshore–offshore movement of sand produced by tidal- and wave-generated currents, there is constant delivery of sand to the inlet and transport of sand away from the inlet produced by the longshore transport system. The discussion below describes the patterns of sand movement at inlets, including how sand is moved past a tidal inlet.

16.7.1   General Sand-Dispersal Trends

The ebb-tidal delta segregates areas of landward versus seaward sediment transport that are controlled primarily by the way water enters and discharges from the inlet as well as the effects of wave-generated currents (Figure 16.19). During the ebbing tidal cycle the tidal flow leaving the backbarrier is constricted at the inlet throat, causing the currents to accelerate in a seaward direction. Once out of the confines of the inlet, the ebb flow expands laterally and the velocity slows. Sediment in the main ebb channel is transported in a net seaward direction and eventually deposited on the terminal lobe due to this decrease in current velocity. One response to this seaward movement of sand is the formation of ebb-oriented sandwaves having heights of 1–2 m.

In the beginning of the flood cycle, the ocean tide rises while water in main ebb channel continues to flow seaward as a result of momentum. Due to this phenomenon, water initially enters the inlet through the marginal flood channels, which are the pathways of least resistance. The flood channels are dominated by landward sediment transport and are floored by flood-oriented bedforms. On both sides of the main ebb

channel, the swash platform is most affected by landward flow produced by the flood-tidal currents and breaking waves. As waves shoal and break, they generate landward flow, which augments the flood-tidal currents but retards the ebb tidal currents. The interaction of these forces acts to transport sediment in net landward direction across the swash platform. In summary, at many inlets there is a general trend of seaward sand transport in the main ebb channel, which is countered by landward sand transport in the marginal flood channels and across the swash platform.

16.7.2   Inlet Sediment Bypassing

Along most open coasts, particularly in coastal plain settings, angular wave approach causes a net movement of sediment along the shore. As we learned in Chapter 13, the net volume of sand transported along the east coast of the United States varies from 100,000 to 200,000 m^3 year^{-1} Thus, there is upward of 200,000 m^3 year^{-1} of sand delivered to tidal inlets along this coast. If the sand reservoirs of these inlets remain approximately constant over the long-term then there must be mechanisms whereby sand moves past tidal inlets and is transferred to the downdrift shoreline. This process is

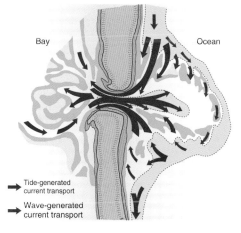

Figure 16.19 Sand dispersal patterns at a tidal inlet reflect the dominance of wave and tidal processes. (*Source:* Modified after Smith (1989).)

called inlet sediment bypassing. There are multiple ways in which inlets bypass sand including: Stable inlet processes; Ebb-tidal delta breaching; and Inlet migration and spit breaching. One of the end products in all the different mechanisms is the landward migration and attachment of large bar complexes to the inlet shoreline.

16.7.2.1 Stable Inlet Processes

This mechanism of sediment bypassing occurs at inlets that do not migrate and whose main ebb channels remain approximately in the same position (Figure 16.20). Sand enters the inlet by: wave action along the beach; flood-tidal and wave-generated currents through the marginal flood channel; and waves breaking across the channel margin linear bars. Most of the sand that is dumped into the main channel is transported seaward by the dominant ebb-tidal currents and deposited on the terminal lobe.

At lower tidal elevations waves breaking on the terminal lobe transport sand along the periphery of the delta toward the landward beaches in much the same way as sand is moved in the surf and breaker zones along beaches. At higher tidal elevations waves breaking over the terminal lobe create swash bars on both sides of the main ebb channel. The swash bars (50–150 m long, 50 m wide) migrate onshore due to the dominance of landward flow across the swash platform. Eventually, they attach to channel margin linear bars forming large bar complexes. Bar complexes tend to parallel the beach and may be more than a kilometer in length. They are fronted by a steep face (25–33°) called a slipface, which may be up to 3 m in height. At mid-tide, pleasure boaters often anchor behind the bars taking advantage of the quiet water and so that swimmers may dive off the bar slipface.

The stacking and coalescing of swash bars to form a bar complex is the result of the bars slowing their onshore migration as they move up the nearshore ramp. As the bars gain a greater intertidal exposure, the wave bores which cause their migration onshore

act over an increasingly shorter period of the tidal cycle. Thus, their rate of movement onshore decreases. The growth of the bar complex is similar to cars on a highway all stacking up when they approach a toll booth.

Eventually the entire bar complex migrates onshore and welds to the upper beach. When a bar complex attaches to the downdrift inlet shoreline, some of this newly accreted sand is then gradually transported by wave action to the downdrift beaches, thus completing the inlet sediment bypassing process. It should be noted that some sand bypasses the inlet independent of the bar complex. In addition, some of the sand comprising the bar re-enters the inlet via the marginal flood channel and along the inlet shoreline.

16.7.2.2 Ebb-tidal Delta Breaching

This means of sediment bypassing occurs at inlets with a stable throat position, but whose main ebb channels migrate through their ebb-tidal deltas like the wag of a dog's tail (Figure 16.21). Sand enters the inlet in the same manner as described above for Stable inlet processes. However, at these inlets the delivery of sediment by longshore transport produces a preferential accumulation of sand on the updrift side of the ebb-tidal delta. The deposition of this sand causes a deflection of the main ebb channel until it nearly parallels the downdrift inlet shoreline. This circuitous configuration of the main channel results in inefficient tidal flow through the inlet, ultimately leading to a breaching of a new channel through the ebb-tidal delta. The process normally occurs during spring tides or periods of storm surge when the tidal prism is very large. In this state the ebb discharge piles up water at the entrance to the inlet where the channel bends toward the downdrift inlet shoreline. This causes some of the tidal waters to exit through the marginal flood channel or flow across low regions on the channel margin linear bar. Gradually over several weeks or convulsively during a single large storm, this process cuts a new channel through the ebb delta thereby providing a more direct pathway for tidal

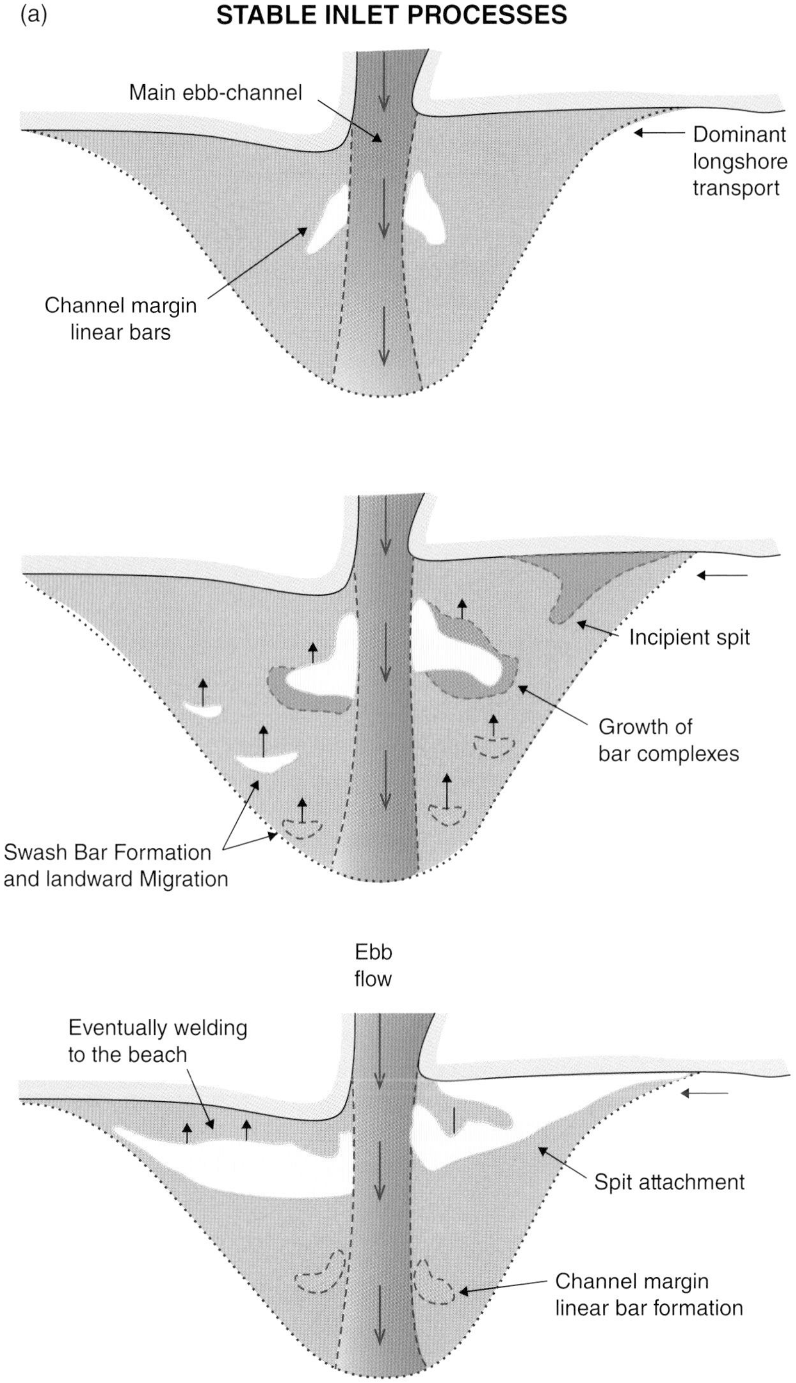

(a)

STABLE INLET PROCESSES

Main ebb-channel

Dominant
longshore
transport

Channel margin
linear bars

Incipient spit

Growth of
bar complexes

Swash Bar Formation
and landward Migration

Ebb
flow

Eventually welding
to the beach

Spit attachment

Channel margin
linear bar formation

Figure 16.20 Inlet sediment bypassing at stable inlets. (a) Model of stable inlet processes. (*Source:* From FitzGerald (1988).) (b) North Inlet along the northern South Carolina coast is an example where sediment bypassing occurs through stable inlet processes. Note the large bars migrating onshore to the downdrift inlet shoreline.

(b)

Figure 16.20 (Continued)

exchange through the inlet. As more and more of the tidal prism is diverted through the new main ebb channel, tidal discharge through the former channel decreases, causing it to fill with sand.

The sand that was once on the updrift side of the ebb-tidal delta and which is now on the downdrift side of the new main channel is moved onshore by wave-generated and flood-tidal currents. Initially, some of this sand aids in the filling of the former channel while the rest forms a large bar complex that eventually migrates onshore and attaches to the downdrift inlet shoreline. The ebb-tidal breaching process results in a large packet of sand bypassing the inlet. Similar to the stable inlets discussed above, some sand bypasses these inlets in a less dramatic fashion, grain by grain on a continual basis.

It is noteworthy that at some tidal inlets the entire main ebb channel is involved in the ebb-tidal delta breaching process, whereas at others just the outer portion of main ebb channel is deflected. In both cases, the end-product of the breaching process is a channel realignment that more efficiently conveys water into and out of the inlet, as well as sand being bypassed in the form of a bar.

One of the largest scale ebb-tidal delta breaching processes takes place at Willapa Bay Inlet on the Oregon coast. This inlet is 11 km wide and more than 12 m deep. Its outer channel is deflected south by a mostly submerged spit that builds 6 km southward from Cape Shoalwater. Every 8–27 years (16 year average) a new channel is breached back to the north straightening the main entrance channel. The submerged shoal that is bypassed moves onshore merging with inner bars. An additional interesting aspect of this cycle is that the breaching process correlates well with El Nino events, which cause water levels on the West Coast to be elevated by 20–30 cm. Higher water levels cause areas within Willapa Bay that are normally above mean high water to be inundated, thereby increasing the tidal prism. In turn, larger tidal prisms lead to stronger tidal flow and greater potential to cut a new channel through the ebb delta.

16.7.2.3   Inlet Migration and Spit Breaching

A final method of inlet sediment bypassing occurs at migrating inlets. In this situation an abundant sand supply and a dominant long-shore transport direction cause spit building at the end of the barrier (Figure 16.22). To accommodate spit construction, the inlet migrates by eroding the downdrift barrier shoreline. Along many coasts as the inlet is displaced further along the downdrift shore-line, the inlet channel to the backbarrier

(a)   **EBB-TIDAL DELTA BRFACHING**

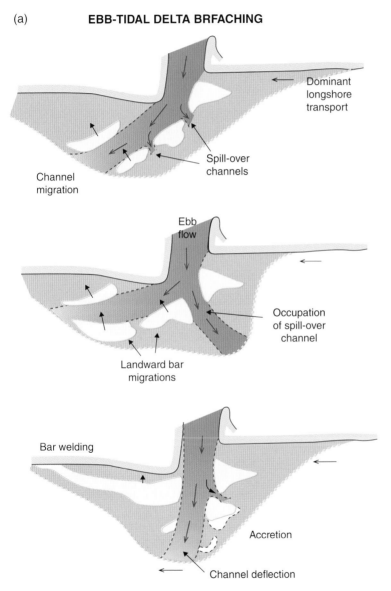

Figure 16.21 Sediment bypassing at inlets whose main ebb channel migrates downdrift in response to wave energy and sand influx via the longshore transport system. (a) Model of inlet sediment bypassing by ebb-tidal delta breaching processes. (*Source:* From FitzGerald (1988).) (b) View of Murrells Inlet, South Carolina showing two large bar complexes welding to the beach after a recent channel was breached through the updrift portion of the ebb-tidal delta.

(b)

Figure 16.21 (Continued)

lengthens, retarding the exchange of water between the ocean and backbarrier. This condition leads to large water-level differences between the ocean and bay, making the barrier highly susceptible to breaching, particularly during storms. Ultimately, when the barrier spit is breached and a new inlet is formed in a hydraulically more favorable position, the tidal prism is diverted to the new inlet and the old inlet closes. When this happens, the sand comprising the ebb-tidal delta of the former inlet is transported onshore by wave action, commonly taking the form of a landward migrating bar complex. It should be noted that when the inlet shifts to a new position along the updrift shoreline a large quantity of sand has effectively bypassed the inlet. The frequency of this inlet sediment bypassing process is dependent on inlet size, rate of migration, storm history and backbarrier dynamics. Nauset Spit along the outer coast of Cape Cod, Massachusetts exhibits a cycle of spit accretion and inlet migration of 10–15 km followed by multiple breachings occurring approximately every 100 years. Kiawah River Inlet along central coast of South Carolina has had a similar history, with at least three periods of southwesterly migration of the inlet of up to 15 km followed by breachings of the spit updrift at about the same position each time during a 150-year period. (See Box 16.1.)

16.7.2.4 Bar Complexes

Depending on the size of the inlet, the rate of sand delivery to the inlet, the effects of storms, and other factors, the entire process of bar formation, its landward migration, and its attachment to the downdrift shoreline may take from 6 to 10 years. The volume of sand bypassed can range from 100,000 to over 1,000,000 m^3. The bulge in the shoreline that is formed by the attachment of a bar complex is gradually eroded and smoothed as sand is dispersed to the downdrift shoreline and transported back toward the inlet.

In some instances, a landward-migrating bar complex forms a salt water pond as the tips of the arcuate bar weld to the beach, stabilizing its onshore movement. Although the general shape of the bar and pond may be

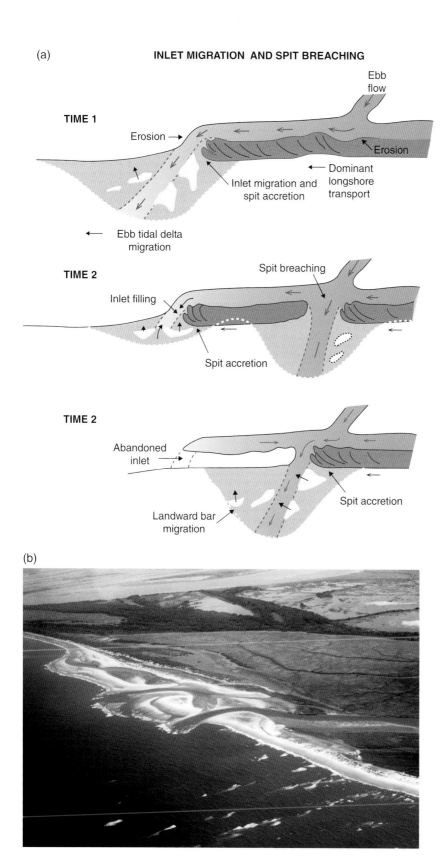

Figure 16.22 Breaching of a spit allows a large quantity of sand to bypass the tidal inlet. (a) Model of inlet migration and spit breaching processes. (*Source:* From FitzGerald (1988).) (b) Small inlet near the mouth of the Santee River, South Carolina illustrating at least two episodes of spit breaching.

Box 16.1 Breaching of Nauset Spit and Formation of New Inlet, Cape Cod, Massachusetts

The Town of Chatham, which is located along the outer coast of Cape Cod, Massachusetts, is protected by a sandy barrier known as Nauset Spit. The ancestors of this town were not foolhardy when they chose to build their homes and establish their community along the glacial uplands across Pleasant Bay. Normally, the northeast storms that wreak havoc along this coast have less effect on the mainland coast here due to the shelter afforded by the barrier. However, all that changed following the northeast storm of January 2, 1987 that breached Nauset Spit, establishing a new opening to Pleasant Bay – New Inlet (Box Figure 16.1.1). The once-idyllic coastal community was now threatened by storm waves, shoreline erosion, shoaling of its navigation channels, closure of its harbors, and tidal inundation of its lowland areas.

The Nauset barrier is part of a spit system that has accreted southward-forming lagoons and bays along the irregular southeast mainland coast of Cape Cod. The chain is broken by several tidal inlets. The sand forming the barrier complex is sourced from eroding glacial bluffs along upper Cape Cod and is transported southward by the dominant northeast wave climate. Prior to the breaching event, Nauset Spit was 14 km long, extending southward from a glacial headland to where it

Box Figure 16.1.1 Aerial photographs of Nauset Spit before New Inlet formed (1985), after the spit was breached (1987), and after a second inlet had formed north of New Inlet in 2007.

Box Figure 16.1.1 (Continued)

overlapped the northern end of Monomoy Island. At this time Pleasant Bay was connected to the open ocean through a long circuitous route of shallow channels, shoals, and a wave-dominated tidal inlet. Scientists studying the long-term history of outer Cape Cod, including Charles McClennen at Colgate University in New York and Graham Giese at Woods Hole Oceanographic Institute on Cape Cod, discovered that the barrier spit system experiences a long-term cycle of growth and decay. They showed that with a frequency of about 100–150 years the barrier maintains a period of southerly accretion followed by a destructional phase whereby the spit becomes segmented and portions of the barrier migrate onshore (Box Figure 16.1.2).

Segmentation of Nauset Spit is related to a gradual reduction in tidal exchange between Pleasant Bay and the ocean, which is caused by a restriction in tidal flow through the existing inlets. This hydraulic inefficiency produces large differences in both tidal range and times of high and low tide on opposite sides of the barrier. These differences lead to certain times of the tidal cycle in which the water level in the ocean is more than a meter higher than in the bay. Under these conditions the barrier is susceptible to breaching, particularly during storms when the storm surge increases the height of the ocean tide level. It should be noted that the periodicity and location of breaching along Nauset Spit is also dependent on the morphology and overall width of the barrier.

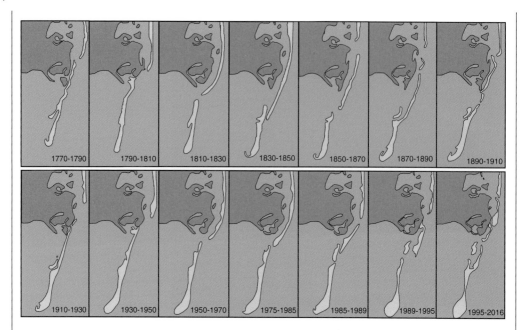

Box Figure 16.1.2 Historical shoreline changes to Nauset spit during the past 200 years. (*Source:* Modified from Giese (1988)).

(a)

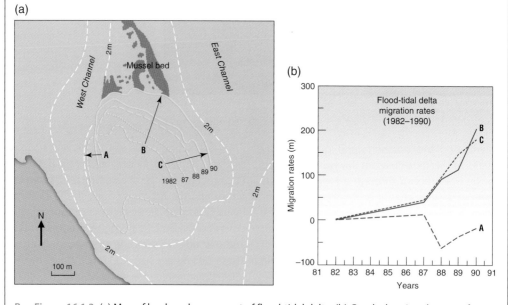

Box Figure 16.1.3 (a) Map of landward movement of flood-tidal delta. (b) Graph showing the rate of growth coinciding with the arrows A–C on the map.

If the spit is wide and has a well-developed frontal dune ridge and secondary dune system, breaching of the barrier is difficult regardless of the potential hydraulic head. In contrast, destruction of the foredune ridge and thinning of the barrier facilitates barrier overwashing, channelization of the return flow, and inlet formation.

On January 2, 1987 high water levels associated with perigean spring tides (see Chapter 7) and a storm surge produced by a strong Nor'easter (see Chapter 5) allowed storm waves to carved away the last vestiges of the frontal dune ridge along central Nauset Spit. As waves continued to overwash the barrier, eventually an overwash channel was created that allowed tidal exchange between the ocean and bay. The day after the storm the channel was several meters wide and about a meter deep. This marked the beginning of a new tidal inlet. As the channel captured an increasingly larger portion of the Pleasant Bay tidal prism, the inlet grew in size from 0.5 km

after two months, to 1.0 km in six months, and by early 1988 New Inlet had reached almost 2.0 km in width.

Opening of New Inlet drastically changed the hydraulic setting and sediment-transport patterns in Pleasant Bay. In the process of enlarging the inlet channel, tidal currents washed much of the sand from the eroding barrier into the bay. Here, the sand was reworked into shoals, bedforms and other deposits. Some of the eroded sand was also transported seaward, forming a large ebb-tidal delta. After the breaching event the tidal range in Pleasant Bay increased by 0.3 m from 1.2 to 1.5 m. The increased tidal fluctuation

(a)

Box Figure 16.1.4 (a) 2003 photograph of the inlet illustrating how ocean waves can propagate through the inlet causing erosion along the landward shoreline. (b) Revetments were constructed to combat this erosion. Location of photo shown in vertical aerial photograph.

(b)

Box Figure 16.1.4 (Continued)

generated stronger tidal currents in the back-barrier channels, which changed the sand dynamics in the bay. The influx of sand and its movement within the bay resulted in the migration of bedforms and shoals, which in turn led to the closure of certain channels and the opening of others. One dramatic example of changes that took place in Pleasant Bay was the movement of a large flood-tidal delta 0.7 km long and 0.5 m wide. Over an eight-year period from 1988 to 1996 the flood delta marched northward into the bay, moving about a half kilometer (Box Figure 16.1.3). As a result of this migration, the access channel to the town's main harbor was temporarily closed and had to be dredged, the buoys in the navigation channel on the west side of delta had to be repositioned, and a major shellfish bed was destroyed as the deltaic sands drowned a large community of mussels.

These were not the only changes to the bay. After inlet formation the inner harbor shoreline landward of the inlet was subjected to wave erosion, endangering millions of dollars-worth of properties (Box Figure 16.1.4). The owners responded by constructing expensive seawalls and revetments, but not before acres of valuable real estate were lost and numerous structures had to relocated. The havoc wreaked by the formation of the new inlet will continue because the new inlet is not stable in its position but is migrating downdrift in response to the dominant southerly longshore transport system. Thus, in the foreseeable future many other erosional and depositional problems will occur as the inlet migrates southward.

modified by overwash and dune-building activity, the overall shoreline morphology is frequently preserved. Lenticular-shaped coastal ponds or marshy swales become diagnostic of bar-migration processes and are common features at many inlets.

16.8 Tidal Inlet Effects on Adjacent Shorelines

Many people wish to live along waterways, particularly at tidal inlets due to their scenic beauty, fishing opportunities and boat access

Figure 16.23 Factors affecting the size and number of tidal inlets along a barrier shoreline. (*Source:* From FitzGerald (1988).)

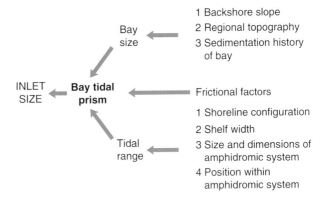

to backbarrier bays and the open ocean. For these reasons, property values are unusually high in the vicinity of tidal inlets and frequently there is considerable demand by the private sector to develop these areas. In conflict with these pressures is the instability of inlet shorelines. In addition to the direct consequences of spit accretion and inlet migration there are the effects of volume changes in the size of ebb-tidal deltas, sand losses to the backbarrier, processes of inlet sediment bypassing and wave sheltering of the ebb-tidal delta shoals. The manner in which these processes affect tidal inlet shorelines is presented below.

16.8.1 Number and Size of Tidal Inlets

The degree to which barrier shorelines are influenced by tidal inlet processes is dependent on their size and number. As the O'Brien Relationship demonstrates, the size or cross-sectional area of an inlet is governed by its tidal prism. This concept can be expanded to include an entire barrier chain in which the size and number of inlets along a chain are primarily dependent on the amount of open water area behind the barrier and the tidal range of the region. In turn, these parameters are a function of other geological and physical oceanographic factors (Figure 16.23). As demonstrated in Chapter 15, wave-dominated coasts tend to have long barrier islands and few tidal inlets and mixed-energy coasts have short, stubby barriers and numerous

tidal inlets. Correspondingly, along the wave-dominated, microtidal coasts of Texas and eastern Florida tidal inlets occur every 40–50 km, whereas along the mixed-energy, mesotidal coasts of Georgia, the East Frisian Islands of Germany, and the Copper River Delta barriers of Alaska, inlets are found every 10–20 km. Presumably, mesotidal conditions produce larger tidal prisms than occur along microtidal coasts, which necessitate more holes in the barrier chain to let the water into and out of the backbarrier. Many coastlines follow this general trend but there are many exceptions due to the influence of sediment supply, large versus small bay areas, and other geological controls. For example, along the central Gulf Coast of Florida the low wave energy of this region, limited sand resources, and large open-water bays produce a coast containing numerous tidal inlets occurring about every 10–20 km.

Along the glaciated coast of southern Massachusetts in Buzzards Bay, the influence of bedrock controls on the size and number of tidal inlets is well illustrated (Figure 16.24). The peninsula and deep embayments of this region are a product of river erosion during the Tertiary (geological time period lasting from 66.4 to 1.6 million years before present) and repeated Pleistocene glaciations. As the valleys of this coast became flooded by rising sea level following deglaciation, erosion of the surrounding glacial deposits produced sediment for the construction of the barriers fronting the embayments. As the Holocene transgression proceeded along the shoreline,

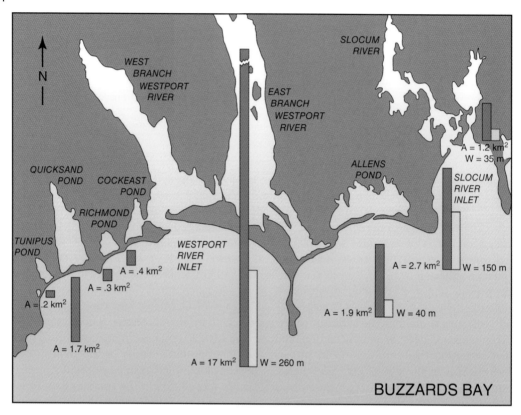

Figure 16.24 Along the transgressive coast of northwestern Buzzards Bay in Massachusetts barriers have built in front of flooded valleys. Rising sea level and a scarcity of sediment has caused the barriers to migrate onshore, decreasing the size of the bays. Reduced bay areas and decreasing tidal prisms has led to closure of several inlets along this coast. Westport River Inlet is the largest inlet in this region and contains the largest bay area. Key: A = Bay area…; W = Inlet width…. (*Source:* From FitzGerald (1996).)

the barriers migrated landward at a faster rate than the bay shorelines were inundated. This resulted in increasingly smaller-sized bays and the gradual closure of many tidal inlets due to decreasing tidal prisms.

16.8.2 Tidal Inlets as Sediment Traps

Tidal inlets not only trap sand temporarily on their ebb-tidal deltas, but they also are responsible for the longer-term loss of sediment moved into the backbarrier. At inlets dominated by flood-tidal currents, sand is continuously transported landward, enlarging flood-tidal deltas and building bars in the tidal creeks. Sand can also be transported into the backbarrier of ebb-dominated tidal

inlets during severe storms (Figure 16.25). During these periods increased wave energy produces greater sand transport to the inlet channel. At the same time the accompanying storm surge increases the water surface slope at the inlet, resulting in stronger than normal flood tidal currents. The strength of the flood currents coupled with the high rate of sand delivery to the inlet results in landward sediment transport into the backbarrier. Along the Malpeque barrier system in the Gulf of St. Lawrence, New Brunswick, it has been determined that over 90% of the sand transferred to the backbarrier took place at tidal inlets and at former inlet locations along the barrier.

Sediment may also be lost at migrating inlets when sand is deposited as channel fill.

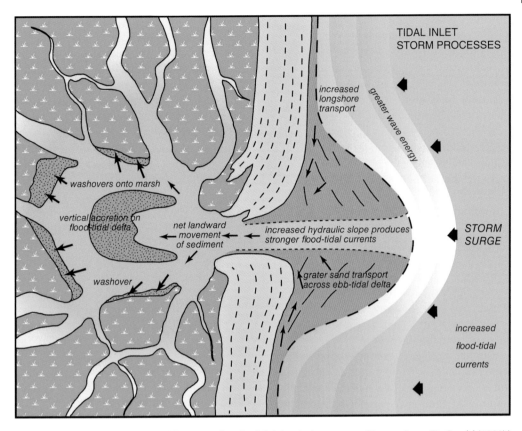

Figure 16.25 Processes of sand addition to a flood-tidal delta during a storm. (*Source:* From FitzGerald (1988).)

If the channel scours below the base of the barrier sands, then the beach sand which fills this channel will not be replaced entirely by the deposits excavated on the eroding portion of the channel. Because up to 40 % of the length of barriers is underlain by tidal inlet fill deposits ranging in thickness from 2 to 10 m, this volume represents a large, long-term loss of sand from the coastal sediment budget. Another major process producing sand loss at migrating inlets is associated with the construction of recurved spits that build into the backbarrier. For example, along the East Frisian Islands recurved spit development has caused the lengthening of barriers along this chain by 3–11 km since 1650. During this stage of barrier evolution the size of the tidal inlets permitted ocean waves to transport large quantities of sand around the end of the barrier, forming recurves that extend far into the backbarrier. Due to the

size of the recurves and the length of barrier extension, this process has been one of the chief natural mechanisms of bay infilling.

16.8.3 Changes in Ebb-Tidal Delta Volume

Ebb-tidal deltas represent huge reservoirs of sand that may be comparable in volume to that of the adjacent barrier islands along mixed-energy coasts (i.e. East and West Frisian Islands, northern Massachusetts, southern New Jersey, Virginia, South Carolina, and Georgia). For instance, the ebb-tidal delta volume of Stono and North Edisto Inlets in South Carolina is $197 \times 10^6 \, \text{m}^3$ and the intervening Seabrook–Kiawah Island barrier complex contains $252 \times 10^6 \, \text{m}^3$ of sand. In this case, the deltas comprise 44 % of the sand in the combined inlet–barrier system. The magnitude of sand contained in ebb-tidal deltas suggests that small changes in their volume dramatically affects the sand supply to the landward shorelines.

To illustrate this concept consider the consequences of changing hydraulic conditions at a tidal inlet along the central coast of Maine. It has been theorized that if a planned hydroelectric power plant is constructed in the Bay of Fundy, tidal ranges in the Gulf of Maine would increase by approximately 30 cm. At the Kennebec River Inlet the larger tidal range would increase the tidal prism by a minimum of 5%. Using the Walton and Adams Relationship, it is calculated that a potential 5% increase in tidal prism would ultimately add over $60 \times 10^6 \, \text{m}^3$ of sand to the Kennebec ebb-tidal delta. Although some of this sand would come from scour of the inlet channel, most of the sand would be eroded from the adjacent beaches, resulting in over 100 m of shoreline recession.

A similar transfer of sand takes place when a new tidal inlet is opened, such as the formation of Ocean City Inlet when Assateague Island, Maryland was breached during the 1933 hurricane. Initially, the inlet was only 3 m deep and 60 m across but quickly widened to 335 m when it was stabilized with jetties in 1935. Since the inlet formed, more than a million cubic meters of sand have been deposited on the ebb-tidal delta (Figure 16.26). Trapping the southerly longshore movement of sand by the north jetty and the growth of the ebb-tidal delta have led to serious erosion along the downdrift beaches. The northern end of Assateague Island has been retreating at an average rate of 11 m per year. The rate of erosion lessened when the ebb tidal delta reached an equilibrium volume and the inlet began to bypass sand.

In contrast to the cases discussed above, the historical decrease in the inlet tidal prisms along the East Frisian Islands has had a beneficial effect on this barrier coast (Figure 16.27) From 1650 to 1960 the reclamation of tidal flats and marshlands bordering the German mainland as well as natural processes, such as the building and landward extension of recurved spits, decreased the size of the backbarrier by 80%. In turn, the reduction in bay area decreased the inlet tidal prisms, which led to smaller sized inlets, longer barrier islands, and smaller ebb-tidal deltas (see Chapter 15). Wave action transported ebb-tidal delta sands onshore as tidal discharge decreased. This process increased the supply of sand to the beaches and aided the lengthening of the barriers.

16.8.4 Wave Sheltering

The shallow character of ebb-tidal deltas provides a natural breakwater for the landward shorelines. This is especially true during lower tidal elevations when most of the wave energy is dissipated along the terminal lobe. During higher tidal stages intertidal and subtidal bars cause waves to break offshore expending much of their energy before reaching the beaches onshore. The sheltering effect is most pronounced along mixed-energy coasts where tidal inlets have well developed ebb-tidal deltas.

The influence of ebb shoals is particularly well-illustrated by the history of Morris Island, South Carolina, which forms the southern border of Charleston Harbor (Figure 16.28). Before human modification, the entrance channel to the harbor paralleled Morris Island and was fronted by an extensive shoal system. The deflected southerly course of the main ebb channel was due to the preferential accumulation of sand on the updrift side (northeast side) of the harbor's ebb-tidal delta caused by the dominant southerly longshore transport of sediment. The shallow and constantly shifting position of the outer portion of the entrance channel made for treacherous navigation into the harbor, resulting in numerous shipwrecks along the outer shoals. In the late 1800s jetties were constructed at the harbor entrance to straighten, deepen, and stabilize the main channel; the project was completed in 1896. During the period prior to jetty construction (1849–1880) Morris Island had been eroding at an average rate of 3.5 m year^{-1}. After the

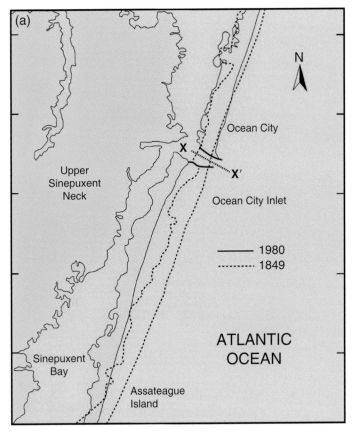

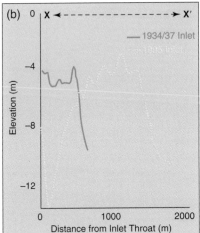

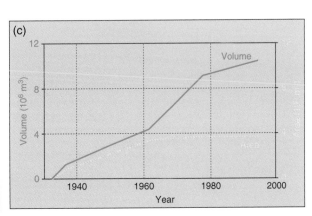

Figure 16.26 Ocean City Inlet was opened along northern Assateague Island, Maryland during the 1933 hurricane. (a) The Army Corps of Engineers then constructed jetties to stabilize the entrance. (b) The ensuing tidal exchange between the ocean and Isle of Wight Bay and Sinepuxent Bay produced a large tidal prism, a deepening of the inlet channel (location of longitudinal section shown in panel A). (c). Since 1933, the volume of the ebb delta has grown to more than a million cubic meters of sand. (*Source:* From Leatherman (1984).) Note in panel B the seaward extension and growth of the ebb delta. As seen in panel A, this sand-trapping has starved the downdrift shoreline of sand, resulting in a landward migration of the northern end of Assateague Island by more than several hundred meters. (*Source:* From Stauble (1997).)

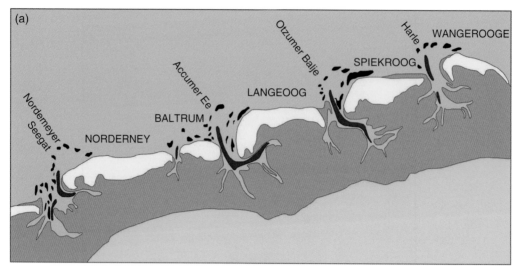

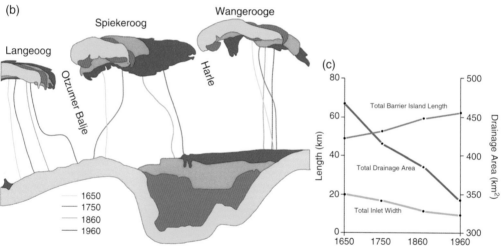

Figure 16.27 (a) Historical morphological changes to the Frisian Islands along the German North Sea coast. (b) From 1650 to 1960 widespread land reclamation along the backside of the barriers and on the mainland significantly decreased the size of the bays behind the tidal inlets. Coincident with a large decrease in drainage area behind Harle Inlet was a narrowing of Harle Inlet and an eastward extension of the Spiekeroog barrier island. (c) Moreover, land reclamation and reduction in bay area along the entire length of the barrier chain decreased the width of the tidal inlets as well as the volume of sand contained in the ebb-tidal deltas. During the 1650–1960 period, sand from the ebb-tidal delta moved onshore, which facilitated the growth the barrier islands.

jetties were in place, the ebb-tidal delta shoals, which bordered the old channel, were cut off from their longshore sand supply. As the shoals eroded and gradually diminished in size, so did the protection they afforded Morris Island, especially during storms. From 1900 to 1973 Morris Island receded 500 m at its northeast end increasing to 1100 m at its southeast end, a rate three times what it had been prior to jetty construction. One dramatic response to this erosion was the detachment of a lighthouse from the southeast end of the island. In 1900 the lighthouse was located 640 m onshore but by 1970 it was sitting in 3 m of water 360 m from the shoreline.

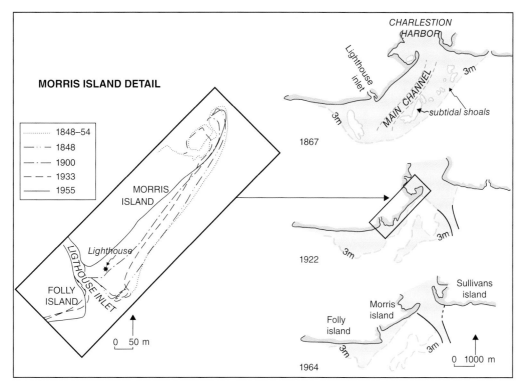

Figure 16.28 Jetty construction at the entrance to Charleston Harbor, South Carolina in the late 1800s significantly diminished the size and extent of sand shoals fronting the harbor, which in turn affected erosional–depositional processes along the landward shorelines. (*Source:* From FitzGerald (1988)).

16.8.5 Effects of Inlet Sediment Bypassing

Tidal inlets interrupt the wave-induced long-shore transport of sediment along the coast, affecting both the supply of sand to the downdrift beaches and the position and mechanisms whereby sand is transferred to the downdrift shorelines. The effects of these processes are exhibited well along the Copper River Delta barriers in the Gulf of Alaska. From east to west along the barrier chain the width of the tidal inlets increases as does the size of the ebb-tidal deltas (Figure 16.29). In this case the width of the inlet can be used as a proxy for the inlet's cross sectional area. These trends reflect an increase in tidal prism along the chain caused by an increase in bay area from east to west while tidal range remains constant. Also quite noticeable along this coast is the greater downdrift offset of the inlet shoreline in an westerly

direction. This morphology is coincident with an increase in the degree of overlap of the ebb-tidal delta along the downdrift inlet shoreline. The offset of the inlet shoreline and bulbous shape of the barriers are produced by sand being trapped at the eastern, updrift end of the barrier. The amount of shoreline progradation is a function of inlet size and extent of its ebb-tidal delta. What we learn from the sedimentation processes along the Copper River Delta barriers is that tidal inlets can impart a very important signature on the form of the barriers.

16.8.5.1 Drumstick Barrier Model

In an investigation of barrier islands shorelines in mixed energy settings throughout the world, Miles Hayes at the University of South Carolina noted that many barriers exhibit a similar shape. Numerous tidal inlets studies by Professor Hayes and his students allowed

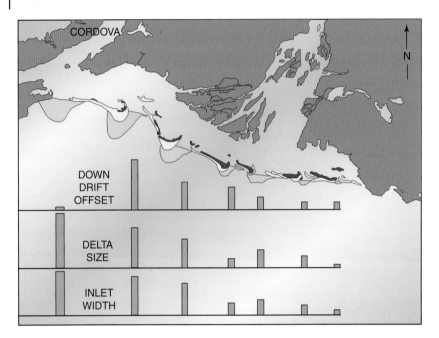

Figure 16.29 The Copper River Delta barrier chain in the Gulf of Alaska is a product of an abundant sand supply from the Copper River and high wave energy. From east to west the extent of open water in the backbarrier increases. This trend leads to larger tidal prisms, larger-sized tidal inlets, and barrier morphology that is strongly influenced by tidal inlet processes. (*Source:* From Hayes (1975).)

him to formulate his drumstick barrier island model (Figure 16.30). In this model the meaty portion of the drumstick barrier is attributed to waves bending around the ebb-tidal delta, producing a reversal in the longshore transport direction. This process reduces the rate at which sediment bypasses the inlet, resulting in a broad zone of sand accumulation along the updrift end of the barrier. The downdrift, or thin part of the drumstick, is formed through spit accretion. Later studies demonstrated that bar complexes migrating onshore from the ebb-tidal delta are an important factor dictating barrier island morphology and the overall erosional–depositional trends, particularly in mixed energy settings. The Copper River barriers conform well to the Hayes drumstick model.

Looking at the East Frisian Islands we see that in addition to drumsticks, barriers can have many other shapes (Figure 16.31). Inlet sediment bypassing along this barrier chain occurs, in part, through the landward migration of large swash bars (>1 km in length) that deliver up to 300,000 m^3 of sand when they weld to the beach. In fact, it is the position where the bar complexes attach to the shoreline that dictates the form of the barrier along this coast. If the ebb-tidal delta greatly overlaps the downdrift barrier, then the bar complexes may build up the barrier shoreline some distance from the tidal inlet. In these cases, humpbacked barriers are developed, such the Norderney or Spiekeroog. If the downdrift barrier is short and the ebb-tidal delta fronts a large portion of the downdrift barrier, such as at the island of Baltrum, the bar complexes weld to the eastern end of the barrier forming downdrift bulbous barriers. Thus, studies of the Friesian Islands demonstrate that inlet processes exert a strong influence on the dispersal of sand along mixed-energy barrier island shorelines and in doing so dictate barrier shape.

16.8.6 Human Influences

Dramatic changes to inlet beaches can also result from human influences including the obvious consequences of jetty construction

DOMINANT LONGSHORE
SEDIMENT TRANSPORT

PREDOMINANT
WAVE APPROACH

SEDIMENT
TRANSPORT
REVERSAL

BARRIER ISLAND
DRUMSTICK MODEL

Figure 16.30 Miles Hayes' drumstick barrier island model explains barrier shape as a function of wave refraction around the ebb-tidal delta trapping sand along the downdrift inlet shoreline. (*Source:* From Hayes and Kana 1978).

that reconfigures an inlet shoreline. By preventing or greatly reducing an inlet's ability to bypass sand, the updrift beach progrades while the downdrift beach, whose sand supply has been diminished or completely cut off, erodes. There can also be more subtle human impacts that can equally affect inlet shorelines, especially those associated with changes in inlet tidal prism, sediment supply, and the longshore transport system. Nowhere are these types of impacts better demonstrated than along the central Gulf Coast of Florida where development has resulted in

the construction of causeways, extensive backbarrier filling and dredging projects, and the building of numerous engineering structures along the coast. Of the 17 inlets that have closed along this coast since the late 1980s, more than half of the closures can be traced to human influences caused primarily by changes in inlet tidal prism. For example, access to several barriers has been achieved through the construction of causeways that extend from the mainland across the shallow bays. Along most of their lengths the causeways are dike-like structures that partition

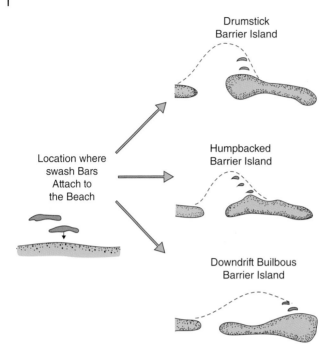

Figure 16.31 The shape of mixed-energy barriers along the Frisian Islands is primarily a function of wave inlet sediment bypassing processes. The bulbous portion of the barrier coincides with the location where sand bypasses the updrift inlet and where bar complexes migrate onshore. (*Source:* From FitzGerald (1988).)

the bays, thereby changing bay areas and inlet tidal prisms. In some instances, tidal prisms were reduced to a critical value causing inlet closure. At these sites the tidal currents were unable to remove the sand dumped into the inlet channel by wave action. Similarly, when the Intracoastal Waterway (a protected inland canal built for barge and boat traffic) was constructed along the central Gulf Coast of Florida in the early 1960s, the dredged waterway served to connect adjacent backbarrier bays, thereby changing the volume of water that was exchanged through the connecting inlets. The Intracoastal Waterway lessened the flow going through some inlets while at the same time increasing the tidal discharge of others. This resulted in the closure of some inlets and the enlargement of others. Improved access to the central Florida barriers led to their development including the formation of marinas and finger canals along the backside of the barriers. These were formed by dredging small canals and then using the dredge spoil to build land peninsulas where there was once just water. As seen in a comparison of historical and present day maps of Boca Ciega Bay, this process has drastically reduced the open-water area, leading to smaller tidal prisms and smaller equilibrium-sized tidal inlets (Figure 16.32). These examples demonstrate that altering the natural system can produce undesired consequences, which emphasizes the need to assessed the potential effects of developmental projects before they are undertaken.

16.9   Summary

As we have seen in this chapter, tidal inlets occur along barrier coasts in coastal plain settings and in other regions where there has been a sufficient supply of sand for barrier spit construction across embayments. Inlet formation today occurs primarily when

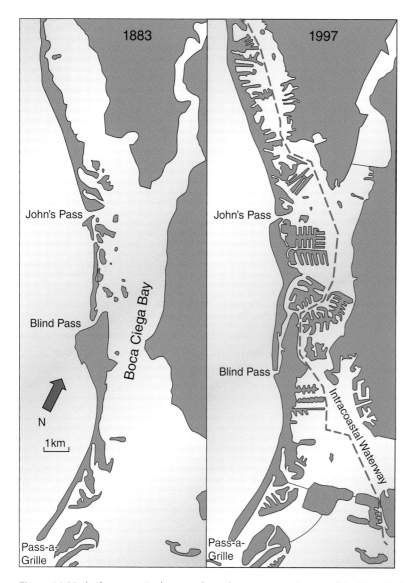

Figure 16.32 Anthropogenic changes along the west-central coast of Florida as indicated by maps of the region in 1883 and 1997. Construction of finger canals and solid causeways have decreased open water areas and reduced water circulation in the bays. This condition has led to smaller bay tidal prisms and reduced sized of associated tidal inlets. (*Source:* From Barnard and Davis (1999).)

narrow, low barriers are breached during severe storms. Tidal inlets are narrowest and deepest at their throat section where tidal currents and potential sediment transport reach their maximums. Many inlets have stabilized next to bedrock outcrops, in former river channels or in resistant sedimentary strata. Migrating inlets are usually shallow and positioned in easily eroded sands. Flood-tidal deltas are horseshoe-shaped shoals situated on the landward side of an inlet and formed from sand entering the inlet channel and being transported into the backbarrier by flood-tidal currents (see Box 16.2). Sand transported seaward by ebb-tidal currents forms arcuate-shaped ebb-tidal deltas.

The morphology of ebb deltas reflects the wave versus tidal energy that shapes them. A direct correspondence exists between an inlet's tidal prism and its throat cross sectional area, and the volume of sand contained in its ebb-tidal delta. These concepts are very useful when planning jetty construction, channel dredging, and sourcing ebb-tidal deltas for beach nourishment programs. There are various mechanisms whereby sand bypasses unmodified tidal inlets. Wave-dominated inlets bypass sand along the periphery of the delta by wave action. At mixed-energy inlets one of the end-products of sediment bypassing is the formation of large bar complexes which migrate onshore and attach to the landward shoreline. Processes of inlet sediment bypassing, volumetric changes of the ebb-tidal delta, sand losses to the backbarrier, the sheltering effect of the ebb-tidal delta, and other inlet processes strongly influence the distribution of sand along inlet beaches. Along mixed-energy coasts tidal inlets may dictate the shape of barrier islands.

Box 16.2 Rapid Changes at Anclote Key, Florida (Box #2)

We generally think that major changes to barrier islands take either a long time to eventuate or are the result of hurricanes or other intense storms. Major changes in the length and morphology of Anclote Key on the Gulf Coast of the Florida peninsula have taken place over only a few decades in the absence of any significant storm. This island, which is the northernmost barrier along this section of coast, is separated from the mainland by a 5-km-wide, shallow expanse of water. Anclote Key is a wave-dominated linear barrier exhibiting recurved beach ridges at both ends. Despite having no nearby barriers, fairly deep, inlet-like channels exist at both ends of the island. Radiocarbon dates of shells and organic material obtained during extensive coring of the island have shown it to be about 1500 years old. The barrier sand rests on a pavement of Miocene limestone that is from 3 to 5 m below mean sea level.

Historical maps and charts of the region indicate that Anclote Key did not change significantly from 1881 to the1960s. Most of the changes that did occur were associated with a modest southward extension of the island. During this time the landward migration and attachment of swash bars to the southern shoreline relegated an old Coast Guard pier to an upland position. The northern end of the island displayed no noticeable change until the early 1960s. Over the next 20 years, spit accretion extended the northern end of Anclote Key by more than a kilometer.

Historical records reveal that the spit system has consistently terminated at the site of a deep tidal channel. Strong flood-tidal currents in this channel (landward-directed currents) along with wave action and a large sediment supply, have produced a long spit that has widened the northern end of the island by about a kilometer (Box Figure 16.2.1).

The growth of a barrier over a 20-year period is in itself remarkable, particularly along a segment of coast that is sediment-starved. Where did all the sand come from? The answer to this question was found in an analysis of aerial photographs taken along this coast during the 1950s (1951 and 1957). Both of these sets of photographs show that sea grass once covered an expansive area of the inner shoreface to within 30 m or so of the shoreline. A 1963 aerial photograph of the same region shows that the vegetation had disappeared. Some process or condition, as yet undetermined, caused the demise of the sea grass, which had been stabilizing the sand substrate. Once the stabilizing effect of the grass was removed, wave action transported the nearshore sand onto the adjacent beaches of Anclote Key. Much of the newly accreted sand was then carried by longshore currents to the northern end of the island forming

Anclote Key, FL, 1979

Box Figure 16.2.1 Aerial photograph of northern Anclote Key, Florida in 1979. (modified from Giese, 1988).

recurved spits. This same phenomenon also occurred a few kilometers south of Anclote Key and was responsible for the construction of long spit on Honeymoon Island as well as the formation of a new barrier island called Three-Rooker Bar.

References

Barnard, P.L. and Davis, R.A. (1999). Anthropogenic versus natural influences on inlet evolution: west-Central Florida. In: *Coastal Sediments '99* (ed. N.C. Kraus and W.G. McDougal), 1498–1504. Reston, VA: A.S.C.E.

FitzGerald, D.M. (1988). Shoreline erosional–depositional processes associated with tidal inlets, hydrodynamics and sediment dynamics of tidal inlets. In: *Hydrodynamics and Sediment Dynamics of Tidal Inlets* (Lecture Notes on Coastal and Estuarine Studies Volume 29) (ed. D.G. Aubrey and L. Weishar), 186–225. Berlin: Springer.

FitzGerald, D.M. (1996). Geomorphic variability and morphologic and sedimentologic controls on tidal inlets. In: *Understanding Physical Processes at Tidal Inlets*, Special Issue #23, Journal of Coastal Research (ed. A.J. Mehta), 47–71. Coastal Education and Research Foundation.

Giese, G.S. (1988). Cyclic behavior of the tidal inlet at Nauset Beach, Chatham, MA. In: *Hydrodynamics and Sediment Dynamics of Tidal Inlets* (ed. D.G. Aubrey and L. Weishar), 269–283. New York: Springer.

Hayes, M.O. (1975). Morphology of sand accumulations in estuaries. In: *Estuarine Research*, vol. 2 (ed. L.E. Cronin), 3–22. New York: Academic Press.

Hayes, M.O. and Kana, T. (1978). *Terrigenous Clastic depositional Environments*. Columbia, SC: Deparment of Geology, University of South Carolina.

Leatherman, S.P. (1984). Shoreline evolution of north Assateague Island, Maryland. *J. Shore Beach* 52 (4): 3–10.

O'Brien, M.P. (1931). Estuary tidal prisms related to entrance areas. *Civ. Eng.* 1: 738–739.

O'Brien, M.P. (1969). Equilibrium flow areas of inlets on sandy coasts. 95: 43–55.

Oertel, G. (1975). Ebb-tidal deltas of Georgia estuaries. In: *Estuarine Research*, vol. 2 (ed. L.E. Cronin), 267–276. New York: Academic Press.

Smith, J.B. (1989). Morphodynamics and stratigraphy of Essex River ebb-tidal delta., Masters Thesis, Boston University, 223 p.

Stauble, D.K. (1997). *Ocean City Inlet, Maryland, and Vicinity Water Resources Study*, 46 p. Baltimore: U.S. Army Corps of Eng.

Walton, T.L., and Adams, W.D. (1976). Capacity of inlet outer bars to store sand. Proc. of 15th Coastal Engineering Conference, ASCE, Honolulu, Hawaii, p. 1919-1937.

Suggested Reading

Aubrey, D.G. and Giese, G.S. (eds.) (1993). *Formation and Evolution of Multiple Tidal Inlets*. Washington, DC: American Geophysical Union.

Aubrey, D.G. and Weishar, L. (eds.) (1988). *Hydrodynamics and Sediment Dynamics of Tidal Inlets, Lecture Notes on Coastal and Estuarine Studies*. New York: Springer.

Boothroyd, J.C. (1985). Tidal inlets andtidal deltas. In: *Coastal Sedimentary Environments* (ed. R.A. Davis), 445–532. New York: Springer.

Bruun, P. (1966). *Tidal Inlets and Littoral Drift*. Amsterdam: North-Holland Publishing.

Bruun, P. and Gerritsen, F. (1960). *Stability of Tidal Inlets*. Amsterdam: North-Holland Publishing.

Cronin, L.E. (ed.) (1978). *Estuarine Research*, vol. 2. New York: Academic Press.

Hayes, M.O. and Kana, T. (1978). *Terrigenous Clastic depositional Environments*. Columbia, SC: Deparment of Geology, University of South Carolina.

17

Glaciated Coasts

17.1   Introduction

Glaciated coasts exhibit diversity in both their types of features and the changing landscape that is unparalleled in geomorphology. The ability of continental glaciers (ice sheets) and valley glaciers to sculpture land surfaces, transport large quantities of rock and sediment, and eventually to deposit these materials in a number of glacial features accounts for this variability. For example, the northern New England coast is mostly rocky with small pocket beaches that range in composition from sand to gravel and even to boulder-sized sediment over distances of less than a kilometer. Interrupting this trend in northern Massachusetts and southern Maine are several extensive barrier systems (5–30 km long) that occur at the mouths of estuaries within arcuate embayments. The existence of these barriers along an otherwise sediment-starved coast is due to the large volumes of sand brought to the coast by rivers following deglaciation. Continuing northward, the central and northeast coast of Maine is rugged and highly irregular with rocky peninsulas giving way to broader embayments and finally to bedrock cliffs. Here glaciers have stripped away most of the sediment overlying the bedrock. This morphology contrasts sharply with that of Cape Cod (Figure 17.1), which has smooth coastlines consisting of mainland beaches, barrier spits, and barrier islands. Cape Cod is composed entirely of glacial sediments that were deposited during the deglaciation of this region approximately 17000–18000 years before present. Not only have the effects of glaciation produced very different sediment abundance along the New England coast, but also glacial processes combined with the pre-existing bedrock geology of the region have created a coast with numerous bays and sounds. In turn, this coastal morphology has resulted in highly varied physical settings including wave-dominated, mixed-energy, and tide-dominated coasts.

The glaciated coast along the tectonically active Gulf of Alaska is even more diverse and more spectacular than that of New England (Figure 17.2). In some locations, such as the Kenai Peninsula, valley glaciers are still found a short distance from the coast, having retreated into the bordering mountain valleys. The deep fjords of this region are a testament to the ability of glaciers to carve coastal landscapes. Along the Alaskan Peninsula, glacial meltwater streams feed sediment into broad embayments forming extensive tidal flats, marshes and barriers spits. During the summer, these streams burgeon with salmon seeking spawning grounds, thereby providing a tasty meal for the waiting brown bears. In yet another site along this active coast, several glaciers east of Prince William Sound have produced a wealth of sediment that has been carried to the coast via a myriad of meltwater streams. The deposition of these sediments and their reworking by waves and tides are responsible for

Beaches and Coasts, Second Edition. Richard A. Davis, Jr. and Duncan M. FitzGerald.
© 2020 John Wiley & Sons Ltd. Published 2020 by John Wiley & Sons Ltd.

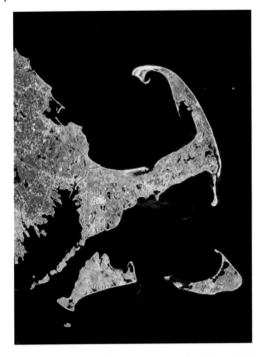

Figure 17.1 Color infrared photograph of Cape Cod and the islands of Nantucket (right) and Martha's Vineyard (left). These regions were built from sediment carried south by glaciers and deposited as end moraines, outwash plains and other sedimentary features.

forming the 80-km long Copper River Delta barrier chain and an expansive backbarrier tidal flat system.

This chapter describes how glacial processes have produced diverse and dramatic landscapes along many high-latitude coasts. The manner in which glaciers excavate bedrock, transport large quantities of sediment and rock, and the processes whereby these materials are deposited are discussed. The effects of sea-level changes associated with the enlargement and melting of continental ice sheets are also explained. Finally, the causes of repeated episodes of glaciation during the Ice Ages (past 2.2–2.4 million years) are explored.

17.2   The World's Glaciers

Glaciers exist on almost every continent of the world, including Africa, where retreating glaciers top portions of Mount Kilimanjaro and Mount Kenya (Figure 17.3). They are not present in Australia but are found in nearby New Zealand. Glaciers occur as narrow ribbons of

Figure 17.2 Aerial view of the northern Alaskan Peninsula. Meltwater from nearby glaciers transports sediment to the coast via a braided stream. A large gravel and sand recurved spit has built from the abundant sediment supply.

(a)

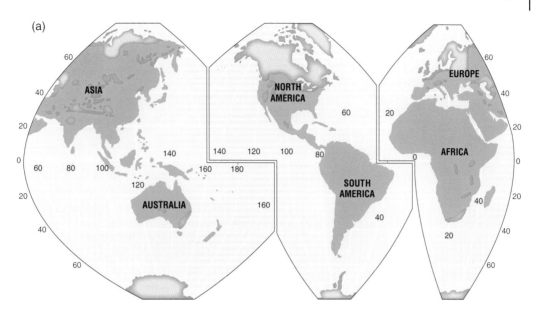

(b)

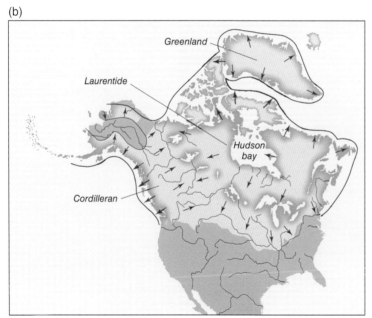

Figure 17.3 Extent of glacial ice during maximum Pleistocene glaciation. (a) Ice covered portions of North America, Europe, and Asia and the high mountains in other regions 18,000 to 20,000 years ago. (b) In North America majors ice sheets included the Laurentide, Greenland, and Cordilleran sheets.

flowing ice in high mountain regions or as thick ice sheets covering vast continental areas. Presently, glaciers cover about 10% of the continental landmass but in the recent geologic past (several times during the past 1.6 million years) they extended over 30% of the land surface. This fact suggests that glacial processes have formed or strongly influenced a large portion of the world's coastlines, a concept that is not widely appreciated.

17.2.1 Glacier Formation

A glacier is defined broadly as a large mass of ice that flows internally. The oldest glacier ice on Earth is a stagnant ice mass in the Dry Valleys region of Antarctica, which was part of an active glacier 14 million years ago. The formation of glaciers is tied closely to certain climatic conditions where cool temperatures and precipitation produce more snow accumulation during the colder months of the year than ablates during the warmer months (Figure 17.4). Ablation is the term given to the collective processes of ice wastage, including melting, sublimation and ice calving into water. Thus, glaciers tend to form in regions of high latitude or high elevation; even in these regions slight changes in either temperature or precipitation can cause glaciers to advance or retreat.

The transformation of snow to ice and then to glacial ice is a progressive process whereby air is gradually forced out to produce a dense mass of interlocking ice crystals (Figure 17.5). Freshly fallen snow is commonly light and fluffy and may be 90% air, although those who live in northern regions and shovel snow during the winter might argue about this presumed weightlessness. As more and more snow accumulates, snow at the base is compacted by the overlying weight, changing the hexagonal snowflake crystals into smaller spherical structures called granular snow (i.e. the type of snow common during spring skiing). With greater weight added by more snowfall, along with some melting and freezing, the granular snow is transformed into a denser, recrystalized granular structure called firn. This is the type of ice comprising old snow banks along the sides of roadways during the close of winter. Under the pressure of more snowfalls and the development of thick firn layers, eventually all the air is

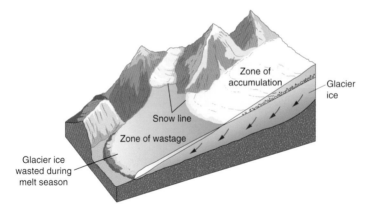

Figure 17.4 Cross-sectional view of a glacier showing the annual accumulation zone and the ablation zone, also called the zone of wastage. The accumulation zone occurs where there is a net gain in ice, whereas the ablation zone experiences greater wastage than ice formation. The zone separating these two regions is called the snow line.

Figure 17.5 Conversion from freshly fallen snow through several states to glacial ice.

expelled (except some minute air bubbles) and a mass of interlocking ice crystals is created. At a depth of about 50 m glacial ice is formed. This ice is highly compact and exhibits a vivid blue color in sharp contrast with the surface ice, which is commonly whitish to gray in color due to the presence of air and sediment.

17.2.2 Glacier Movement

Flow is a characteristic of all glaciers. Movement by glaciers is achieved by two major mechanisms: Plastic Flow and Basal Slip (Figure 17.6). Most of our experience with ice leads us to believe that ice is brittle and will shatter when a force is applied to it, as, for instance, when dropping an ice cube on the floor and seeing it break into many pieces. However, at a thickness of 50 m or more ice behaves as a plastic material and can be deformed. Under these conditions the glacial ice below 50 m will flow downslope under the influence of gravity. Differential stresses in the overlying brittle ice may produce deep cracks in the ice surface called crevasses.

Basal slip is an equally important means of producing movement in glaciers. In this process water acts as a lubricant and reduces the friction between the base of the ice and the underlying bedrock or sediment surface.

Figure 17.6 Cross-section of a glacier illustrating ice movement involving two methods. (a) The first is by plastic flow below a depth of about 50 m. (b) The second process is by sliding along the bottom, known as basal slip.

(a)

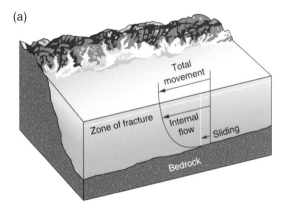

(b)

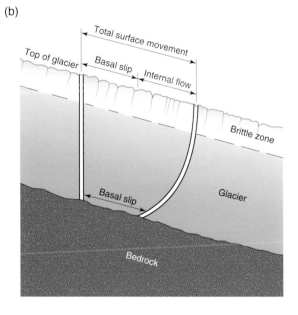

Basal slip allows the entire glacier to slide downslope along a layer, or in some cases a thin film, of water. Formation of meltwater at the base of the glacier is caused by several different mechanisms including frictional heating, which is produced when the flowing glacier comes in contact with the bedrock or sediment surface. Heat rising from within the Earth's interior may also contribute to the warming and melting of ice at the base of the glacier. Meltwater is also formed beneath the glacier due to pressure exerted by the thickness and weight of the overlying ice. Finally, meltwater formed at the surface of the glacier may eventually end up at the base of the glacier by descending through the glacier via a giant well-like vertical to near vertical shafts called a moulins.

17.2.3 Distribution and Types of Glaciers

17.2.3.1 Alpine Glaciers

In mountainous regions where winters are long with abundant snowfall and summers are cool and short-lived, conditions are perfect for forming alpine glaciers, or valley glaciers as they are also called (Figure 17.7).

Named for the Alps where these glaciers are common, alpine glaciers occur throughout the world in all major mountain belts. They originate in highest parts of mountains and flow down former river valleys under the force of gravity. Alpine glaciers in Alaska, cover an area equivalent to half the size of New England (75,000 km^2). Most of these glaciers occur in southern Alaska, including many that have formed in mountain ranges along the Gulf of Alaska. Alpine glaciers are responsible for carving coastal landforms in some locations and for delivering large quantities of sediment to others.

17.2.3.2 Ice Sheets

Unlike alpine glaciers which are confined to mountain valleys, ice sheets stretch across millions of square kilometers, reaching several kilometers in thickness. Because of their great surface area, they are also called continental glaciers. Although ice sheets once occupied vast regions of North America, Europe, and Asia as recently as 18,000 years ago, there are only two ice sheets left today, one in each of the two hemispheres. Ice covers over 80% of Greenland (1.7 million square kilometers), the world's largest island.

Figure 17.7 Aerial view of an alpine glacier terminating in a proglacial lake.

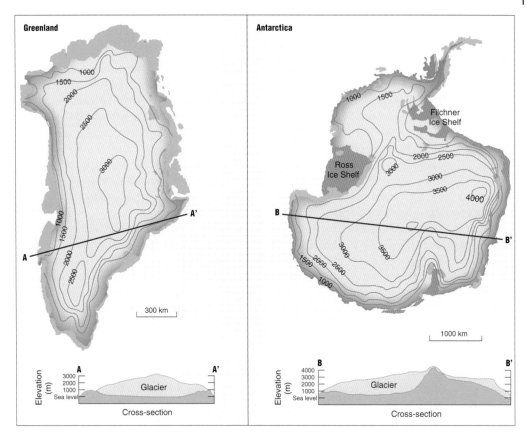

Figure 17.8 The Greenland and Antarctica ice sheets have a combined area of almost 16 million square kilometers, which is slightly less than the area of South America.

The central portion of the ice sheet is 3000 m thick and thins toward the coast, producing a lens-shaped ice mass (Figure 17.8). The thicker ice sheet in the interior of the island causes the ice to radiate outward and flow toward the coast. Here it is met by rugged mountain systems that fringe much of the Greenland coast, interrupting its passage to the sea. The coastal mountains act as dams, causing the ice to bulge and build pressure behind them. Ridge systems dissect the ice sheet while mountain passes allow individual lobes of ice to extend toward the sea. These glaciers are called outlet glaciers and are similar in appearance to alpine glaciers. Flow rates of the main ice sheet are on the order of 40–120 m per year, in contrast to outlet glaciers, which may speed along by as much as a meter per day.

The Antarctic ice sheet is a many times larger than the one in Greenland (1.6 million square kilometers) having an area of just over 14 million square kilometers, which is about 1.5 times the size of the contiguous United States. Antarctica is composed of the large East Antarctic Ice Sheet (EAIS) and the smaller West Antarctic Ice Sheet (WAIS). The EAIS has a maximum thickness of about 4200 m and overlies mostly a bedrock basement (Figure 17.8). The smaller WAIS is mostly marine-based and is susceptible to melting beneath the ice, which thins the ice causing eventual flotation and large-scale calving (see Box 17.1). In several locations along the coast, the ice sheet extends across large embayments forming ice shelves. These are regions where the ice thins and is no longer in contact with the land surface, but

Box 17.1 The Fate of the Ice Sheets

With recent warnings of increasing global warmth, it is sometimes difficult to remember that the Earth is still held firmly in the grips of a great ice age that began about 40 million years ago and intensified dramatically over the last three million years (Ma). Thirty million cubic kilometers of ice still reside in Antarctica (Box Figure 17.1.1) with an additional three million cubic kilometers of ice in Greenland. Visitors to Antarctica see the world much as it was 20,000 years ago, when similar sheets of ice, up to 4 km thick, covered much of the northern hemisphere. Over at least the last 2.5 Ma, great ice sheets have expanded and contracted in concert with changes in the geometry of the Earth's orbit reaching similar maximum dimensions every 100,000 years or so. During each of these glacial maximums, eustatic sea level falls due to water being temporarily stored in the ice sheets – only to be returned to the oceans, sometimes catastrophically, when the ice sheets return to near their present dimensions.

If all of the ice in both Greenland and Antarctica today were to melt, global eustatic sea level would rise by approximately 66 m (216 feet). Under a projected global warming scenario of 2-3 °C resulting from a doubling of atmospheric CO_2, we now ask the question: will this ice melt with future warming? Or, is the remaining ice in the polar ice sheets stable and well-positioned to endure the worst-case scenario of greenhouse-induced global warmth? Furthermore, if the ice is at risk of collapse, how quickly might it disappear?

In order to predict the ice sheets' response to a warmer world, we first must recognize the different geographic and physical characteristics of the individual ice sheets. The Greenland ice sheet, which if melted would contribute approximately 6 m of sea level rise, is located in the relatively warmer northern hemisphere and is believed to be vulnerable to a nearly complete disintegration, mostly from surface melting and accelerated flow, under a 2-3 °C

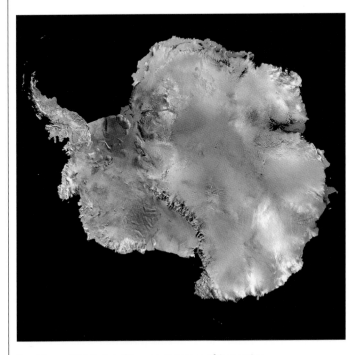

Box Figure 17.1.1 Satellite mosaic image of Antarctica.

increase in global temperature. The Antarctic ice sheet is not everywhere equally stable and will have a more complex response to warming. The west Antarctic ice sheet (WAIS) is grounded mostly below sea level and would contribute approximately 3 m of sea level rise if melted. Because its base is vulnerable to melting in a warmer ocean, the WAIS is considered by most researchers to be at risk of near total collapse under a warming scenario of only 2-3 °C. On the other hand, the east Antarctic ice sheet (EAIS) is grounded predominately above sea level where it is less susceptible to ocean warming. Is the EAIS still vulnerable to future global warming? The answer to this question is highly relevant. If the EAIS - the true sleeping giant of the ice sheets - were to melt completely, it would raise eustatic sea level by a massive 57 m.

One key to unlocking the mystery of the future behavior of the East Antarctic Ice Sheet comes from examining the past behavior of the ice sheet during periods of prior global warmth – particularly those that match or exceed predicted warming from greenhouse emissions. One such time interval is the early-to-mid-Pliocene, 3–4 million years ago. During this time period, global temperatures were 2-3 °C warmer than today, Greenland was predominantly ice free, and the WAIS was mostly non-existent. However, a decades-long debate has surrounded the stability of the EAIS during this period. Central to the debate are the presence of Pliocene-aged marine diatoms located in glacial deposits in a mountain range that crosses Antarctica (the Transantarctic Mountains). One theory to explain the placement of marine diatoms in interior Antarctica suggests that EAIS was almost completely melted during the early-to-mid-Pliocene which allowed for the existence of marine seaways crossing the South Pole. Opposing theories suggest that the marine diatoms were simply transported to the interior by wind blowing in from the distant open ocean and point to other evidence to suggest the EAIS was essentially the same size in the early-to-mid-Pliocene as it

is today. The most recent evidence suggests that the real answer probably lays somewhere in the middle of these two extremes, but that the majority of the EAIS is stable in the face of a moderate amount of warming.

Overall, the above discussion suggests that 5-15 m of eustatic sea level rise from melting ice sheets is theoretically inevitable under a global warming scenario of only 2-3 °C. However, even more uncertain and arguably more important than the eventual fate of the ice sheets is rate at which they may decline. A sea level rise of 5-15 m occurring over a few decades would have a vastly different human impact than the same sea level increase stretched over many millennia. Up until relatively recently, most researchers believed that the decline of the ice sheets would be fairly gradual and partially offset by increased snowfall (and resulting ice accumulation) in warming world. From satellite observations, we know that the Greenland ice sheet has already reduced in mass at an average pace of 280 gigatons per year over the last 15 years and that WAIS has lost an average of 125 gigatons of ice per year over the same period. This rate of ice loss, if continued at its current pace, would result in only a few centimeters of eustatic sea level rise by the year 2100. However, ice sheet behavior is known to be highly nonlinear and several recent observations and models hint that a rapid acceleration of ice loss could be in store for the near future. These include the disintegration and destabilization of ice shelves that buttress the flow if ice on WAIS, the retreat of submarine grounding lines for both WAIS and Greenland, the acceleration of ice flow velocity on all the ice sheets, and the discovery of deep marine channels that may be capable of bringing in warm ocean water to the base of critical portions of the EAIS. Considering all of the evidence, the current conservative estimate of sea level rise - which includes both the effect of melting ice sheets and the thermal expansion of seawater - is between 40 and 100 cm by the year 2100.

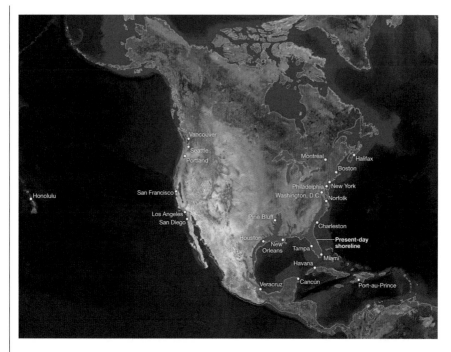

Box Figure 17.1.2 Hypothetical shoreline of North America 18,000 years ago during glacial maximum and after ice sheets melt. (*Source:* From E. J. Tarbuck & F. K. Lutgens, 2002, Earth, New York, Prentice Hall.)

Wherever the truth lies, one thing is certain. Ice contained in the great ice sheets has the potential to completely change the world as we know it. Most of the world's largest cities would be lost if this ice should melt (Box Figure 17.1.2). It will melt sometime. The question is when.

Sean L. Mackay
Polar Scientist
Earth and Environment Department
Boston University

rather floats above the sea floor. The ice shelves, of which the Ross and Filchner are the largest, are fed by ice flowing from the Antarctic interior.

17.3 Pleistocene Glaciation

17.3.1 Introduction

The cycles of glaciation which began approximately 2.4 million years ago marked one of the most dramatic periods of change in the Earth's recent history. During this time, fluctuations in the worldwide climate caused periodic advances of huge ice sheets in high-latitude regions followed by a general retreat of the glacial ice (Figure 17.9).

This period is commonly referred to as the Ice Ages. Through bedrock excavation and sediment deposition, glaciers significantly altered the landscape of large sections of North America, northern Europe and Siberia, and lesser areas in the southern hemisphere. The vacillating extent of the ice sheet and shifting climatic conditions led to widespread changes in patterns of vegetation as well as the types of animals dwelling in northern regions. Glaciers created the Great Lakes, as well as Lake Winnipeg, Great Slave Lake, Great Bear Lake, and numerous other large and small lakes. The advance and retreat of the ice sheets is tied very closely to sea-level changes. The precipitation that falls on ice sheets causing their growth ultimately originates from

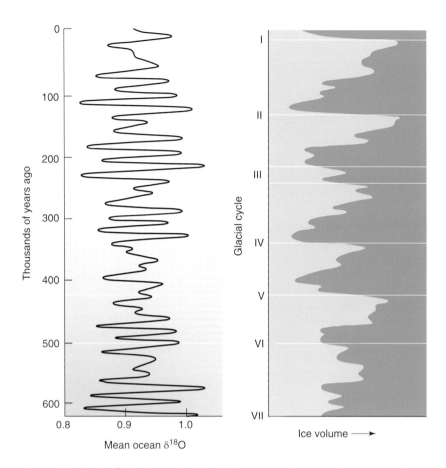

Figure 17.9 Cycles of glaciation coincide with long-term changes in the intensity of summer sunshine in northern latitudes, which are driven by variations in Earth's orbital characteristics (*Source:* From Broeker and Denton, *Scientific American*, January 1990).

water that is evaporated from the ocean surface. Thus, when an ice sheet enlarges, it means that more and more water from the ocean is being stored in the form of glacial ice and that sea level will correspondingly drop. During the most recent glacial maxima, sea level was lowered by at least 120 m.

The drastic changes such as ice sheet movement and sea-level fluctuations had a major impact on many of the world's coastal regions. In high latitudes, glacial action resulted in various types of erosional and depositional coastlines. In low latitudes, not affected by ice, the rise and fall of sea level produced numerous features that give clues to the timing and magnitude of the Ice Ages.

17.3.2   Defining the Pleistocene

The Ice Ages are intimately associated with the geologic time period known as the Pleistocene Epoch. Charles Lyell, a British geologist, originally defined the beginning of the Pleistocene. He based his designation on fossil-bearing sedimentary rocks in Italy, which have been dated at 1.65 million years old. During the past 700,000 years the glacial–interglacial cycles lasted approximately 100,000 years. Furthermore, analyses of fossils from deep-sea sediment cores reveal that the Earth has experienced as many as 20 episodes of glaciation. Consequently, it is now accepted that although the lower

Pleistocene boundary dates at 1.65 million years, the Ice Ages began 2.2–2.4 million years ago.

17.3.3   Causes of the Ice Ages

In the mid-1800s Louis Agassiz, a Swiss scientist, became a convert to the idea that large masses of glacial ice once covered much of Europe and extensive parts of North America. He eventually became the chief spokesman of the glaciation theory and through his studies and many lectures is credited with establishing glaciation as a major geological event. General acceptance of the glacial theory among scientists spawned numerous hypotheses to explain the cause of the Ice Ages. These ideas are still being debated today. Any satisfactory theory must account for the following:

- Although glaciations have occurred in the geologic past, they have not been a common geological phenomenon.
- During most of the Earth's history the climate was warmer than it is today. Beginning about 65 million years ago global temperatures began cooling, eventually leading to the Ice Ages that commenced about 2.2–2.4 million years ago.
- During the Ice Ages there has been a repeated succession of ice sheet growth followed by ice retreat coinciding with variations in global temperatures of about 5 °C.
- Periods of glaciation and interglacial climates occurred at approximately the same times in both the northern and southern hemispheres.

17.3.3.1   Effects of Plate Tectonics

When climatic conditions are favorable, ice sheets form in polar regions and advance to the mid-latitudes. Of course, this process can only occur if landmasses are present in the high and mid-latitudes. This requirement may explain why periods of glaciation have been rare events in geological history. Extensive glaciation occurred approximately 600 and 250 million years ago. In the latter case, the landmasses had assembled over the South Pole in the super-continent of Pangaea (see Chapter 2). Glaciation ceased in these areas after the breakup of Pangaea as the continents moved to more equatorial regions. Thus, it is apparent that the rarity of glacial episodes throughout Earth's history is because there have been few instances in which the continents have been in polar positions when climatic conditions were conducive for snow accumulation and ice sheet formation (Figure 17.10).

17.3.3.2   Carbon Dioxide Abundance

As already mentioned, the Earth was considerably warmer in the geologic past and has cooled by as much as 10–15 °C during the past 65 million years. Coincident with this cooling trend has been a dramatic decrease in the amount of carbon dioxide (CO_2) in the atmosphere, falling to a quarter of its level since the Cretaceous. Because CO_2 is an important greenhouse gas, a decrease in its abundance causes less trapping of solar and re-radiated radiation by the atmosphere, resulting in cooler climates. Because the amount of CO_2 in the oceans is about 60 times greater than that found in the atmosphere, it would appear that the oceans must play a strong role in controlling atmospheric CO_2 and Earth temperature.

17.3.3.3   Milankovitch Climatic Cycles

Although plate tectonics explains well the long periods between glaciations (measured in hundreds of millions of years) that have characterized most of Earth's history, the repeated glacial and interglacial climates that have occurred during the Ice Ages require a different mechanism. Plates move too slowly to account for the waxing and waning of ice sheets over periods of 100,000 years and less. In the early 1900s Milutin Milankovitch, a Serbian mathematician, calculated seasonal changes in radiation received at various latitudes during the past 600,000 years. He linked cyclic variations in the Earth's orbital characteristics to changes in climatic conditions, principally the Earth's surface temperature. These variations are called

(a)

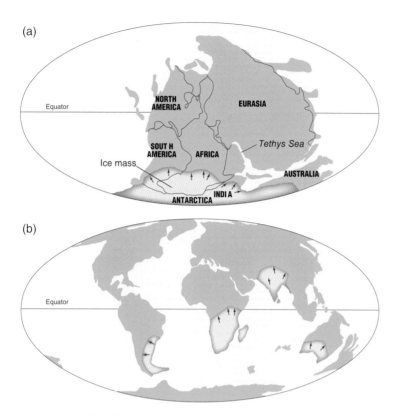

(b)

Figure 17.10 Glaciations occur when climatic conditions are favorable and when continental landmasses are situated in polar regions. (a) View of Pangaea and the ice sheet that covered the Antarctic and surrounding region 300 million years ago. (b) Present day position of the continents and aerial extent of the former glaciated terrain.

Milankovitch cycles (Figure 17.11) and are defined below:

1) Eccentricity – The Earth's present orbit about the Sun is elliptical but at other times it is almost circular. The span of time for the orbit to cycle from its most elliptical to near-circular and back to elliptical is approximately 96,000 years. The amount of eccentricity, which is a measure of how elliptical the orbit is, dictates changes in the distance between the Earth and Sun over the course of a year. Presently, the Earth is closest to the Sun when the northern hemisphere is experiencing winter and the southern hemisphere has its summer and furthest away from the Sun during summer in the northern hemisphere. The opposite was true approximately 50,000 years ago.

2) Obliquity – The Earth's axis of rotation is inclined 23.5° with respect to a line drawn perpendicular to a plane containing the Earth's orbit. This is commonly referred to as the Earth's tilt. Every 41,000 years the Earth's tilt cycles between a minimum value of 21.8° and a maximum of 24.4°. The greater the tilt the more pronounced are the seasons.

3) Precession – Presently, the Earth's axis of rotation points toward the North Star. However, in 11,500 years the axis will point to the star of Vega, and 11,500 years after that it will once again be directed toward the North Star. This 23,000-year cycle describes a circular precession of the axis of rotation. It can be likened to a spinning top that slows down and begins to wobble. The wobble is the precession

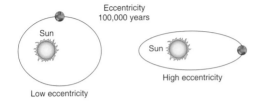

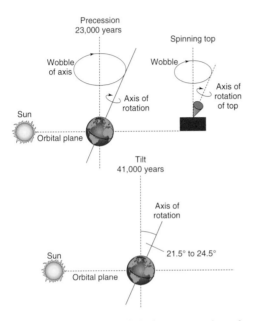

Figure 17.11 Milankovitch Cycles are a product of variations in Earth's orbital behavior. (a) Orbital elements include Eccentricity, Precession, and Tilt. (from Press and Siever 1998)). (b) The cyclicity of orbital variations is responsible for the intensity of the seasons and are linked to climatic changes (*Source*: From Graedel and Crutzen (1993)).

of the spin axis. In the present configuration, summer in the northern hemisphere occurs when the Earth is inclined toward the Sun. It is tilted away from the Sun during winter. In 11,500 years the tilt will have precessed 180° such that the present summer months will become winter months and vice versa.

Milankovitch showed that the interaction of these three cycles did not change the total amount of solar radiation reaching the Earth; rather it affected the contrast in the seasons. For example, if summers were cooler, then glaciers might be expected to enlarge due to less melting of ice while at the same time slightly warmer winters may actually increase snowfall. This is an oversimplified scenario but it does demonstrate the type of climatic changes that are induced by the Milankovitch cycles. Present thought is that Milankovitch cycles are somehow propagated through changes in atmospheric and ocean circulation and these broad conveyer belts of heat and cold control global climatic fluctuations. (See also Chapter 4, Section 4.5.1 and Box 17.20.)

17.3.4 The Late Pleistocene

Waxing and waning of ice sheets characterized the Pleistocene Epoch throughout the world. In North America the last of these major glaciations is referred to as the Wisconsin Ice Age, named after the state in which the deposits left behind by the ice sheet are easily studied. During the beginning of the Wisconsin (70,000–90,000 years ago) the ice sheets began expanding, reaching their maximum southern extent about 20,000–18,000 years before present. The margin of the ice sheets is defined in many regions by particular types of glacial deposits, providing an ideal means of mapping the limit of the ice. Ice sheets covered all of Canada and the mountainous areas of Alaska. In western Canada coalescing glaciers flowed to the Pacific. An expansive ice sheet, called the Laurentide Ice Sheet, was centered over Hudson Bay in eastern Canada and flowed outward in all directions. In Hudson Bay the ice was almost 4000 m thick and in New England it reached more than 2000 m in thickness. The Laurentide Ice Sheet extended eastward to the Atlantic Ocean and as far south as central Illinois and Indiana. Ice sheets also covered much of northern Europe and parts of Siberia. The only vestige of the Laurentide ice sheet today is the Barnes ice cap on Baffin Island in northeastern Canada. The end of the Pleistocene coincided with a period of abrupt global warming and rapid retreat of ice sheets in the northern and southern hemispheres. It should be noted

Figure 17.12 Erosion by glaciers occurs through ice wedging and plucking and by abrasion.

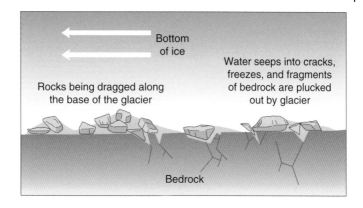

that this and other sudden shifts in the Earth's climatic patterns are not easily explained by Milankovitch cycles and it is probable that other mechanisms are responsible, such as changes in ocean circulation. By definition, the Pleistocene ended 10,000 years ago when the Holocene commenced. The beginning of the Holocene marks a period of rapid warming in North America and Europe, as indicated by pollen records.

17.4   Glacial Effects on Coastlines

The growth and decay of ice sheets dramatically affects the morphology of coastlines due to the ability of glaciers to carve into bedrock, strip away loose materials overlying bedrock, deposit large quantities of sediment and change the level of the world's oceans. Direct effects of glaciation include those due to glacial erosional and depositional processes and those resulting from elevation changes of the coast associated with ice loading and unloading the land. Indirect effects of glaciation are linked to sea-level fluctuations produced by volume changes of the ice sheets.

17.4.1   General Erosional Processes

Erosional processes by glaciers have been responsible for the formation of the Matterhorn in the Swiss Alps, Half Dome in

Yosemite National Park in California and the deepwater port of New York City. Erosion by glaciers occurs by several processes, the most important of which are ice wedging, plucking and abrasion (Figure 17.12). Ice wedging occurs when meltwater flows into the cracks and crevices of the bedrock underlying the ice. As the meltwater refreezes, the expansion of water to ice applies appreciable pressure on the sides of the crack. As this process is repeated over and over again, pieces of the bedrock, large and small are wedged free. The process is especially prevalent where the original bedrock surface is highly fractured and meltwater can readily penetrate numerous cracks. The excavation process (plucking) is completed when the pieces of rock are quarried from the bedrock and incorporated into the ice. This is achieved through material freezing to the base of the glacier. The plastic nature of glaciers at depth also allows larger rocks and boulders to be enveloped by ice. Once the material is incorporated within the ice, it flows with the glacier and is transported toward the ice margin. Along its journey, the rock fragments carried at the base of the ice abrade the underlying bedrock. Just as the sediment that is transported by the Colorado River accelerates the cutting of the Grand Canyon, the rock material carried by a glaciar significantly increases the erosion process. The grooves and scratches cut into bedrock surfaces, called striations, are evidence of glacial abrasion (Figure 17.13). Scientists use the orientations of these linear features to determine the flow direction of the ice.

Figure 17.13 View of glacial striations on bedrock along the coast of Maine. Striations are oriented left to right.

Glacial
excavation

Figure 17.14 Fjords are flooded coastal mountain valleys that are created when alpine glaciers excavate bedrock valleys below sea level.

17.4.2 Fjords

17.4.2.1 Formation

The erosional coastal landscapes exposed by the retreating ice at the end of the Pleistocene exhibited considerable variability. Their form was dependent on the original topography and bedrock structure of the region. Along mountainous coastlines in high-latitude regions, alpine glaciers deepened existing mountain valleys producing U-shaped glacial troughs. In these settings the ends of the valleys were often eroded well below sea level. When the glaciers retreated from the valleys and sea level rose following deglaciation, the glacial troughs were flooded creating fjords (Figure 17.14). These features are found throughout the high-latitude coastlines of the world including the Scandinavian countries, Iceland, Greenland, eastern and western Canada, Alaska, New Zealand, and Chile. Fjords are commonly spectacular features with steep, cliff-like sides and winding valleys that follow the original sinuous mountain valleys from which they were formed. They can be hundreds of meters deep; some reaching more than 1000 m in depth. This means that creation of the fjords is attributable to more than just rising sea level resulting from melting ice sheets. In most instances, the depth of a fjord is chiefly a function of the amount of bedrock excavation that was accomplished by the alpine glaciers.

17.4.2.2 Kenai Peninsula, Alaska
The Kenai Peninsula is a 250 km-long landmass that juts southwestward from southern Alaska. It is separated from the Alaskan Peninsula by the long, narrow, macrotidal embayment of Cook Inlet; Anchorage is located in the upper reaches of Cook Inlet. The mountainous landscape of this region is a product of tectonic processes associated with the subduction of the Pacific Plate beneath North America. During numerous episodes of alpine glaciation, ice flowed southeastward from the Kenai Mountains and excavated a series of deep fjords along the open coast, many of them more than 100 m deep (Figure 17.15). Steep-walled bedrock slopes and cliffs that rise several hundred meters high characterize much of the seaward shoreline of the fjords. In contrast, the heads of many fjords contain outwash fans and fan deltas that are actively prograding into the deep water. Braided streams draining meltwater and sediment from retreating glaciers produce these sedimentary features and resulting low topography. During the 1964 Good Friday Earthquake this region was downwarped a maximum of 2.3 m below sea level. One of the major effects of this event was the formation of drowned forests at the heads of some of the fjords and in other embayments along the Kenai coast.

17.4.3 Rocky Coasts

17.4.3.1 Formation
The effects of continental glaciers are strikingly different from those of alpine glaciers. Whereas alpine glacial processes tend to erode and deepen mountain valleys, thereby accentuating the rugged terrain, continental glaciers tend to reduce the relief along a coast, although they may produce a highly irregular coast. Continental glaciers are too thick to be confined to valleys but rather spread over entire landscapes, including low mountain systems. Along many high-latitude, glaciated coasts in North America, Europe and elsewhere, the major effect of the Pleistocene ice sheets was to strip away the sediment cover and excavate several meters of the underlying bedrock. Although the glaciers left behind a thin layer of undifferentiated sediment, called till, these coastal regions are normally very rocky with numerous embayments, bedrock promontories, and islands. Beaches and barriers are uncommon along these types of coastlines and where they do occur, their sediment supplies are close by.

17.4.3.2 Northern New England
The coast of northern New England illustrates well the general effects of the Pleistocene ice sheets. From northern Massachusetts to northeastern Maine the coast exhibits a wide range of morphologies that are a function of isolated sediment sources, a highly variable bedrock fabric, and a land area that has been inundated by the sea during the past 11,000 years. The bedrock imprint on this region is particularly important, producing cliff coasts in extreme eastern Maine, broad deep embayments with numerous islands and peninsulas in central Maine, and finally a straighter coast along southern Maine, New Hampshire and northern Massachusetts that contains bedrock promontories separated by gently curved embayments. The rocky nature of this coast is a direct consequence of glacial erosion including the removal of sediment that had

(a)

(b)

Figure 17.15 Views of two of the many fjords found along the Kenai Peninsula of Alaska.

once covered the bedrock basement. Along much of this coast the only depositional landforms are small pocket beaches and barriers in protected embayments that have developed from the reworking of local glacial deposits (Figure 17.16). The numerous bedrock islands that characterize this coast have also led to the development of tombolos and cuspate spits. The composition of these depositional features is usually sand and gravel, reflecting the mixed sediment of the glacial sources.

Exceptions to the general trends cited above occur in regions where inland deposits

Figure 17.16 Pocket beach located along a rocky sediment-starved coast. Sediment is derived from the reworking of local thin till deposits.

of sand have been delivered to the coast in large quantities. In the lowlands of Maine, as in many other areas of New England, there are immense sand and gravel deposits that were produced as the ice sheet retreated northward from this region (discussed below). During and following deglaciation, these deposits were excavated by tributaries and trunk streams of major river systems and brought to the coast. This movement of sand down rivers, although less active, continues to the present time. The sediment deposited at the mouths of these rivers has been redistributed alongshore as well as onshore by wind, waves, and tides to form sandy beaches and barriers chains, some of which reach 30 km in length. Most of these barrier systems are located within arcuate embayments, and individual barriers are usually anchored to bedrock promontories or glacial headlands. Likewise, tidal inlets along these chains have stabilized next to bedrock outcrops or glacial deposits (Figure 17.17). Depositional features in this region are usually isolated and directly linked to nearby sediment sources such as rivers.

17.4.4 General Depositional Processes

Glaciers not only carve mountain valleys and strip away the sediment cover from vast areas; they are also responsible for widespread sediment deposition. The rock and sediment that are removed and transported away by glaciers from one location are eventually deposited at another site. The large boulders that are moved by glaciers and laid down far from their origin are called erratics (Figure 17.18). They are commonly the size of a car; however, the largest one in North America, the Okotoks of southern Alberta, Canada, is larger than a two-story house. By matching the rock type of erratics to the bedrock from which they were derived, glacial geologists are able to determine the direction of ice flow and how far the erratics have traveled. For ice sheets, this distance may be as little as a few kilometers or in some cases as much as 1000 km. Erratics litter the landscape of New England, portions of the Mid-West, and many areas within Canada and northern Europe. Louis Agassiz used

Figure 17.17 Along glaciated coasts tidal inlets are commonly anchored next to bedrock exposures or till headlands (*Source:* Westport River Inlet, Massachusetts).

Figure 17.18 Large erratic along the Cape Cod shoreline.

these features as the primary evidence to advance his glaciation theory.

Glacial material comprises two major categories of sediment: Till- sediment deposited directly by the ice, and Stratified Sediment – layered sediments deposited by glacial meltwater (Figure 17.19). Glaciers carry a variety of sediment sizes from clay-sized material to large boulders. When a glacier retreats, the sediment it carries melts out from the ice and is deposited in an unsorted, chaotic mass called till. Anyone who has dug a hole in till to plant a bush or excavate a trench is familiar with its bouldery composition. In the farmlands that stretch across glaciated areas the boundaries of their fields are commonly outlined by stone walls. The rocks making up these walls, large and small, were placed there by farmers who wished to rid their land of the boulders that obstructed the cultivation of their fields.

Unlike till, stratified drift is deposited by flowing water derived from the melting ice. Because the energy needed to transport different sized sediment is directly related to the speed of the flow, water-laid sediments tend to be sorted and are deposited in layers. In stratified drift these layers commonly consist of sand and fine gravel; however, if the current is very strong, coarse gravel layers can form, or if the current flows into a standing body of water, layers of silt and clay may develop.

Figure 17.19 Glacial sediment consists of till, undifferentiated sediment deposited by the glacier, and stratified drift, layered sand and gravel deposited by meltwater (*Source:* From Miles Hayes).

17.4.5   Depositional Landforms

Along glaciated coasts, there are numerous examples where glacial deposits formed the initial shoreline and although coastal processes have subsequently modified these deposits, the original glacial features are still recognizable. These deposits may be large (>100 km) or small (<5 km) and may be composed of till or stratified drift, or in the case of Cape Cod, Massachusetts and Long Island, New York they consist of both types of deposits. The major types of glacial deposits along coasts include **end moraines**, **outwash plains**, and **drumlins**.

17.4.5.1   End Moraines

As discussed earlier, ice sheets flow outward from interior regions of ice accumulation. Glaciers continue to advance as long as more ice is formed during the winter than melts during the summer. For ice sheets, the distance separating the area of net ice formation from the region of ice melting along the margin of the glacier may be over 1000 km. If the amount of ice that forms in the interior of an ice sheet equals the amount that is lost through melting and sublimation (process whereby ice moves directly from solid to gaseous state), then the ice marginal position will remain stationary. Under these conditions the ice sheet, acting like a conveyor belt, continuously transports sediment to the terminus of the glacier where it melts out from the ice and is deposited. As this process continues through time, the accumulating sediment forms a ridge of till called an end moraine (Figure 17.20). The longer the ice terminus remains in the same position, the greater the amount of sediment that is delivered to the ice front and the larger the end moraine becomes. Because the conditions forming end moraines are so variable, their size and extent range widely from prominent ridges 20–50 m in height and 100 km long to those that are only few meters high and extend discontinuously along the former ice margin. End moraines are useful to scientists studying past glaciations because they mark the furthest advance of an ice sheet as well as its recessional positions where the ice front remained stationary for relatively short periods of time before continuing to retreat. In the midwest of the US, a series of end moraines hundreds of kilometers long outline the southern borders of the Great Lakes.

Figure 17.20 End moraines are formed when the margin of a glacier remains in the same position for a period of time. Under these conditions the flowing ice piles sediment at its terminus forming a ridge of till called an end moraine (*Source:* From Miles Hayes).

17.4.5.2 Outwash Plains

As an end moraine is created at the terminus of a stationary ice sheet, the melting ice produces torrents of water choked with sediment. On warm summer days water flows everywhere from underneath, over the surface, and through tunnels within the glacier. Meltwater discharging from the ice front forms a broad network of shallow streams whose channels regularly divide and rejoin in a braid-like pattern. Within a short distance of the ice, much of the bedload transported by these braided streams is deposited due to the gentle slope and resulting decrease in current velocity. Meltwater streams sort the glacial sediment, leaving behind the largest-sized material near or within the ice front. Sand and gravel are deposited in layers beyond the ice terminus and in time build an expansive outwash plain (Figure 17.21). The finest sediment (clay and silt) is transported out of the system and often into the ocean or a lake. Thus, outwash plains are large sandy-gravel regions (some more than 100 km long and an area greater than 1000 km^2) with low topography consisting of stratified drift.

Examples include most of Cape Cod, Massachusetts and Long Island, New York as well as the Skiederarsandur in southeast Iceland. Outwash plains are also formed at the end of some valley glaciers.

17.4.5.3 Drumlins

One of the most distinctive glacial depositional features is a drumlin. Drumlins are teardrop-shaped accumulations of till ranging from 15 to 50 m in height and 0.5–2 km in length (Figure 17.22). In some instances they may also be composed of highly contorted stratified drift while others may be cored by bedrock. The blunt, steep side of a drumlin faces the direction from which the glacier advanced. and the down-glacier end is streamlined, with a gentler slope. This geometry is reminiscent of a spoon turned upside-down. Because drumlins are elongated parallel to the flow direction, they are used by glaciologists to study patterns of ice sheet movement. Drumlins usually occur in significant numbers. In upstate New York over 10,000 drumlins exist in an area of approximately 12,000 km^2. A few hundred are located in the

(a)

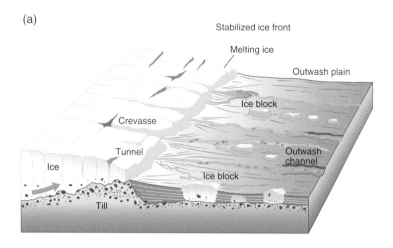

(b)

Figure 17.21 An outwash plain consists of layers of sand and gravel (stratified drift). (a) They are formed through deposition by glacial meltwater streams. (*Source:* From Strahler (1966).) (b) The braided streams in this photograph are most active during the summer when nearby glaciers discharge abundant meltwater (*Source:* From Dave Marchant).

immediate vicinity of Boston, Massachusetts, including a drumlin where the famous Revolutionary War battle of Bunker Hill was fought in 1775. An interesting note of history is that the Bunker Hill Monument commemorating the battle stands on the drumlin of Breeds Hill, the actual battle was fought on the adjacent drumlin of Bunker Hill. The fact that

(a)

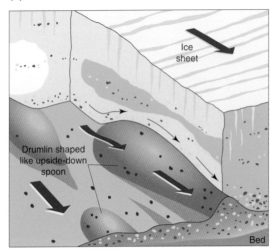

Spoon

(b)

Figure 17.22 Drumlins are composed of till, and usually occur in groups. (a) They are formed beneath the ice and have the form of an upside down spoon. (*Source:* From (Chernicoff (1995).) (b) Erosion of a drumlin in Nova Scotia has left behind a boulder-retreat lag, which can be used to trace the former extent of the landform.

drumlins always occur in clusters with similar geometry and orientations leads scientists to believe that they represent an equilibrium bed configuration in which the flowing ice molds the underlying till. Other drumlins may be a product of glacial erosion. In addition to the drumlin fields of New York and New England others are found in the United Kingdom, Ireland, eastern Canada, Wisconsin, Michigan, and western Washington.

17.5 Examples of Glaciated Coastlines

17.5.1 Cape Cod

17.5.1.1 Formation

During the summer, one of the most popular vacation spots in New England is Cape Cod, Massachusetts, including the two islands of Martha's Vineyard and Nantucket that lie directly off the Cape's southern shore. Cape Cod is shaped like an extended bent arm with a curled-fingered hand pointing northward. The Cape and Islands were formed during the retreat of the Laurentide Ice Sheet, which reached its maximum extent approximately 20,000–18,000 years ago (Figure 17.23). These landmasses consist almost entirely of unconsolidated glacial sediments (about 90 % by area). Exceptions are the beaches, barriers, marshes and tidal flats that outline the present Cape and even these features were formed from reworked glacial sediment. Although some large erratics exist on the Cape, there are no bedrock exposures.

To discuss the glacial origin of Cape Cod and the Islands we must go back to an interglacial period (between glaciations) when the climate was the same or slightly warmer than it is today, about 130,000–100,000 years ago. At this time there was no Cape Cod. Following this period, temperatures cooled and the Laurentide Ice Sheet grew is size, expanding outward from the Hudson Bay region. It reached its maximum southern extent in the Great Lakes region about 18,000 years ago and in eastern United States perhaps a little earlier (~20,000 years ago). In the vicinity of southeastern New England, the terminus of the ice sheet consisted of several large lobes that coincide with the present-day locations of Nantucket, Martha's Vineyard, Block Island off the Rhode Island coast, and Long Island in New York. It is the overall position, configuration, and dynamics of these ice lobes that determined the shape and location of the Cape and islands to the south and west.

Cape Cod and the Islands developed during at least two periods when the margin of the Laurentide Ice Sheet stabilized and large quantities of sediment were deposited at these ice terminuses. At the ice sheet's southernmost position, an end moraine formed corresponding to the northern third of Martha's Vineyard. This is a hilly, bouldery region and is relatively elevated compared with the rest of the island. A similar but less extensive moraine exists on Nantucket. During the same time as the moraines were formed, broad, sandy outwash plains were deposited by meltwater streams draining water and sediment from the ice terminus. These sand plains define the southern border of both islands. Following this period of ice-front stability, the ice sheet retreated 50 km northward and stabilized again. In this position, the Sandwich and Buzzards Bay Moraines and a number of outwash plains were formed that together constitute the southern extent of Cape Cod. The forearm of the Cape developed next after the ice sheet withdrew further northward. At this time, the lobate nature of the ice front left a large low area between the Cape Cod Bay Lobe, filling much of Cape Cod Bay, and the South Channel Lobe, which was located east of the present Cape. This low area was filled with stratified drift (outwash) produced by meltwater braided streams that flowed westward from the South Channel Ice Sheet. With this final stage of glacial deposition completed, the general form of Cape Cod and the Islands was achieved.

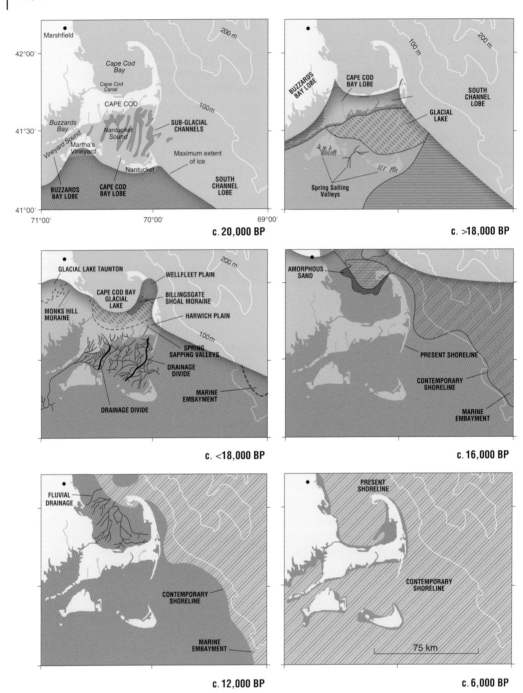

Figure 17.23 Sequential diagrams depicting the formation of Cape Cod and the islands of Martha's Vineyard and Nantucket. They consist of sediment that was carried south and deposited by the Laurentide Ice Sheet in the form of moraines, outwash plains, and other glacial deposits. (*Source:* From Uchupi et al. (1996).)

17.5.1.2 Modification

Cape Cod and the Islands have undergone many important modifications during and after the period that ice withdrew from New England. During the period of ice retreat, at least two large glacial lakes occupied what is now Nantucket Sound and Cape Cod Bay. The first of these lakes to develop, Lake Nantucket Sound, was dammed by the glacial deposits to the west and south including Martha's Vineyard and Nantucket and by ice to the north and east. Meltwater discharging from the ice terminus filled the lake. The unconsolidated, porous nature of the deposits that make up Martha's Vineyard and Nantucket led to water from the glacial lake being piped southward through these sediments where it eventually flowed out along the surface. This process gradually formed channels that ate their way northward across the sand plains as sand and fine gravel were eroded at the heads of these channels and transported southward. These channels are called groundwater-sapping channels and can be seen on a small-scale on a beach at lowtide where groundwater leaks out along the beachface (Figure 17.24). Along Nantucket and Martha's Vineyard groundwater-sapping processes have formed a series of semi-parallel channels that have since been inundated with seawater due to rising sea level. Exactly the same process was responsible for groundwater-sapping channel development along the southern shore of Cape Cod when a large glacial lake occupied Cape Cod Bay. Many of these elongated lagoons have become important harbors along this part of the Cape.

The other major modification of Cape Cod and the islands has been a general smoothing of the shorelines through erosional and depositional processes. The initial form of Cape Cod left by the retreating ice sheet was very different from what it appears today. Immediately following deglaciation, sea level was as much as 120 m lower than the present position and thus the deposits comprising Cape Cod and the Islands were simply regions of somewhat higher elevation. It

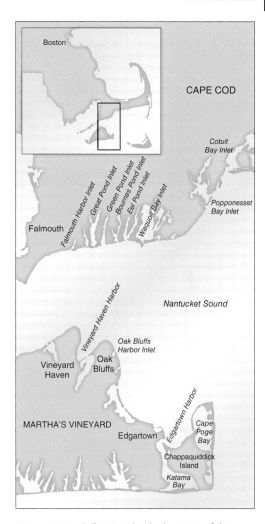

Figure 17.24 Following the deglaciation of this region, groundwater-sapping channels developed in the outwash plains of Cape Cod and Martha's Vineyard. As the ice retreated northward, large lakes formed in what are now Nantucket Sound and, later, Cape Cod Bay. Water from these lakes flowed southward through the outwash plain sediments to the lower sea level. This process created sapping channels, which elongated by headward erosion. When sea level rose to its present position, the channels were flooded, forming the highly indented lagoons along the southern shore of Cape Cod and the Islands.

wasn't until eustatic sea level rose to within 7–8 m of where it is today, that the general configuration of Cape Cod became recognizable. The initial outline of the Cape was very

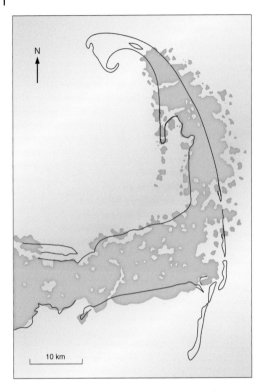

Figure 17.25 The present smooth outline of Cape Cod is a product of wave erosion and the construction of spits in front of embayments. (*Source:* From Strahler [1966]).

irregular with numerous promontories, embayments and islands (Figure 17.25). However, wave action gradually eroded headlands and small islands, liberating sediment that was transported along shore, eventually building spits across embayments. Examples of this process are seen in Cape Cod Bay, where Sandy Neck has accreted in front of Barnstable Bay, and along the outer Cape, where Nauset Spit has formed in front of Pleasant Bay. The widespread glacial cliffs that occur along Cape Cod and the Islands are evidence of the erosional process that has smoothed the coast. Some of the cliffs are more than 30 m high, such as those along the open-ocean coast of northeastern Cape Cod. Much of the sediment that has been eroded from these cliffs has been transported northward forming the extensive spit of the Provincelands.

17.5.2 Drumlin Coasts

In some glaciated terrains, notably the Eastern Shore of Nova Scotia, Clew Bay on the central west coast of Ireland, and Massachusetts Bay, drumlins and accretionary landforms that have developed from reworked glacial sediment dominate the coast. In Massachusetts Bay for example, drumlins exert a strong imprint on the entire landscape. All of the islands, save a few bedrock ledges, are drumlins and the northeast and southeast borders of Boston Harbor consist of drumlins and reworked drumlin deposits. D.W. Johnson, one of the first marine geologists to study glaciated coasts, presented an evolutionary model for this shoreline in 1910 (Figure 17.26). Similar to the scheme presented for the Eastern Shore of Nova Scotia (see Chapter 15, Section 15.8.1), the outer coastline of Boston Harbor represents the end-product of sand and gravel eroded from offshore drumlins and moved onshore where it formed beaches and barriers. Additional sediment was derived from the erosion of onshore drumlins and transported alongshore. When sediment is abundant, drumlins form pinning points for barrier and spit development. As sediment supplies begin to wane and the shoreline recedes, perhaps in response to sea-level rise, the drumlin anchor points erode and contribute new sediment to the system. With continued sea-level rise, eventually the entire barrier complex may narrow and become low enough such that it migrates onshore through rollover processes until stabilizing next to landward drumlins. Alternatively, the sediment is moved onshore in the form of a subtidal sand sheet, feeding sediment to spit systems that build from nearby drumlins. Intertidal and subtidal boulder pavements identify offshore drumlins that have been exhausted of their sediment sources. These boulder accumulations are lag deposits and represent the sediment that was too coarse to be transported by storm waves. In some instances, the boulder pavements may be large enough such that

Stage 1

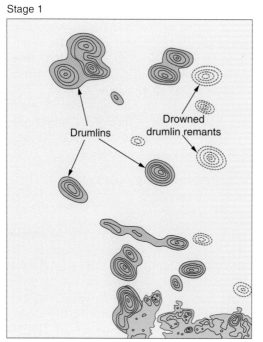

Stage 2

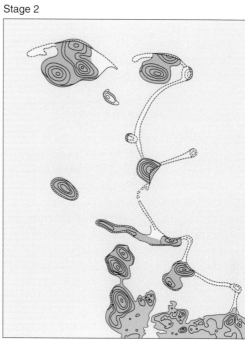

Stage 3

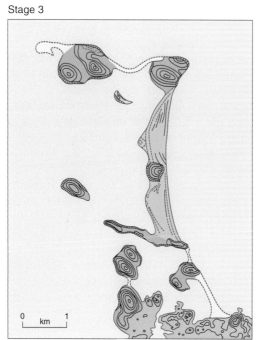

Stage 4

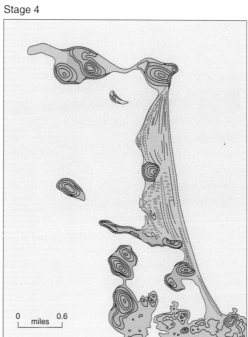

Figure 17.26 Evolutionary model of Nantasket Beach, Massachusetts as envisioned by D.W. Johnson (1925).

they influence the distribution of wave energy and create shadow zones landward of the former drumlins. It is not uncommon to find gravel tombolos or subtidal bars developing in these types of regions.

17.5.3 Sand and Gravel Beaches

Beaches along glaciated coasts come in every size, shape, and composition. This diversity is unique to glaciated coasts and is attributable to the isolated nature and highly variable composition of the sedimentary deposits that occur along these coasts. The deposits left behind by glaciers may consist of stratified sands, stratified gravel and sand, boulder tills, sandy tills, and other types of sediment. It is customary to find glacial deposits having very different compositions occurring in close proximity. When waves and currents rework these deposits, they form very different kinds of beaches. In addition, because glaciated coasts are usually irregular, containing headlands and embayments, there is little exchange of sediment between embayments and therefore little mixing of sediment. Thus, it is not uncommon to find a pocket sandy beach or barrier within a kilometer of a boulder or cobble beach.

Sandy-gravel beaches are common along glaciated coasts, particularly where the beach material comes from erosion of nearby till deposits such as drumlins or moraines (Figure 17.27). The exact make-up of these beaches is dependent on the composition of the glacial deposit. Gravel beaches often have multiple ridges or berms, and their number tends to increase with increasing supply of gravel, tidal range and exposure to storm waves. Each ridge is related to a particular magnitude of storm and tide level. The highest ridge is formed by the highest-elevation storm wave event and usually coincides with spring tide conditions. Each successively lower ridge corresponds to a decreasing magnitude of storm and/or tide level.

Individual clasts making up gravel beaches have a variety of shapes and forms; however, they can be grouped into four general form classes. Using the Zingg Diagram, which is based on ratios of the short, intermediate,

and long-axes of a clast, the gravel forms can be separated into rollers (shaped like a rolling pin), disks (the best type of "skipping stone"), spheres, and finally blades which have three different dimensions and look like "bricks" (Figure 17.28). It is common to find disks high along the elevated portions of gravel beaches. Because of their form, they have a relatively large surface:mass ratio for their size and are easily transported to the upper beach by storm waves.

17.5.4 Uplifted Coasts

During the maximum extent of the Wisconsin glaciation (late Pleistocene), the Laurentide Ice Sheet varied from 1 km to more than 3 km in thickness. Remember from the discussion on Plate Tectonics (Chapter 2) that lithospheric plates are floating on the semi-plastic region of the mantle called the asthenosphere. Thus, as the ice sheet grew in size and volume, the weight of the ice gradually depressed the lithosphere and the land surface sank. Depression of the lithosphere was accomplished by the semi-plastic asthenosphere flowing outward from underneath the portion of the continent that was covered by the ice sheet. Conversely, when the Laurentide Ice Sheet melted away and the weight of the ice was removed from the lithosphere, the land surface rebounded. This re-equilibration process that produces uplift is called glacial rebound.

The effects of glacial rebound are readily seen along the Hudson Bay and Baltic Sea shorelines as both these areas were the centers of large ice sheets during the late Pleistocene. The Laurentide ice sheet was almost 3000 m thick in Hudson Bay region and since its disappearance, the shoreline rimming the bay has rebounded almost 300 m. Scientists estimate that another 100 m of uplift is likely, with the highest rebound occurring where the ice was thickest (Figure 17.29). The effect of this uplift has been the formation of over a hundred raised shorelines that are arranged in stair-step fashion extending from the present bay shoreline to several kilometers inland. These features consist of sand and gravel beach ridges and intervening swales

(a)

(b)

Figure 17.27 Gravel beaches are common along glaciated coasts due to the prevalence of till deposits along these shores. (a) Popplestone beach consisting of well-rounded boulders in Magnolia, MA and (b) Jasper Beach in Machiasport, Maine. The red color comes from a form of quartz called jasper, which contains iron inclusions.

with a relief of 1–3 m. The beach ridges are a product of storm waves reworking glacial and beach sediments piling up gravel and sand along the rear of the beach. Glacial rebound eventually lifts the ridge displacing it vertically and horizontally away from the shoreline. Thus, it is preserved from any further modification by storm waves. Similar processes of glacial rebound and formation of raised shorelines have occurred along the Scandinavian Baltic coast. Present rates of uplift in this region are on the order of 6–9 mm per year.

(a)

dS: short diameter
dI: intermediate diameter
dL: long diameter

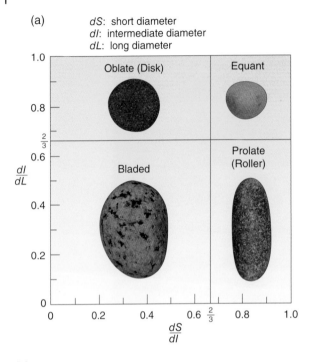

(b)

Figure 17.28 Wave abrasion shapes pebbles and cobbles along gravel beaches. (a) Gravel clasts can be divided into four major types using the Zingg (1935) classification. These classes are based on the relative dimensions of their short, intermediate, and long axes. (b) Examples of a disk, sphere, roller, and blade.

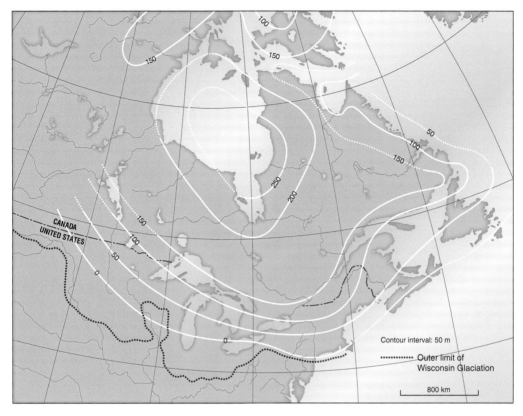

Figure 17.29 Isostatic rebound in northeast North America resulting from retreat of the Laurentide ice sheet. (*Source:* From P.B. King (1965).)

17.5.5 Drowned River Valleys

In addition to the direct effects of glaciers, the waxing and waning of ice sheets caused the world's ocean levels to fall and rise numerous times during the Pleistocene Epoch. For example, during the last glaciation (Wisconsin glaciation) sea level dropped about 120 m and then rose back again to its present level. When sea level falls, rivers extend their courses across the exposed continental shelves to the new, seaward shoreline. During this process the original mouth of the river deepens and widens as the channel seeks to establish a new equilibrium with the lowered sea level. When the ice sheets melt, water is returned to the ocean basin and rising sea level floods the enlarged valleys forming drowned river valleys (Figure 17.30). Delaware and Chesapeake Bays are large-scale examples of these drowned valleys (Figure 17.31).

17.6 Summary

This chapter emphasized the diversity of glaciated coasts and the wide range of glacial processes that produce these varied coastlines. The deep fjords of Alaska, Scandinavia, and many other high-latitude regions demonstrate the ability of alpine glaciers to sculpt and excavate coastal mountain valleys. The erosional effects of ice sheets are more subdued than alpine glaciers but more far-reaching, due to their size, which can extend across entire continents. The rocky coastlines of New England, northern Europe, and elsewhere attest to the ability of ice sheets to strip away sediment covers exposing the underlying bedrock. While glaciers are impressive eroding agents, they also are important vehicles of sediment delivery to coasts. Long Island, Cape Cod, and several regions in Alaska are examples where

Time 1. Highstand

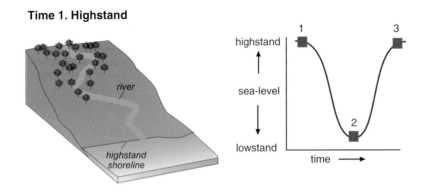

Time 2. Sea-level fall

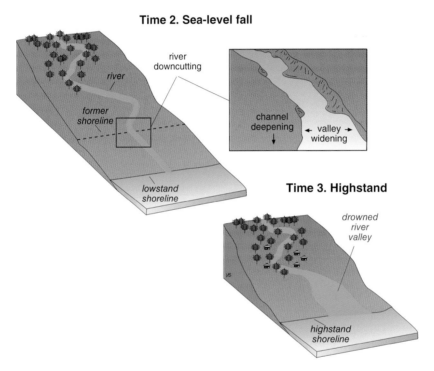

Time 3. Highstand

Figure 17.30 Model of drowned river valley formation.

glacial processes have essentially built the coastline. These deposits consist of end moraines that are composed of till, and outwash plains that are made up of stratified drift. Drumlins are a special type of glacial deposit forming till headlands and intervening spits such as those along the Eastern Shore of Nova Scotia and the Boston Harbor region. Beach and barriers along glaciated coast tend to be discontinuous with a wide range of compositions due to highly variable and often isolated glacial sediment supplies. Sand and gravel beaches are common.

The growth and decay of ice sheets have also produced changes in sea level and position of the shoreline. During the most recent glaciation water removed from ocean basins and stored in huge ice sheets lowered sea level by 120 m. Through this process of sea level lowering and rising back again, which occurred many times throughout the Pleistocene, drowned river valleys were formed along many coastal plain settings of the world. The enlargement and retreat of ice sheets were also responsible for loading and unloading the Earth's crust. In the Hudson

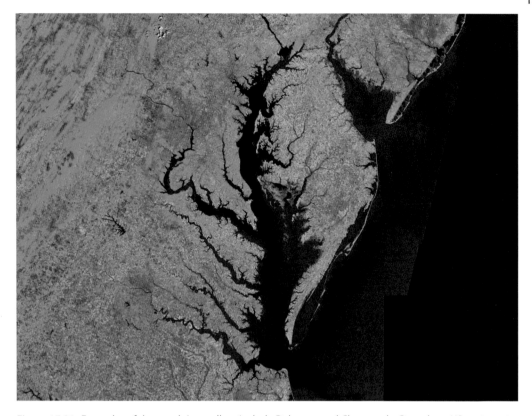

Figure 17.31 Examples of drowned river valleys include Delaware and Chesapeake Bays along US east coast.

Bay region of Canada and in the Scandinavian Baltic Sea, which were sites of thick ice accumulation during maximum glaciation, melting of the ice sheet and rebound of these coasts have produced an extensive set of raised beaches, rimming the shoreline.

Formation of ice sheets occurs only when plate tectonics have arranged the continents in such way that landmasses are situated in polar regions. The onset of the Ice Ages began after the Earth underwent long-term cooling during the Cenozoic (past 65 million years) when average temperature decreased by about 5 °C. The periodicity of glaciations during the past two million years appears to be linked to Milankovitch cycles, which are produced by variations in the Earth's orbital characteristics. These in turn affect the temperature contrast of the seasons. Present thought is that Milankovitch Cycles are somehow linked to changes in atmospheric and oceanic circulation and these broad conveyer belts of heat and cold control global climatic fluctuations and the waxing and waning of ice sheets.

References

Broeker, W.S. and Denton, G.H. (1990). *Scientific American* 48–56.

Chernicoff, S. (1995). *Geology*. New York: Woth Publishers.

Graedel, T.E. and Crutzen, P.F. (1993). *Atmospheric Change*. San Francisco, New York: W.H. Freeman and Co.

Johnson, D.W. (1925). *The New England-Acadian Shoreline*. New York: John Wiley and Sons.

King, P.B. (1965). Tectonics of quaternary time in middle North America. In: *The Quaternary of the United States* (ed. H.E. Wright and D.G. Frey). Princeton, NJ: Princeton University Press.

Press, F. and Siever, R. (1998). *Understanding Earth*. San Francisco: W.H. Freeman and Co.

Strahler, A.N. (1966). *Geologist's View of Cape Cod*. Orleans, MA: Parnassus Imprints.

Tarbuck, E.J. and Lutgens, F.K. (2002). *Earth*. New York: Prentice Hall.

Uchupi, E., Giese, G.S., Aubrey, D.G., and Kim, D.J. (1996). *The Late Quaternary Construction of Cape Cod, Massachusetts: A Reconsideration of the W.M. Davis Model*, Special Paper 309. Geological Society of America.

Zingg, T. (1935). Beitrage zur Schotteranalyse. *Schweizerische Mineralogische und Petrographische Mitteilungen* 15: 39–140.

Suggested Readings

Bennett, M.R. and Glasser, N.F. (1997). *Glacial Geology: Ice Sheets and Landforms*. New York: John Wiley and Sons.

Coates, D.R. (ed.) (1974). *Glacial Geomorphology*. Binghamton, NY: State University of New York Press.

Drewy, D.J. (1986). *Glacial Geologic Processes*. London: Edward Arnold.

FitzGerald, D.M. and Rosen, P.S. (eds.) (1986). *Glaciated Coasts*. New York: Academic Press.

Hambrey, M. and Alean, J. (1992). *Glaciers*. Cambridge: Cambridge University Press. 208 pages.

Johnson, D.W. (1919). *Shore Processes and Shoreline Development*. New York: Hafner Publishing.

Patterson, W.S.B. (1981). *The Physics of Glaciers*. Oxford, UK: Pergamon Press.

Ruddiman, W.F. (2000). *Earth's Climate Past and Future*. San Francisco: W.H. Freeman.

Sharp, R.P. (1989). *Living Ice: Understanding Glaciers and Glaciation*. Cambridge: Cambridge University Press.

Shepard, F.P. and Wanless, H.R. (1971). *Our Changing Coastlines*. New York: McGraw-Hill.

Sugden, D.E. and John, B.S. (1976). *Glaciers and Landscape*. New York: John Wiley and Sons.

18

Rocky Coasts

18.1　Introduction

Rocky coasts offer some of the most striking ocean vistas in the world due to the exquisite beauty produced by waves crashing against their jagged shores (Figure 18.1). These coasts are typically associated with cliffs and other erosional landforms such as stacks, arches, and caves. Along many active continental margins, rocky coasts are bordered by majestic mountainous hinterlands. In contrast, low plains or hilly regions border the rocky shores of some mid-ocean islands. Despite the high wave energy and apparent harsh environment of rocky coasts, their intertidal zones commonly teem with life and are far more productive biologically than sandy shorelines.

Most people are unaware of the extent of rocky shorelines because many coastal cities, major coastal population centers, and even vacation sites are located in coastal plain settings, at the mouths of estuaries, and in other lowland areas. It is estimated that 75% of the world's shorelines are rocky, which includes beaches that are backed by bedrock cliffs or rocky uplands. Although rocky coasts are not confined to a single type of geological setting, they are more common along tectonically active coasts than they are on passive margins. For example, rocky coasts comprise much of the west coasts of North and South America, whereas beaches and barriers characterize most of the non-glaciated east coast of North America. Still, there are many sites throughout the world where rocky and rugged coasts are found along passive margins such as South Africa, parts of Argentina, eastern Canada, southern Australia, and sections of northwest Europe.

Rocky coasts display a wide range of morphologies because they are composed of different types of rocks and have formed in a variety of geologic settings. For example, the Bahamian coast consists mostly of limestone shorelines that are derived from coral, shells and coralline algae. In southern California and Victoria, Australia, cliff coasts have developed from the recession of relatively weak sedimentary rocks including sandstones, siltstone and mudstones. Marine erosion of volcanic rocks along the south coast of the island of Hawaii has produced dramatic cliffs. Differential weathering of folded metamorphic rocks intruded by granitic batholiths form the indented/island coast of central Maine. Tectonic processes control the rugged mountainous coasts of Alaska and Chile. Although erosional processes dominate most rocky coasts, depositional landforms, including pocket beaches situated between bedrock headlands and even barrier spits downdrift of river mouths and estuaries are common in these regions.

This chapter will explore the extent and variability of rocky coasts around the world. It will be shown that the morphology of these coasts is a function of the structure and type of bedrock as well as the physical, chemical and biological processes operating on the coast. The many different erosional features common to these coasts will be discussed in terms of the processes by which they have formed.

Beaches and Coasts, Second Edition. Richard A. Davis, Jr. and Duncan M. FitzGerald.
© 2020 John Wiley & Sons Ltd. Published 2020 by John Wiley & Sons Ltd.

Figure 18.1 The Indian Ocean coast of South Africa, where wave heights are frequently in excess of 3 m.

18.2 Types and Distribution

Three-quarters of the world's coasts are rocky and have a lack of sediment accumulating on them. They also occur where the original sediment cover has been eroded away or where the sediment has been turned to rock through cementation. The most extensive rocky coasts are associated with mountainous regions and glaciated areas. Other shorter and/or discontinuous sections of rocky coasts occur in volcanic regions and in a variety of other geological settings.

18.2.1 Tectonic Settings

The tectonic setting of the continental margin has an overriding influence on whether a coast is rocky or if it contains beaches, barriers, tidal flats, river deltas, or other types of sediment accumulation forms. As we learned in Chapter 2, the convergence of oceanic and continental plates produces high-relief, continental borderlands such as the Andes mountains along Peru and Chile or the North America Cordillera.[1] The rocky and rugged landscape of these mountains usually extends to the sea, giving the coast a rugged appearance as well (Figure 18.2). In addition, the mountain chains of these continental margins act as dams that prevent rivers from delivering sediment from continental interiors to the coast. The rivers draining mountains along collision margins are typically short and usually transport coarse-grained sediment to the coast. Further, due to their small drainage areas these rivers discharge relatively small quantities of sand and mud, particularly in comparison to rivers along trailing-edge coasts. Waves tend to be large along collision coasts because the steep continental shelves of these margins dampen little of the deepwater wave energy. During storms, high-energy waves remove large quantities of sand from beaches and transport it along shore and offshore. Along the California coast beach sands are permanently lost to the deep ocean basin due to the proximity of submarine canyons to the shoreline (<1 km) resulting from the narrow continental shelf. These canyons capture the longshore transport of sand at their heads and carry it to the deep ocean out of the coastal system. For all these reasons, collision coasts tend to be rocky, containing few depositional features. Because of their relative youth, neotrailing edge coasts such as the Arabian coast along the Red Sea are also rugged and mostly rocky. There has been insufficient time to develop a mature continental margin with a wide shallow shelf to dissipate waves.

18.2.2 Glaciated Regions

During the Pleistocene Epoch, continental ice sheets covered approximately 30% of the world's land surface. This means that the

---

1 The North American Cordillera is a mountain system comprising the west coast of the United States, Canada and Alaska.

Figure 18.2 Subduction of the Pacific Plate beneath North America has produced the rugged mountainous coastline of Alaska, where earthquakes and volcanic eruptions are common geologic processes.

morphology of much of the world's coastline has been strongly imprinted by glacial processes. Although there are large sections of coastline that consist entirely of glacial sediment or reworked glacial deposits, such as Cape Cod, Massachusetts, Long Island, New York, extensive areas along the Great Lakes, and portions of the eastern shore of Nova Scotia (see Chapter 17), most glaciated coasts have widespread bedrock exposures. In mountainous regions, coastlines take the form of bedrock headlands with intervening, deep, flooded valleys. The fjords of Scandinavia, Iceland, Chile and Western Canada, are examples of where alpine glaciers carved and deepened existing mountain valleys (Figure 18.3a). At the head of fjords in the Kenai Peninsula of Alaska and along much of Greenland, vestiges of the valley glaciers that formed these twisting, water-filled valleys are still present. Repeated advance and retreat of great ice sheets stripped away the sediment cover from the land, exposing the underlying bedrock along many other glaciated coasts. These coastlines are commonly embayed to deeply indented and contain numerous islands and

bedrock ledges (Figure 18.3b). Beaches and barriers along these coasts are usually scarce and where they do exist, they are adjacent to rivers, estuaries, or nearby isolated glacial deposits.

18.2.3 Other Bedrock Coasts

Volcanic coasts occur where hot spot activity in the mantle has produced island chains such as those found in the Pacific Ocean, including the Hawaiian and Marshall Islands (Figure 18.4). Outpourings of lava and welded tuff have also formed portions of rocky coasts along island arcs in the Caribbean, and the northern and western Pacific. In tropical regions, the seaward extension of these volcanic coasts forms the platforms upon which luxuriant coral reefs have developed. In some places, sinking volcanic islands have become coral atolls.

Another variety of rocky coast is formed from the shells of dead marine organisms. This type of coast is most common in low-latitude regions including the Caribbean and Mediterranean Seas, where calcium carbonate skeletal material is produced in

(a)

(b)

Figure 18.3 Glaciated coasts exhibit a wide diversity, but are dominated by a rocky landscape. (a) View of a fjord on the coast of Norway. (b) Repeated Pleistocene glaciations have stripped much of the sediment from the Maine coast leaving it rocky and irregular with numerous bedrock islands but few beaches.

high abundance in coastal waters. High rates of shell production may also occur in cooler, higher-latitude regions such as the south coast of Australia and South Africa, where input of other types of land-derived sediment is absent. Many of these rocky carbonate coasts were created during the Pleistocene when sea level fell and onshore winds blew carbonate sand onshore, building dunes and beaches. This sediment was

Figure 18.4 Volcanic island chains such as the Hawaiian Islands form extensive rocky coasts in the Pacific Ocean and other parts of the world. Rapid physical and chemical breakdown of volcanic rock produces sediment that forms pocket beaches.

turned to rock by a process called lithification. Individual grains of sand are welded together when ocean spray, rainwater and groundwater percolate through the sand partially dissolving calcium carbonate and precipitating it as cement. The high temperatures of these coasts produce rapid evaporation of wetted surfaces, which accelerates the lithification process. Dunes that are converted to rock are called eolianites. Wide exposures of this rock with well-developed dune layering can be seen throughout Bermuda (Figure 18.5), several Caribbean islands, along the Yucatan coast of Mexico, and southeastern Australia.

Similar chemical processes in tropical settings produce rocky outcrops along the shore called beachrock (Figure 18.6). In regions with high evaporation rates concentrations of various salts increase in the seawater. Salt spray and tidal inundation constantly bathe the beach with this high-saline water and partial evaporation of the water and interaction with groundwater cause the precipitation of calcium carbonate. This process cements sand grains producing beachrock. The rapidity at which it operates often creates beachrock containing

modern day trash, including such articles such as cans and bottles.

18.3 Erosional Processes

Unlike sandy shorelines, where the effects of a winter storm or passage of a hurricane are readily visible, changes along a rocky coast occur very slowly and may be imperceptible over human life-spans. The present-day appearance of rocky coasts is the product of physical, chemical, and biological processes that have been operating over thousands of years. The rate at which these processes act to weather the shoreline is a function of rock type and its structure, wave intensity, vegetation, climate and numerous other factors. In most cases, physical processes operate over a larger scale and at faster rate than do either chemical or biological agents.

18.3.1 Physical Processes

Wave-Induced Erosion
The processes involved in eroding bedrock shorelines are numerous but the most visible and important agent is wave action. Breaking and shoaling waves are responsible for several interactive processes including wave hammer, air compression, quarrying and abrasion. The first of these, wave or water hammer, occurs when a wave directly impacts a cliff face or sloping rock exposure exerting a tremendous hydraulic force against the rock, particularly if the waves are large. As reported in numerous storm accounts, waves have been responsible for tearing the tops of seawalls and throwing huge five-ton granite blocks tens of meters across roadways. However, in most cases it is not a single wave that finally causes a rock face to fail, but the accumulated effects of tens to hundreds or even thousands of years of wave action.

In some instances, breaking waves trap air between the water and the cliff face. As the

Figure 18.5 Differential weathering of carbonate rock along sections of the Bermuda coast results in the formation of caves and embayments.

Figure 18.6 View of beachrock that has formed along the shoreline of Kubbar, an island off the coast of Kuwait in Arabian Gulf. In this region both terrigenous and biogenic sands are welded together by carbonate cement.

wave collapses, extremely high pressures are instantaneously produced as the air is compressed on the rock surface (Figure 18.7). This process is particularly important when air pockets are compressed into the crevices of rocks, leading to the enlargement of the crack and ultimately to a shattering of the rock. Quarrying is the removal of pieces of rock ranging from small grains to large blocks from a cliff face or bedrock exposure that

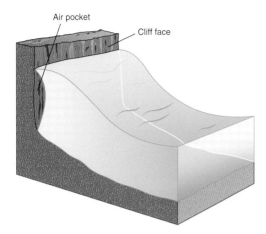

Figure 18.7 Air is instantaneously compressed against a cliff when waves break. This compression of air can lead to the enlargement of cracks and failure of the rock face.

have been loosened or separated from the parent rock by biological, chemical, or other physical processes.

One of the most important mechanisms of physical erosion takes place when sand, granules, and larger-sized gravel are entrained by waves and washed, rolled, and scraped across the rock surface. Abrasion also includes the more violent process of large waves picking up pebbles, cobbles, and boulders and propelling them against the rock surface. The widespread cliffs along the British Isles and along the Alaska Peninsula are believed to have formed due to abrasion by gravel (Figure 18.8). However, there are numerous other coastal sites around the world, including parts of New Zealand, much of Tasmania in Australia, and the Bay of Fundy in Eastern Canada, where cliff shorelines are fronted by sandy beaches. Thus, we can conclude that sea cliffs can be produced when the rock face is abraded by sand or any other available hard particles in conjunction with other wave processes.

The rate at which abrasion takes place is related to wave energy, composition of the rock, and the type and abundance of abrading agents (gravel, sand, or other particles). The process of abrasion tends to produce a smoother rock surface than does quarrying. Exceptions occur when there are variable weaknesses in the bedrock caused by non-uniform rock type (such as mafic dike cutting through a metamorphic rock) or structure (presence of cracks or joints) resulting in differential abrasion rates. In these settings, abrasion locally produces grooves, ridges,

Figure 18.8 Vertical sea cliffs along the southwestern coast of Victoria, Australia.

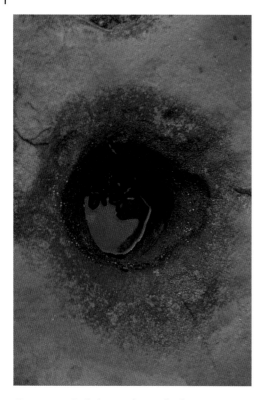

Figure 18.9 Potholes are the result of wave action swirling gravel in a depression. The hole deepens and widens as the rock is preferentially abraded.

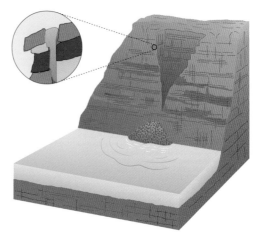

Figure 18.10 Frost wedging occurs when water freezes in the cracks and joints of rocks. The expansion caused by ice formation dislodges rock from the cliff face.

and other surface irregularities (Figure 18.9). Under special circumstances potholes measuring tens of centimeters deep and wide are formed. These rounded depressions are the result of pebbles and cobbles being swirled in the slight concavities of the rock by passing waves. With time, the gravel tools abrade and enlarge some of the depressions to become potholes.

Effect of Freeze and Thaw

There is another less visible physical process that contributes to erosion along rocky coasts in northern latitudes. This process involves the expansion and contraction of water as it changes from a liquid to a solid state. During moderate winter days melting snow may flow into the cracks, crevices, and joints of rocks only to freeze that night as temperatures fall. Because water expands by 9 % when it freezes,

ice formation can exert great pressures in confined spaces (Figure 18.10). In high latitudes, freeze and thaw cycles can occur on a daily basis especially in the intertidal zone where temperatures are governed by the rise and fall of the tide. In addition to melting ice and snow, water can come from rain and groundwater. Although seawater may also freeze, the ice that forms from seawater is relatively soft and therefore is less capable of causing frost-wedging than freshwater ice. It should be emphasized that it is the repeated cycles of expanding ice applying force on the walls of the rock followed by the release of this pressure when the ice melts that eventually causes the rock to fail and break. The effectiveness of freeze and thaw on the erosion process is dependent on how well the rock is fractured (number and size of cracks) and its porosity. Rocks that are thinly layered, such as schists, or highly porous, such as sandstones, are much more susceptible to freeze and thaw weathering than are massive rocks such as granite or basalt. Another important weathering process contributing to rock failure is the growth of other substances in the cracks of rocks, including calcite, halite (rock salt), and clays. Calcite and halite expand and contract due to temperature changes and some clay minerals do likewise when exposed to water. That latter process can be particularly effective in the upper-intertidal zone where these minerals are continuously being wetted and dried during the tidal cycle.

18.3.2 Biological Processes

Bioerosion is a more subtle process than the mechanical wearing-away of rocks, but in tropical regions where carbonate rocks commonly dominate it is an important means of sculpturing the coastal landscape. There are numerous plants and animals that remove the substrate in search of food or shelter or both. The most effective bio-eroder of calcium carbonate rock is the microscopic blue-green algae that penetrate into the rock by as much as 1 mm. A microscope reveals that as many as million algae may colonize a square centimeter of rock surface. These organisms are able to bore into the rock by dissolving away the calcium carbonate. They break down the structure

of the rock, causing individual grains to wash away by wave action. Other rock borers include certain sponges, worms, bivalves and echinoderms.

The presence of algae as well as fungi and lichens in the intertidal and supratidal zones promotes additional erosion by larger grazing organisms, which seek out algae, bacteria and other encrusting organisms as sources of food. Marine invertebrates including snails, limpets, sea urchins, and chitons abrade the rock surface as they feed on microflora. This process is not restricted to carbonate environments alone; the Channel Islands off the southern California coast are a good example. Along the shores of San Nicholas Island, located 160 km west of Los Angeles, mudstones and sandstones are home to a large community of sea urchins and abalone (Figure 18.11). These organisms have fashioned homes by boring into the intertidal sedimentary rocks accelerating the physical and chemical erosion of this rocky coast. The end-result of the bioerosion is a honeycombed outcrop consisting of hollows and ridges.

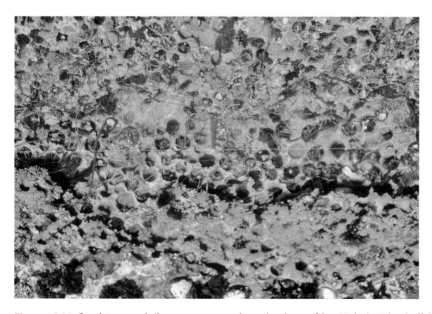

Figure 18.11 Sandstone and siltstone outcrops along the shore of San Nicholas Island off the California coast are home to sea urchins, abalone and other intertidal fauna. These organisms bore into the rock, finding shelter from the high wave energy and protection from predators.

18.3.3 Chemical Processes

The chemical breakdown of rock in the coastal zone is a very slow process due to the slow rate at which rocks react chemically with water. The effects are difficult to notice over human life-spans. However, it is a pervasive process that affects all rocky coasts from the subtidal to supratidal zone. Climate and rock type are the primary factors controlling the rate of chemical weathering. Because chemical reactions involve water and usually proceed more rapidly as temperature increases, hot and humid tropical climates experience much greater chemical weathering than arid temperate coasts or hot desert regions. Similarly, limestone is relatively soluble and so those coasts may erode at a rate of $1\,mm\,year^{-1}$, whereas coasts composed of quartzite are essentially stable because quartz is almost inert under surface conditions.

Chemical weathering of rocks encompasses many different chemical reactions including solution, which is the dissolution of different minerals into water, and hydrolysis, an important reaction that converts feldspars to clays. Because feldspar is a major constituent of many igneous (granite), metamorphic (gneiss), and sedimentary (sandstone) rocks, hydrolysis is a significant chemical weathering process that accounts for much of the world's mud. Oxidation is a chemical reaction between iron and the oxygen in water whereby "rust" is formed. These reactions take from tens of thousands of years for limestone to be dissolved to millions of years for feldspar to be transformed to clay. Because of these slow rates, physical erosion processes usually obscure the effects of chemical weathering.

18.4 Factors Affecting Rates of Erosion

Rocky coasts look pretty much the same visit after visit because the processes that change them work very slowly. Characteristics such as rock type and degree of fracturing, exposure to wave energy, tidal range, climate, relative sea-level changes, and other factors affect the rate of erosion and ultimately, the morphology of the coast.

Rock type

The influence of rock type and structure is particularly well illustrated by comparing parts of the west and east coasts of the United States. In southern California waves have cut into relatively soft sandstones, mudstones and other flat-lying sedimentary rocks producing sea cliffs along much of the coast. In contrast, the metamorphic rocks and igneous intrusives (granites) of the central Maine coast are tough rocks that have resisted erosion. Here, the coastline reflects the structure of the rocks and patterns of differential erosion. The folded metamorphic rocks are turned on end, producing a jagged, indented shoreline. Interrupting this coastal landscape are granite outcrops that, due to their size and resistance to erosion, have formed headlands and islands (Figure 18.12). Likewise, less resistant portions of the sandstone cliffs along the coast of Victoria near Port Campbell, Australia have produced large reentrants with spectacular interior pocket beaches (Figure 18.13).

Degree of fracturing

Fractures include the cracks, joints and faults of a rock outcrop. Their number and trend are significant because erosion is usually concentrated at these sites. Fractures increase the area exposed to various weathering agents. Joints and faults can be indirectly responsible for the dissection of some rocky coasts and the formation of stacks, arches, and other features. Preferential weathering at joints and faults may also produce deep crevasses and the development of narrow embayments.

Wave energy

Wave energy is important because it controls the intensity of physical erosion and the removal of debris that is produced during

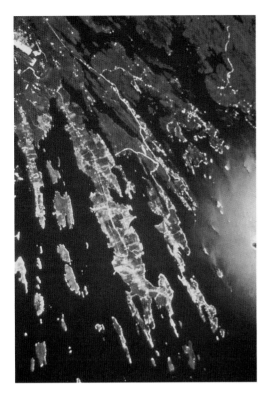

Figure 18.12 False-color infrared photograph of the central peninsula coast of Maine. The highly indented nature of this coast is a product of differential erosion. The ridge and valley morphology has been accentuated by several episodes of Pleistocene glaciation.

the weathering and abrasion processes. Researchers have shown that waves exert their greatest force at a position just above mean high water. This is because water levels have a maximum duration at mean high tide and waves break for an extended period at this elevation. Secondly, waves are larger at high tide than mid- or low tide due to the deeper water and lower frictional resistance generated by the rocky intertidal zone. This condition is apparent when looking at seawalls and noting that the greatest abrasion to the wall occurs near the high-tide line. This level is also where chemical weathering processes are at a maximum. Although waves are even larger and reach higher elevations during storms, they have less effect due to their infrequent nature.

Waves and their attendant processes will continue to erode a rocky shore as long as the sediment that is produced during the weathering process is removed. For example, if a cliff erodes and produces enough sediment to form a wide beach, then the cliff effectively becomes protected from further wave action and erosion (Figure 18.14). Also, if the rocky coast is barren of sediment, there is no material to provide for abrading the rock surface.

Tidal range
Rocky coasts are found in a wide range of tidal settings from the microtidal coasts in the Mediterranean (TR ~ 1 m) to some of the highest tidal range coasts of the world (TR > 10 m) including the Bay of Fundy and the eastern coast of the Bay of Saint-Malo in France. Along most rocky shores the direct influence of tidal forces is minimal. However, tides do control the amount of time that waves erode different elevations along a rocky coast. Along steep or vertical cliffs wave action is focused in a relatively narrow zone, particularly along microtidal coasts. In contrast, along macrotidal settings the large excursion of the tide disperses wave energy over a relatively wide zone.

Climate
Climate influences the rate and type of weathering processes. Tropical areas tend to undergo greater chemical weathering whereas temperate and polar regions experience more frost weathering. Climate also controls the patterns of storms and prevailing winds and therefore the deepwater wave energy along a coast. Generally, temperate regions have greater wave energy and physical processes operate more energetically than in tropical areas, where average wave energies are lower.

Relative sea level
This factor has an obvious effect on the development of marine terraces as they are a product of relative sea-level changes. They form when a coast rises tectonically and marine platforms are lifted from the water. They may also develop when eustatic sea level drops.

Figure 18.13 Formation of large reentrants along the coast of Victoria in Australia is the result of wave-induced preferential erosion of less resistant portions of the sandstone cliff. *Source:* https://en.wikipedia.org/wiki/Port_Campbell_National_Park#/media/File:A191,_Port_Campbell_National_Park,_Australia,_cove_from_helicopter,_2007.JPG © Brian W. Schaller / Licensed under FAL 1.3.

Figure 18.14 Photo of 300-m high cliffs in Resurrection Bay near Seward, Alaska. Note the talus slope at right side of photo that protects the onshore cliff.

18.5 Morphology

Rocky coasts exhibit a wide variety of morphologies because there are so many factors that have influenced their development. One characteristic that they share is an overall scarcity of sediment. Thus, they are associated with tectonically active coasts and are also common in former glaciated settings where the sediment cover was removed by the ice. In addition, low rocky coasts occur in tropical areas where sediment has become lithified by chemical processes. There are too many types of rocky coasts to fully treat their range of morphologies. Instead, the major classes of rocky coasts are presented below as well as some of their striking features.

18.5.1 Sea Cliffs

Cliffs have many different profiles and heights, reflecting their composition, structure and weathering processes. There are the spectacular vertical cliffs at the entrance to Resurrection Bay leading to the deepwater port of Seward, Alaska (Figure 18.15). These rise 200 m vertically from a deep glacially-cut through before ascending more gradually into mountains with elevations in excess of a kilometer. Other substantial vertical cliffs (height > 100 m) include the White Cliffs of Dover in England (Figure 18.16) the Cliffs of Moher (Figure 18.17), and the extensive cliffs along the Nullarbor Plain of South Australia. The fact that these cliffs have similar morphologies despite differing compositions and structure indicates that their development has been dominated by the same efficacy of erosional processes. In other cases, weaknesses in the bedrock

surface provide conditions where wave attack can produce preferential erosion. An examples occurs along Giant's Causeway in Northern Ireland, where the coast is composed of large vertical columns of polygonally jointed rock that formed by cooling of shallow intrusive basalt (Figure 18.18). Fractures between the columns provide locations where waves cause individual columns to separate and fail.

In some regions, cliffs are not vertical: sections of southern California, portions of southwestern Wales, and much of the Alaskan Peninsula. Regional geology and intensity of various weathering processes control the steepness and profile of cliffs. When sea cliffs consist of relatively unfractured, massive rocks such as basalt, granite or quartzite (metamorphosed sandstone), erosion proceeds in a uniform manner. However, when sea cliffs are formed through the erosion of horizontally-layered sedimentary rocks, there are commonly large differences in how individual layers react to weathering and abrasion processes. Under these conditions the cliff erodes unevenly producing steps, notches, and other irregularities. Various patterns of joints, faults and

Figure 18.15 The vertical sheeted dike complex in Resurrection Bay near Seward, Alaska has produced a face which, in places, rises over 800 m above the water surface. These rocks (gabbro, a type of igneous rock) were formed at a mid-ocean ridge system during production of oceanic crust. The ocean crust was subsequently raised to the surface by tectonism – the subduction of the Pacific Plate beneath North America.

Figure 18.16 The White Cliffs of Dover, England are composed chiefly of coccolithophores, a planktonic alga with a calcium carbonate exoskeleton.

Figure 18.17 The Cliffs of Moher in southwestern Ireland rise more than 200 m above the sea.

Figure 18.18 Giant's Causeway in Northern Ireland is known as one the most spectacular coastal outcrops of columnar-jointed basalt in the world.

folds may also result in jagged slopes. Vertical sea cliffs and very steep slopes are created when the rocks are homogeneous and marine weathering processes clearly dominate subaerial erosion. In these situations marine agents cut the slope back at a faster rate than the upland can be eroded to form an incline. When wave abrasion undercuts the base of a sea cliff causing eventual failure of the upper slope, a steep or vertical face is usually produced. Conversely, as erosion of the surface of the slope increases relative to marine erosion, sea cliffs become more gently inclined.

Surface processes coupled with marine erosion have created a spectacular coastal landscape along the southern shore of the island of Hawaii. Here, surface and groundwater flows have cut large V-shaped channels into lava flows. Deep chasms have developed where large amounts of water have flowed to the coast. Along other sections of these sea cliffs, smaller drainage systems have formed dramatic hanging valleys, some with 300 m-high waterfalls (Figure 18.19).

The height of sea cliffs appears to be controlled by wave energy. On a worldwide basis there is a correspondence between cliff height and latitude. Most high-relief cliffs occur in mid-latitudes where waves are relatively large. Modest and low cliffs are found in low and high latitude regions where waves tend to be small. In polar regions, coasts are protected by sea ice during part of the year. Because wave energy is proportional to the square of wave height, large waves exert a much greater force on the face of a cliff than do small waves. Abrasion rates are also greater in regimes of high wave energy due to the large amounts of sediment entrained by large waves. The greatest rates of abrasion coincide with regions of high waves and the presence of pebble, cobble, and small boulder abrading agents. In low latitudes, rock cliffs are limited in height because they have eroded into beach rock and lithified dunes. The complexities of coastal geology around the world, and the highly varied marine erosional processes that operate along coasts have produced many exceptions to the generalized trends presented here.

Figure 18.19 Hanging valleys along the coast of Hawaii are the sites of waterfalls more than 200 m high. The cliffs are probably the result of a huge landslide in which a massive block of volcanic rock slid into the abyss. The valleys were cut by groundwater sapping processes.

18.5.2 Horizontal Erosional Landforms

Platforms and Benches

Wave erosion along rocky coasts not only creates vertical to steeply inclined landforms; horizontal features such as platforms (also called *benches* by early researchers) also form during the process (Figure 18.20). Shore platforms are the flat bedrock ledges that border the base of sea cliffs. They vary in width from a few meters to more than a kilometer. Where platforms have formed on flat-lying sedimentary rocks their surfaces are relatively smooth. When they develop in regions of variable geology, such as highly jointed igneous rocks or on dipping metamorphic or sedimentary rocks, they commonly exhibit irregular surfaces with relief up to meter. Differences in resistance to weathering processes produce grooves, gullies, knobs and other features.

Platforms may be almost horizontal to seaward, dipping with slopes ranging from a few degrees to more than 30 degrees. In some areas, particularly where abrasion is the dominant erosion process, a correlation exists between tidal range and gradient of the platform, with increasing tidal range producing steeper slopes. Tidal range controls the distribution of wave energy over the platform and to some extent, the wave erosional processes. Throughout the tidal cycle waves shoal and break along the seaward edge of the platform. However, only at near high tide levels do waves directly affect the landward, highest portion of the platform. Thus, the greatest rates of vertical erosion occur at high tide, when wave processes are most effective.

Slightly elevated platforms are created in weak sections of rock cliff that erode at a faster rate than the rock below it. These benches are most common in areas of horizontally layered sedimentary rock, such as bedded limestone, sandstone and mudstone, and occur at elevations above mean high water. They are usually fairly narrow and are connected to platforms by a step.

There are platforms in southern California and elsewhere in the world that extend to

(a)

(b)

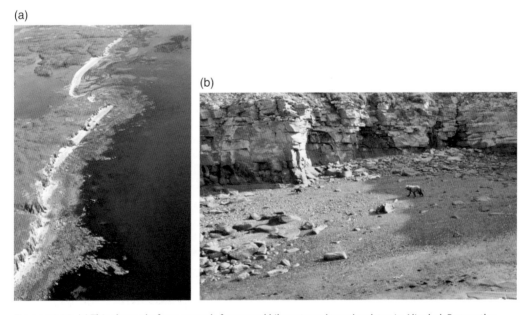

Figure 18.20 (a) This shore platform extends for several kilometers along the shore in Alinchak Bay on the Alaskan Peninsula. Platforms develop through wave abrasion and other erosion processes, producing an intertidal, sub-horizontal bedrock surface that widens as the abutting cliff retreats. (b) Close-up view of the retreating cliff illustrating that the upper beach contains thin beach deposits. Note the brown bear for scale.

depths greater than 10 m. Because it has been shown that wave erosion is essentially inactive below 10 m, the deeper portions of these platforms are viewed as relict and likely were drowned by rising sea-level. Cliff retreat is slow and rates are measured in terms of centimeters per year along most rocky coasts, meaning that platforms were created very slowly as well. This suggests that platform development requires a stable or near stable sea-level position over an extended period of time. The widest shore platforms occur in regions of moderate to high wave energy where rock cliffs are highly susceptible to erosion, such as those that are composed of weakly cemented sedimentary rocks. Resistant rocks, such as quartzite, are commonly associated with narrow, steep platforms.

The formation of platforms involves wave processes and a type of chemical weathering called water-layer weathering. Wave-eroded platforms are closely related to sites of cliff retreat and encompass wave abrasion and quarrying. These processes are most energetic in areas of moderate to high wave energy where abrading tools are abundant. Wave energy must be sufficient to remove the debris that is produced during the erosion process, otherwise the platform becomes insulated from further weathering processes.

Water-layer weathering includes all processes resulting from the wetting and drying of the bedrock including salt crystallization. The process is most operative in areas of horizontally bedded, highly permeable rocks in tropical regions with low tidal ranges such as Hawaii, Australia, the Caribbean or the Mediterranean coasts. High temperatures favor evaporation and chemical weathering. This type of platform is common along carbonate coasts and can be identified by small shallow-water pools that occur near mean high water. Granitic rocks along the coast of Maine may exhibit these same shallow pools of water at elevations from one to several meters above mean high water. These pools contain granules of quartz and feldspar that have weathered directly from the granite during the water-layer erosion process.

Marine Terraces

Marine terraces are platforms that have been displaced from their original position by tectonic forces, sea-level changes, or both. They are located above sea level where the land is rising and below sea level where it is sinking (Figure 18.21). Along the coast of Southern California there are at least six marine terraces in the Palos Verdes region and as many as ten on the islands of San Clemente and Santa Cruz. Extensive terrace development also exists along the tectonic coasts of New Zealand, New Guinea, the Alaska Peninsula and along the Pacific side of Central and South America. Steep slopes separate individual terraces. Elevation difference between successive terraces is a function of the uplift rate along the coast. Marine terraces are best developed in areas of moderate to high wave energy along rising coasts that are composed of easily eroded, horizontally stratified, sedimentary rocks. It should be noted that marine terraces also form along unconsolidated sediment coasts.

The formation of marine terraces requires periods of sea level stability in an overall regime of tectonic activity (Figure 18.22). Most scientists now agree that terraces represent platforms raised from the water during the Pleistocene Epoch. Remember from Chapter 15, that during the Pleistocene the advance and retreat of continental ice sheets produced coincident changes in sea level by as much as 120 m. It is believed that most platforms were formed while sea level stood at a highstand position (approximately present-day sea level) that lasted perhaps thousands of years. They were raised during the interval of time after the highstand when sea level fell to a lowstand position and then rose back to approximately the present sea level. This process was repeated during each period of glaciation producing a series of increasingly higher marine terraces that record the times of Pleistocene highstands. Knowing the elevation difference between successive terraces and determining their age of formation using various dating techniques, provides a means of calculating the rate of tectonic uplift for that particular coast.

(a)

(b)

Figure 18.21 (a) Marine terrace development along the southwestern coast of Victoria, Australia, and (b) in Bodega Bay along the central California coast. Note that the marine stacks in Bodega Bay are morphologically similar to the rock outcrops atop the marine terrace. Indeed, the rock outcrops were marine stacks when the shore platform was developing. *Source:* 123RF/ bennymarty. Reproduced with permission of 123RF.

18.5.3 Sea Stacks, Arches, and Erosional Features

One of the striking features of rocky coasts is the erosional remnants that are left behind as cliffs retreat. Through differential weathering of bedrock, landforms such as sea stacks and arches are left sitting atop shore platforms. In other regions preferential erosion along cliffs produces notches and caves. The 12 Apostles along southwestern Victoria, Australia are

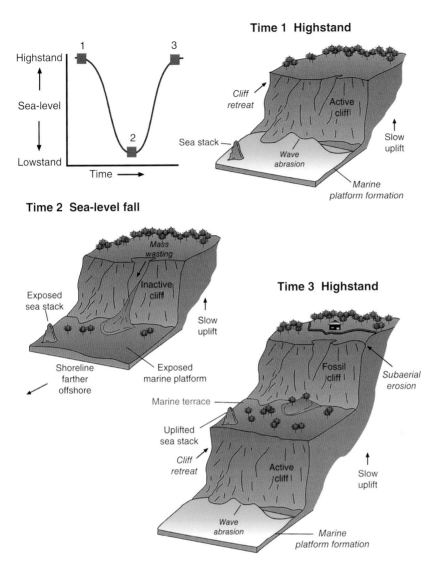

Figure 18.22 Marine terraces form along coasts undergoing uplift. Time 1. terrace development begins during a sea level highstand with the formation of a marine platform produced by wave abrasion. Time 2. The marine platform is subsequently abandoned when sea level falls at the onset of a global glaciation event. Time 3. During the period of time in which sea level falls to a lowstand and then rises back to a highstand position, the marine platform is uplifted above the highstand position, becoming a marine terrace. Idealized sea-level curve shows timing of marine terrace development.

one of the most famous set of stacks in the world and are used widely to depict rocky coasts (Figure 18.23). Some stacks, such as those along the Bay of Fundy shoreline of New Brunswick, Canada, are large and tall enough to have trees growing on top of them. Stacks exhibit many different forms but commonly are taller than they are wide and are generally lower in elevation than the adjacent cliffs. Arches are equally impressive due to their unusual likeness to bridges that go nowhere.

Formation of these features is the result of differential marine erosion superimposed on specific types of bedrock geology. The development of a single stack is illustrated well

Figure 18.23 The Twelve Apostles are sea stacks that have formed from flat-lying limestone rock (Miocene age) in Port Campbell National Park along the coast of Victoria, Australia. *Source:* https://commons.wikimedia.org/wiki/File:Apostles_3_GOR.JPG. Licensed under CC BY SA 3.0.

where waves have eroded ash, welded tuff, and other rocks comprising the exterior of a volcano while its neck remains as a pinnacle. This is because the neck is composed of relatively strong basalt that resists erosion. Multiple stacks are generally related to horizontally layered sedimentary rocks, such as sandstones, limestones and conglomerates that are highly jointed. These giant cracks are weaknesses in the rock that promote erosion due to increased surface area and greater exposure to marine weathering processes. As erosion proceeds, joints become indentations, then deep chasms, and with further wave attack finally, embayments. Eventually, the headland is detached from the mainland and a stack is formed. Thus, stacks can be created if they are composed of rock that is more resistant than the cliff rock. Alternatively, stacks can form due to structural weaknesses of the cliff rock, which makes them more susceptible to differential erosion and retreat.

Arches are commonly found near sea stacks and are usually associated with horizontally layered sedimentary rocks. They develop after the cliff rock has been dissected and narrow headlands or elongated stacks have formed. The next stage in development occurs when wave action preferentially erodes the middle or lower section of the headland or stack. The "tunnel" that develops is a product of wave abrasion and other processes that concentrate their erosion at the high water line and attack at both sides of the landform. As the tunnel widens, a sea arch is created. Eventually, continued erosion causes the bridge to collapse into the sea and two stacks are produced. London Bridge was the name given to well-known sea arch located along the southwestern coast of Victoria, Australia (Figure 18.24). In February 1989 the arch suddenly collapsed while two tourists were investigating its seaward end. Lucky for them they were not in the middle when the rocks gave way; they were later rescued by a helicopter and taken to shore. All in all they had a memorable day.

Along many rocky coasts the pounding of waves at the base of sea cliffs undercuts the

(a)

(b)

Figure 18.24 (a) London Bridge was a famous sea arch along the southwestern coast of Victoria, Australia. (b) It collapsed suddenly in 1989, stranding two tourists who were later rescued by helicopter. *Source:* Courtesy of Andrew Short.

rock producing a notch (Figure 18.25). In addition to abrasion and other wave processes, erosion is also attributed to freeze and thaw cycles in northern latitudes and salt crystal formation and other chemical and biological processes in tropical regions. Notches occur predominantly just below the mean high water line but may be found at

Figure 18.25 Notches usually develop near the high-tide line because wave energy is concentrated in this zone.

lower elevations, particularly along coasts with large tidal ranges such as in the Bay of Fundy. Notches rarely extend more than several meters deep into the cliff face because as the overhang increases, support of the overlying rock diminishes. Eventually, this leads to failure of the cliff and the production of debris on the shore platform. Stacks can also become notched giving them a mushroom appearance.

Notches can be transformed into caves under special geological conditions. This usually occurs where the bedrock is locally weak and susceptible to erosion. One of the authors will not forget walking along the shoreline of Puale Bay on the Alaska Peninsula and coming to a small headland that jutted into the bay preventing further travel. We climbed several meters up the rock face finding a 5 m-wide cave that had been excavated into a conglomerate. The stronger, more resistant sandstones and mudstones formed the boundaries of the opening. The cave, which extended about 15 m into the cliff, broadened toward the rear and seemed to be lit by sunlight. Sure enough there was second passageway to the cave that opened to the other side of the headland.

Figure 18.26 Waterfalls are common features along tectonic coasts where the rate of uplift is greater than the rate at which the stream can downcut.

Figure 18.27 Sea arch created in layered rocks due to jointing patterns and slight difference in rock type.

As we approached this opening a waterfall that rained down from the overhead cliff top obscured our view of the bay (Figure 18.26). In fact, this was one of many spectacular landforms along this rocky coast which featured shore platforms and marine terraces, numerous waterfalls cascading to beaches below, and a wide variety of stacks, arches and notches (Figure 18.27). The stacks of this coast are favorite nesting sites of bald eagles due the protection afforded by the isolation of the stacks and the proximity to food sources.

18.6 Summary

Rocky coasts exist in a wide variety of settings, but are most common along collision coasts due to the high relief and low sediment supplies. They also occur in glaciated regions where sediment has been stripped from the bedrock. In tropical regions, rocky shores are found where sand has been lithified by calcium carbonate cement. Rocky coasts are resistant to erosion and retreat imperceptibly. They are slowly modified by a combination of physical, chemical, and biological processes. In areas of moderate to high wave energy where gravel and/or other sediment are present, wave abrasion is the dominant eroding agent. High-latitude coasts are affected by freeze and thaw whereas in tropical areas chemical weathering and bioerosion tend to be important erosional processes.

Cliffs are common features of retreating coasts and are often fronted by shore platforms. These are best developed in regions where the bedrock consists of horizontally-layered, easily-eroded sedimentary rocks. Along tectonic coasts undergoing uplift, shore platforms have been raised forming marine terraces. Downwarped coasts may have subtidal terraces. Erosional remnants, including stacks and arches, as well as notches and caves, are found along coasts that experience differential erosion of the retreating cliffs. This can be a product of differences in bedrock composition or structural weaknesses.

Suggested Reading

Bird, E.C.F. (1993). *Submerging Coasts.* Chichester, UK: Wiley.

Bradley, W.C. and Griggs, G.B. (1976). Form, genesis, and deformation of some Central California wave-cut platforms. *Bull. Geol. Soc. Am.* 87: 433–449.

Carter, R.W.G. and Woodroffe, C.D. (eds.) (1994). *Coastal Evolution.* Cambridge: Cambridge University Press.

Davies, J.L. (1980). *Geographical Variation in Coastal Development*, 212 pp. Harlow, UK: Longman.

Easterbrook, D.J. (1999). *Surface Processes and Landforms.* Upper Saddle River, NJ: Prentice Hall.

Griggs, G.B. and Savoy, L.E. (1985). *Living with the California Coast.* Durham, NC: Duke University Press.

Inman, D.L. and Nordstrom, C.E. (1971). On the tectonic and morphologic classification of coasts. *J. Geol.* 79: 1–21.

Johnson, D.W. (1919). *Shore Processes and Shoreline Development.* New York: Hafner Publishing.

Shepard, F.P. and Wanless, H.R. (1971). *Our Changing Coastlines.* New York: McGraw-Hill.

Trenhaile, A.S. (1987). *The Geomorphology of Rock Coasts.* Oxford, UK: Oxford University Press.

Index

Beaches and Coasts, Second Edition. Richard A. Davis, Jr. and Duncan M. FitzGerald.
© 2020 John Wiley & Sons Ltd. Published 2020 by John Wiley & Sons Ltd.

Index

c

California 7
California Current 97
Cape Cod 454, 472–4
    formation 476–80
    modification 479–80
capillary waves 133
carbonate mud 235, 245
carbonates 3
causeways 447, *449*
centrifugal force 154–7, 175
    channel margin linear bars
        413, 415, 428–9
    cliffs 469, 480, 496–511
Chesapeake Bay 211
clay minerals 61, 216
climate change 76, 78, 69
coastal construction 333
coastal invasion 10
coastal marsh 281
coastal sediment supply 47–50
coastal tourism 10
collision coasts 2, 36–9, 42,
        45, **47**, 51, 72, 360, 364–5,
        490, 511
    continental 36–7, 39
    island arc 36, 39, 42
Columbia River 180
continental drainage
        45–9, 363
continental drift 16–17
continental margins 31–4
    rises 32–3
    shelves 32–9
    slopes 32–6
continent–continent plate
        convergence 27–8
convection cells 30, 98, 100
Coorong, Australia 243
Copper River Delta 181
coppice mounds 344
Coriolis effect 75, 98,
        100–2, 110, 112, 162–3,
        165, 168, 175
Corps of Engineers 295
CRAB 139
crevasse splay 185, 187, 286
crustal movement 15
cumulative curve 58

cyanobacteria 237
cyclones 96–7, 101–3,
        105–10, **112**, 113, 125
    formation of 105–6
    cyclonic systems 101, 125

d

Dalrymple, Robert A. 66
Davis, Robert 108
Daytona Beach, Florida **111**,
        336, 337
Dean, Robert G. 66
de Beaumont, Elie 377
delta classification 190, 191
delta front 86
delta marshes 196
delta plain 184, 185, 199, 200
deltas 4, 5
depositional coasts **47**, 96, 394
desert 498
dikes 398, 495, 501
desiccation features 255, 268
distributaries 184
distributary mouth bar *402*
Dolan, Robert 108
dredge and fill 304
drowned river valleys *46*,
        377, 417–19, 485–7
drumlins 391, 401, *418*,
        473–82, 485
Duck Pier 138, 139
dune dynamics 349
dune erosion 350
dune migration 347
dune ridge 344, 345
dune walkovers 353, 355

e

earthquakes 15, 21–2, 25–6,
        28–31, 34, 36, 38–42,
        44–5, **47**, 50, 52, 469, *491*
east coast of the United States
        35, 45–6, 107, 167, 363–4,
        380–1, 415, 418, 427
East Friesian Islands 383,
        395, 397, 439, 441–2, 446
ebb shield 410, 412–13
ebb spit 410, 412–13
El Nino 69

end moraine *454*, 473–4,
        477, 486
engineers, coastal 423, 426
eolianites 493
erratics 471, 477
estuarine hydrology 306
estuary 5
    definition 207
    types 208

f

Fairweather fault 40
feldspars 20, 360, 498, 505
fencing 384, 394, 395
Ferrel cell 100
Field Research Facility 138,
        139
Figure Eight Island, North
        Carolina 359
Fisher, John 379
Fisk, H. N. 178
fjords *16*, 38, *42*, 174, 203, 204,
        453, 468–70, 485, 491–2
flaser bedding 260–3
flood channel 410, 412–13,
        427–8
flood ramp 410, 412–13
flood plain 186
Florida Bay 300
Florida Keys 110, 119
foreshore zone 312
frontal systems 96–108

g

Galapagos Islands 302
Galveston Sea Wall 9
Galveston, Texas 95, **111**,
        127, 359
Ganges–Brahmaputra 95
Georgia Bight 396, *400*
ghost crab 326, 329
ghost shrimp 319
Gilbert, G. K. 177, 377
glacial cycles 83
glaciated coast 2, 170, 359,
        374, 391, 399, 400, 419, *421*,
        439, 453, 469, 472–3, 477,
        480, 482–3, 485–6, 491–2
    alpine 458–9, 468–9